Bivectors and Waves in Mechanics and Optics

APPLIED MATHEMATICS AND MATHEMATICAL COMPUTATION

Editors

R.J. Knops, K.W. Morton

Text and monographs at graduate and research level covering a wide variety of topics of current research interest in modern and traditional applied mathematics, in numerical analysis, and computation.

1 Introduction to the Thermodynamics of Solids *J.L. Ericksen* (1991)

2 Order Stars *A. Iserles and S.P. Norsett* (1991)

3 Material Inhomogeneities in Elasticity *G. Maugin* (1993)

4 Bivectors and Waves in Mechanics and Optics *Ph. Boulanger and M. Hayes* (1993)

5. Mathematical Modelling of Inelastic Deformation *J.F. Besseling and E. van der Geissen* (1993)

(Full details concerning this series, and more information on titles in preparation are available from the publisher)

Bivectors and Waves in Mechanics and Optics

Ph. BOULANGER

Departement de Mathematique
Universite Libre de Bruxelles
Belgium

and

M. HAYES

Mathematical Physics Department
University College Dublin
Ireland

CRC Press
Taylor & Francis Group
Boca Raton London New York

CRC Press is an imprint of the
Taylor & Francis Group, an informa business
A CHAPMAN & HALL BOOK

CRC Press
Taylor & Francis Group
6000 Broken Sound Parkway NW, Suite 300
Boca Raton, FL 33487-2742

ISBN 13: 978-0-412-46460-7 (hbk)

Visit the Taylor & Francis Web site at
http://www.taylorandfrancis.com

and the CRC Press Web site at
http://www.crcpress.com

To Violeta and Colette

Contents

Preface

The first systematic study of bivectors, or complex vectors, was by J.W. Gibbs (1881). In 1881 and 1884 he published, privately, a 73-page pamphlet *Elements of Vector Analysis*, of which the final seven pages are devoted to bivectors. With each bivector Gibbs associated an ellipse, called its 'directional ellipse' and was thus able to give a geometrical interpretation to the dot product of two bivectors being zero. From notes of Gibbs' lectures it is clear that he made use of bivectors in his studies of electromagnetic waves.

Previously, W.R. Hamilton (1853) made passing reference to bivectors, in the context of biquaternions, in his *Lectures on Quaternions*, but he did not develop the theory. His colleague, James MacCullagh (1847), when working on electromagnetic waves, derived a result (MacCullagh's theorem, Chapter 2) central to the development of bivectors, though not in the context of bivectors. Hamilton (Halberstam and Ingram, 1967, p. 142) described MacCullagh's theorem as 'a remarkable use of the symbol $\sqrt{-1}$'.

Recently, Synge (1964) gave a systematic treatment of the algebra of bivectors, but, apparently unaware of Gibbs' work, did not give geometrical significance to his results.

Bivectors occur naturally in the description of elliptically polarized homogeneous and inhomogeneous plane waves. The description of a homogeneous plane wave generally involves a vector – the unit vector along the propagation direction, and a bivector – the complex amplitude of the wave. The ellipse associated with the amplitude bivector is the polarization ellipse of the wave. For instance, the Jones vectors which are used in optics for the description of elliptical polarization are amplitude bivectors.

Inhomogeneous plane waves, sometimes called 'evanescent waves', are those waves for which the planes of constant phase are different from the planes of constant amplitude. They are described in terms of two bivectors – the complex amplitude and the complex slowness.

The real part of the slowness bivector characterizes the propagation direction and the phase speed, whilst its imaginary part gives the attenuation direction and attenuation coefficient. The use of bivectors and of their associated ellipses is essential for the presentation of a systematic method (the 'directional ellipse' method introduced in Chapter 6) for deriving all possible inhomogeneous plane wave solutions in a given context.

The purpose of this book is to give an extensive treatment of the properties of bivectors both algebraic and geometrical, and to show how these may be applied to the theory of homogeneous and inhomogeneous plane waves. Nowadays, there is, unfortunately, little classical euclidean geometry in undergraduate mathematics courses. For that reason, some basic material on ellipses and ellipsoids has been included here.

Because elliptically polarized transverse waves are so important, one chapter is devoted to their description. Use of bivectors makes the treatment direct and simple and leads naturally to the parameters introduced by Stokes (1852). The links with Jones vectors are identified and properties of the Poincaré sphere are noted.

In order that the material be self-contained, Chapter 8 is devoted to considerations of energy flux for trains of homogeneous and inhomogeneous plane waves. For homogeneous waves the systems considered are linear and conservative. The link between the mean energy flux vector and group velocity for homogeneous plane waves is established. For inhomogeneous plane waves the systems considered are linear, conservative and such that the propagation condition depends only upon the slowness bivector and is independent of the frequency of the wave train. In this case some general results are obtained.

Having laid the foundations, the final three chapters are devoted to applications – to electromagnetic waves in crystals, to waves in linearized elasticity and to waves in viscous fluids.

For each chapter, we have included many exercises with answers. They have a double purpose. On the one hand, they may be used as a working test for students. On the other hand, many of them present further useful properties which are used and referred to afterwards.

The material of this book is suitable for senior undergraduate and first year graduate students. We hope that it may also prove useful for researchers interested in homogeneous and inhomogeneous plane waves.

We are grateful to a number of people – Robin Knops for his encouragement; for their continued support and patience, to our colleagues, in UCD: Joe Pulé, John Kennedy, Ted Cox and Dermot McCrea, and in ULB: Christine Demol, Jules Leroy and Georges Mayné. We particularly thank Christine Demol for her help in drawing the figures. Also, Professors I. Shih-Liu and J.A. Salvador from the Universidade Federal do Rio de Janeiro are gratefully acknowledged for having invited one of us (Ph.B.) to deliver lectures on some of the topics covered by this book. Finally, we wish to thank the secretaries Bridget Mangan (UCD) and Nicole Aelst (ULB) for their careful typing of the manuscript.

Neither of us could have managed without the patience and support of our families for which we are very grateful.

Ph. Boulanger
M. Hayes

1

The ellipse

Because ellipses and bivectors are very closely related we begin by drawing together their basic properties. We consider pairs of conjugate radii of ellipses and present results in terms of them.

1.1 The ellipse

Referred to rectangular cartesian axes Ox and Oy, the equation of the ellipse centred at O, with major semi-axis of length a along Ox and minor semi-axis of length b along Oy is

$$\frac{x^2}{a^2}+\frac{y^2}{b^2}=1. \tag{1.1.1}$$

The coordinates of a generic point P on the ellipse may be given in terms of a parameter θ through

$$x=a\cos\theta, \quad y=b\sin\theta. \tag{1.1.2}$$

Clearly, this satisfies (1.1.1). The angle θ may be interpreted through use of the circle centred on O with the major axis as diameter which is called the 'auxiliary circle' (Figure 1.1). If a radius OQ of the auxiliary circle makes an angle θ with Ox, then Q has coordinates $(a\cos\theta, a\sin\theta)$ and the perpendicular QM onto Ox intersects the ellipse at P since $OM=a\cos\theta$. Thus if the position vector of P is denoted by $\boldsymbol{r}$, then

$$\boldsymbol{r}=\boldsymbol{OP}=a\cos\theta\mathbf{i}+b\sin\theta\mathbf{j}, \tag{1.1.3}$$

where $\mathbf{i}$ and $\mathbf{j}$ are unit vectors along Ox and Oy respectively.

If R is the point on the auxiliary circle, so that $R\hat{O}Q=\frac{1}{2}\pi$, then R has coordinates $(a\cos(\theta+\frac{1}{2}\pi),\ a\sin(\theta+\frac{1}{2}\pi))$. The perpendicular RS onto Ox intersects the ellipse at T which has coordinates

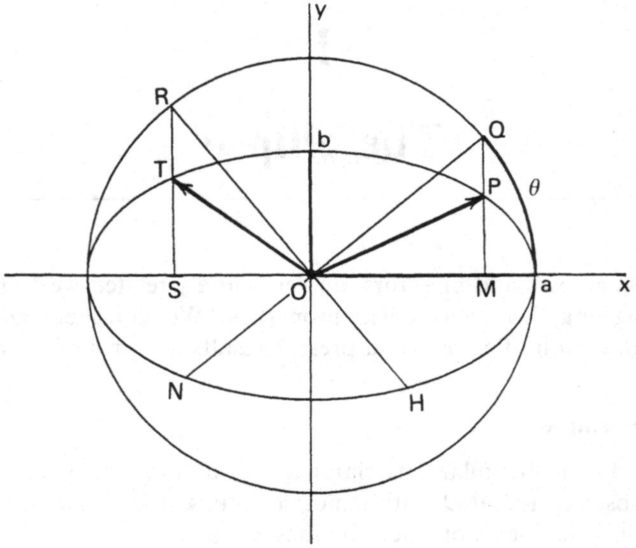

Figure 1.1 *Ellipse and conjugate radii.*

$(a\cos(\theta + \frac{1}{2}\pi),\ b\sin(\theta + \frac{1}{2}\pi))$. Hence

$$\boldsymbol{OT} = -a\sin\theta\mathbf{i} + b\cos\theta\mathbf{j}. \qquad (1.1.4)$$

Two radii OP and OT so related are said to be 'conjugate' (Figure 1.1). PN and TH are conjugate diameters. Any pair of directions parallel to a pair of conjugate diameters are said to be 'conjugate directions'.

Using (1.1.3) and (1.1.4) two invariant properties of pairs of conjugate radii may be easily derived. We have

$$(OP)^2 + (OT)^2 = a^2 + b^2, \qquad (1.1.5)$$

so that the sum of the squares of pairs of conjugate radii of an ellipse is constant. Also

$$|\boldsymbol{OP} \times \boldsymbol{OT}| = ab, \qquad (1.1.6)$$

so that the areas of the parallelograms formed by pairs of conjugate radii of the ellipse are all equal to the area of the rectangle on the principal semi-axes.

It is now shown that the tangent to the ellipse at P is parallel to

OT and the tangent at T is parallel to OP. This provides a procedure alternative to the construction based on Figure 1.1 for the determination of pairs of conjugate radii.

The tangent to the ellipse at P is along $\mathrm{d}\boldsymbol{r}/\mathrm{d}\theta$. It is parallel to $\boldsymbol{OT}$ because

$$\frac{\mathrm{d}\boldsymbol{r}}{\mathrm{d}\theta} = -a\sin\theta\mathbf{i} + b\cos\theta\mathbf{j} = \boldsymbol{OT}. \tag{1.1.7}$$

The slope of the tangent at P is $-a^{-1}b\cot\theta$ which is also the slope of OT; and the slope of the tangent at T is $-a^{-1}b\cot(\theta+\frac{1}{2}\pi) = a^{-1}b\tan\theta$ which is also the slope of OP. Note that the product of these slopes is $-b^2a^{-2}$.

The conjugate diameters have the further property that any chord of the ellipse parallel to one diameter is bisected by its conjugate diameter; conversely if a diameter bisects a chord parallel to another diameter, then the two diameters are conjugate. This may be proved as follows.

Let $\boldsymbol{r}_1 = a\cos\theta_1\mathbf{i} + b\sin\theta_1\mathbf{j}$ and $\boldsymbol{r}_2 = a\cos\theta_2\mathbf{i} + b\sin\theta_2\mathbf{j}$ be any two radii of the ellipse (1.1.3). The equation of a chord parallel to $\boldsymbol{r}_2$ is $\boldsymbol{r} = k\boldsymbol{r}_1 + \mu\boldsymbol{r}_2$. Here μ is a parameter, k is a number in the range $0<k<1$ so that $\boldsymbol{r} = k\boldsymbol{r}_1$ is the point of intersection with the radius $\boldsymbol{r}_1$. The chord meets the ellipse (1.1.3) where for some θ,

$$a\cos\theta\mathbf{i} + b\sin\theta\mathbf{j} = k[a\cos\theta_1\mathbf{i} + b\sin\theta_1\mathbf{j}] + \mu[a\cos\theta_2\mathbf{i} + b\sin\theta_2\mathbf{j}].$$

Thus

$$\cos\theta = k\cos\theta_1 + \mu\cos\theta_2,$$
$$\sin\theta = k\sin\theta_1 + \mu\sin\theta_2.$$

Hence, eliminating θ, we obtain the quadratic for μ:

$$\mu^2 + 2k\mu\cos(\theta_1 - \theta_2) + k^2 - 1 = 0.$$

Now, if $\boldsymbol{r}_1$ and $\boldsymbol{r}_2$ are conjugate, then $\cos(\theta_1-\theta_2)=0$ and hence the roots for μ are equal and opposite so that $\boldsymbol{r} = k\boldsymbol{r}_1$ is the mid point of the chord. Conversely, if $\boldsymbol{r} = k\boldsymbol{r}_1$ is the mid point, then the roots for μ must be equal and opposite and hence $\cos(\theta_1-\theta_2)=0$, so that $\theta_1 = \theta_2 \pm \frac{1}{2}\pi$, the condition that $\boldsymbol{r}_1$ and $\boldsymbol{r}_2$ be conjugate radii.

Pairs of conjugate directions may be obtained easily as follows (Figure 1.2). Let BOC be any diameter and let A be any point on the ellipse. Then the chords AB and AC are in conjugate directions.

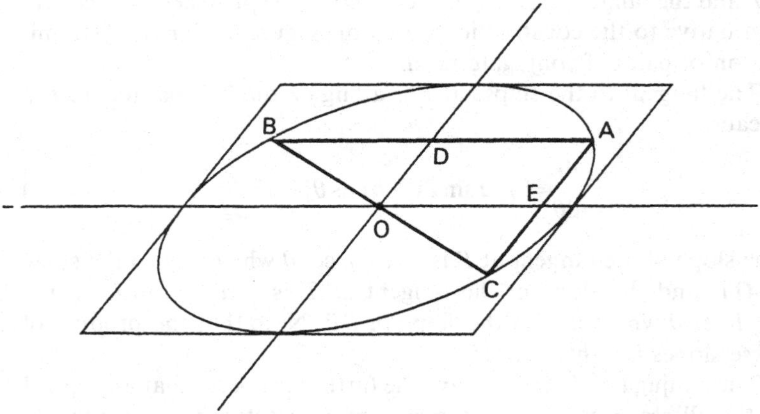

Figure 1.2 *Construction of conjugate directions. BOC is any diameter, A any point on the ellipse. BA and CA are along conjugate directions.*

For if D and E are the mid points of AB and AC respectively, then since O is the mid point of BC, it follows that DO is parallel to AC and bisects BC and AB, and EO is parallel to AB and bisects AC and BC. Thus AB and AC are in conjugate directions.

Exercise 1.1

Prove that if the product of the slopes of two radii is $-b^2a^{-2}$, then these radii are conjugate.

1.2 Equation of an ellipse referred to axes along conjugate diameters

Let axes Ox' and Oy' be taken along a pair of conjugate radii OM and ON respectively. Let $\mathbf{i}'$ and $\mathbf{j}'$ be unit vectors along Ox' and Oy' respectively so that $\boldsymbol{r}$ is given by $\boldsymbol{r} = x'\mathbf{i}' + y'\mathbf{j}'$. Now the equation of the ellipse must have the form

$$a'x'^2 + b'y'^2 + h'x'y' + p'x' + q'y' = 1. \qquad (1.2.1)$$

If a point with coordinates (x', y') is on the ellipse then the points with coordinates $(x', -y')$, $(-x', y')$ and $(-x', -y')$ must also be on it. Using this information in (1.2.1) it follows that $h' = p' = q' = 0$.

Hence the equation of an ellipse referred to (oblique) axes along a pair of conjugate diameters may be written

$$\frac{x'^2}{c^2} + \frac{y'^2}{d^2} = 1, \tag{1.2.2}$$

which has precisely the same form as (1.1.1). Clearly c and d are the lengths of the conjugate radii OM and ON respectively. Thus, an ellipse is uniquely determined by giving a pair of conjugate radii OM and ON: its equation referred to oblique axes along these radii is (1.2.2).

When an ellipse is referred to oblique axes of centre 0 but not along conjugate directions, its equation is of the form (1.2.1) with $p' = q' = 0$ but $h' \neq 0$.

1.3 Equation of an ellipse with given conjugate radii

Here we show how to obtain the vectorial, parametric and cartesian forms of the equation of an ellipse with given conjugate radii OM, ON. Let

$$\boldsymbol{c} = \boldsymbol{OM} = c\mathbf{i}', \quad \boldsymbol{d} = \boldsymbol{ON} = d\mathbf{j}'. \tag{1.3.1}$$

1.3.1 Parametric form

From (1.2.2) a generic point on this ellipse has coordinates

$$x' = c\cos\phi, \quad y' = d\sin\phi, \tag{1.3.2}$$

in terms of a parameter ϕ. Then its position vector is

$$\begin{aligned} \boldsymbol{r} &= x'\mathbf{i}' + y'\mathbf{j}' \\ &= c\cos\phi\,\mathbf{i}' + d\sin\phi\,\mathbf{j}', \end{aligned}$$

or

$$\boldsymbol{r} = \cos\phi\,\boldsymbol{c} + \sin\phi\,\boldsymbol{d}. \tag{1.3.3}$$

In Figure 1.3, $\phi = 0$ at M, $\phi = \frac{1}{2}\pi$ at N. This is the vector parametric form of the equation of the ellipse which has $\boldsymbol{c}$ and $\boldsymbol{d}$ as a pair of conjugate radii. Note that $\mathrm{d}\boldsymbol{r}/\mathrm{d}\phi$, which is along the tangent, is given by

$$\frac{\mathrm{d}\boldsymbol{r}}{\mathrm{d}\phi} = -\sin\phi\,\boldsymbol{c} + \cos\phi\,\boldsymbol{d}, \tag{1.3.4}$$

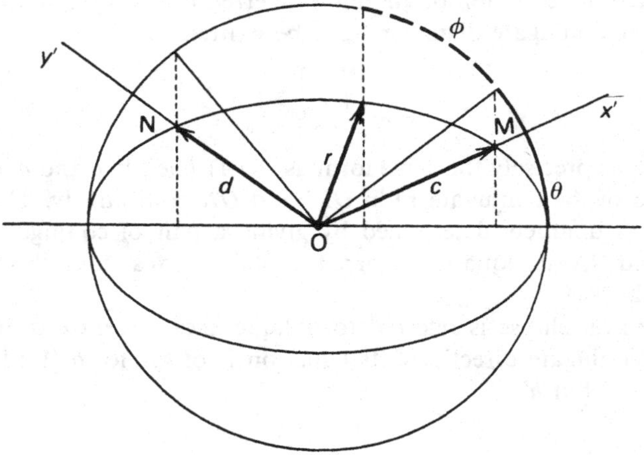

Figure 1.3 *Interpretation of the parameter* ϕ.

so that the tangent at M where $\phi = 0$ is parallel to $\boldsymbol{d}$ and the tangent at N where $\phi = \frac{1}{2}\pi$ is parallel to $-\boldsymbol{c}$. Generally, the tangent at $\boldsymbol{r}(\phi)$ is parallel to the radius $\boldsymbol{r}(\phi + \frac{1}{2}\pi) = \cos(\phi + \frac{1}{2}\pi)\boldsymbol{c} + \sin(\phi + \frac{1}{2}\pi)\boldsymbol{d} = -\sin\phi\boldsymbol{c} + \cos\phi\boldsymbol{d}$. Thus $\boldsymbol{r}(\phi)$ and $\boldsymbol{r}(\phi + \frac{1}{2}\pi)$ form a pair of conjugate radii.

1.3.2 Cartesian form

Let $\boldsymbol{c}^*$ and $\boldsymbol{d}^*$ be the pair of vectors coplanar with $\boldsymbol{c}$ and $\boldsymbol{d}$ and reciprocal to them. By definition, $\boldsymbol{c}^*$ and $\boldsymbol{d}^*$ satisfy

$$\boldsymbol{c}^*\cdot\boldsymbol{c} = \boldsymbol{d}^*\cdot\boldsymbol{d} = 1, \quad \boldsymbol{c}^*\cdot\boldsymbol{d} = \boldsymbol{d}^*\cdot\boldsymbol{c} = 0. \tag{1.3.5}$$

From these it follows that

$$\begin{gathered}\boldsymbol{c}^* = -\kappa(\boldsymbol{c}\times\boldsymbol{d})\times\boldsymbol{d}, \quad \boldsymbol{d}^* = \kappa(\boldsymbol{c}\times\boldsymbol{d})\times\boldsymbol{c},\\ \boldsymbol{c}^*\cdot\boldsymbol{c}^* = \kappa\boldsymbol{d}\cdot\boldsymbol{d}, \quad \boldsymbol{d}^*\cdot\boldsymbol{d}^* = \kappa\boldsymbol{c}\cdot\boldsymbol{c}, \quad \boldsymbol{c}^*\cdot\boldsymbol{d}^* = -\kappa\boldsymbol{c}\cdot\boldsymbol{d},\\ (\boldsymbol{c}^*\cdot\boldsymbol{c}^* - \boldsymbol{d}^*\cdot\boldsymbol{d}^*)(\boldsymbol{c}^*\cdot\boldsymbol{d}^*)^{-1} = (\boldsymbol{c}\cdot\boldsymbol{c} - \boldsymbol{d}\cdot\boldsymbol{d})(\boldsymbol{c}\cdot\boldsymbol{d})^{-1},\\ \kappa^{-1} = (\boldsymbol{c}\times\boldsymbol{d})\cdot(\boldsymbol{c}\times\boldsymbol{d}) = |\boldsymbol{c}\times\boldsymbol{d}|^2.\end{gathered} \tag{1.3.6}$$

Using (1.1.6) note that $\kappa^{-1} = a^2b^2$, where a, b are the principal semi-axes of the ellipse which has $\boldsymbol{c}$ and $\boldsymbol{d}$ as conjugate radii.

Now from (1.3.3),

$$\cos\phi = \boldsymbol{r}\cdot\boldsymbol{c}^*, \quad \sin\phi = \boldsymbol{r}\cdot\boldsymbol{d}^*, \tag{1.3.7}$$

and thus

$$(\boldsymbol{r}\cdot\boldsymbol{c}^*)^2 + (\boldsymbol{r}\cdot\boldsymbol{d}^*)^2 = 1. \tag{1.3.8}$$

This is the cartesian equation of the ellipse. It may be written as

$$\boldsymbol{r}^{\mathrm{T}}\boldsymbol{\alpha}\boldsymbol{r} = 1, \quad x_i\alpha_{ij}x_j = 1, \quad (i, j = 1, 2), \tag{1.3.9}$$

where

$$\boldsymbol{\alpha} = \boldsymbol{c}^* \otimes \boldsymbol{c}^* + \boldsymbol{d}^* \otimes \boldsymbol{d}^*. \tag{1.3.10}$$

Thus if the equation of the ellipse is given in parametric form by (1.3.3), then its cartesian form (1.3.9) may be immediately written down.

We note that

$$\boldsymbol{c}^* \otimes \boldsymbol{c} + \boldsymbol{d}^* \otimes \boldsymbol{d} = \mathbf{1} = \boldsymbol{c} \otimes \boldsymbol{c}^* + \boldsymbol{d} \otimes \boldsymbol{d}^*, \tag{1.3.11}$$

where $\mathbf{1}$ is the 2×2 unit matrix. For, if $\boldsymbol{w}$ is any vector in the plane, so that $\boldsymbol{w} = \sigma\boldsymbol{c}^* + \varepsilon\boldsymbol{d}^*$, for some σ and ε, then $(\boldsymbol{c}^* \otimes \boldsymbol{c} + \boldsymbol{d}^* \otimes \boldsymbol{d})\boldsymbol{w} = \sigma\boldsymbol{c}^*(\boldsymbol{c}\cdot\boldsymbol{c}^*) + \varepsilon\boldsymbol{d}^*(\boldsymbol{d}\cdot\boldsymbol{d}^*) = \boldsymbol{w}$. The result follows because this is valid for all $\boldsymbol{w}$.

Example 1.1

Determine the parametric and cartesian equations of the ellipse which has $\boldsymbol{c} = 2\mathbf{i} + 3\mathbf{j}$, $\boldsymbol{d} = -6\mathbf{i} + 5\mathbf{j}$ as conjugate radii.

From (1.3.3) we immediately obtain the parametric equations

$$x = 2\cos\phi - 6\sin\phi, \quad y = 3\cos\phi + 5\sin\phi.$$

Using (1.3.6) and (1.3.10) we obtain

$$28\boldsymbol{c}^* = 5\mathbf{i} + 6\mathbf{j}, \quad 28\boldsymbol{d}^* = -3\mathbf{i} + 2\mathbf{j},$$
$$(28)^2\boldsymbol{\alpha} = 34\mathbf{i}\otimes\mathbf{i} + 40\mathbf{j}\otimes\mathbf{j} + 24(\mathbf{i}\otimes\mathbf{j} + \mathbf{j}\otimes\mathbf{i}),$$

so that the cartesian equation of the ellipse is

$$(5x + 6y)^2 + (-3x + 2y)^2 = (28)^2,$$

or equivalently

$$34x^2 + 40y^2 + 48xy = (28)^2.$$

Example 1.2 The harmonic oscillator

Consider the ordinary differential equation

$$\frac{d^2\boldsymbol{r}}{dt^2} = -\omega^2\boldsymbol{r}, \tag{1.3.12}$$

where $\boldsymbol{r}$ is the position vector, t is time and ω is a constant. This equation describes the motion of a particle of mass m subject to a restoring force $-k\boldsymbol{r}$, with $\omega^2 = km^{-1}$. If the initial position and velocity of the particle are

$$\boldsymbol{r}(0) = \boldsymbol{c},$$

$$\frac{d\boldsymbol{r}}{dt}(0) = \omega\boldsymbol{d}, \tag{1.3.13}$$

then the solution of (1.3.12), subject to (1.3.13), is

$$\boldsymbol{r} = \boldsymbol{c}\cos\omega t + \boldsymbol{d}\sin\omega t. \tag{1.3.14}$$

This is the parametric equation of an ellipse in the plane of $\boldsymbol{c}$ and $\boldsymbol{d}$, with $\boldsymbol{c}$ and $\boldsymbol{d}$ as a pair of conjugate radii. The period of the oscillation is $2\pi\omega^{-1}$.

If $\boldsymbol{c}\cdot\boldsymbol{d} = 0$, then the ellipse has principal axes along $\boldsymbol{c}$ and $\boldsymbol{d}$.

If $|\boldsymbol{c}| = |\boldsymbol{d}|$ and also $\boldsymbol{c}\cdot\boldsymbol{d} = 0$, then (1.3.14) describes a circle of radius $|\boldsymbol{c}|$.

If the initial velocity is along $\boldsymbol{c}$, so that $\boldsymbol{d} = \boldsymbol{c}$, then $\boldsymbol{r}$ is given by

$$\begin{aligned}\boldsymbol{r} &= \boldsymbol{c}\cos\omega t + \boldsymbol{c}\sin\omega t \\ &= \sqrt{2}\boldsymbol{c}\sin(\omega t + \tfrac{1}{4}\pi).\end{aligned}$$

Then the particle continues to move in simple harmonic motion along $\boldsymbol{c}$ with amplitude $2^{1/2}|\boldsymbol{c}|$.

1.4 Conjugate directions

Now we derive, in terms of the matrix $\boldsymbol{\alpha}$, a necessary and sufficient condition that two vectors $\boldsymbol{u}$ and $\boldsymbol{v}$ be along conjugate directions of the ellipse (1.3.3).

Assume that $\boldsymbol{u}$ and $\boldsymbol{v}$ are along conjugate directions of the ellipse (1.3.3) for some angle ϕ. For some scalar β and value ϕ, we must have

$$\beta\boldsymbol{u} = \boldsymbol{c}\cos\phi + \boldsymbol{d}\sin\phi. \tag{1.4.1}$$

Also since $\boldsymbol{v}$ is along a radius conjugate to $\beta\boldsymbol{u}$, we must have, for some scalar γ,

$$\gamma\boldsymbol{v} = -\boldsymbol{c}\sin\phi + \boldsymbol{d}\cos\phi. \tag{1.4.2}$$

Thus

$$\begin{aligned}\cos\phi &= \beta(\boldsymbol{u}\cdot\boldsymbol{c}^*) = \gamma(\boldsymbol{v}\cdot\boldsymbol{d}^*),\\ \sin\phi &= \beta(\boldsymbol{u}\cdot\boldsymbol{d}^*) = -\gamma(\boldsymbol{v}\cdot\boldsymbol{c}^*),\end{aligned} \tag{1.4.3}$$

and hence

$$(\boldsymbol{u}\cdot\boldsymbol{c}^*)(\boldsymbol{v}\cdot\boldsymbol{c}^*) + (\boldsymbol{u}\cdot\boldsymbol{d}^*)(\boldsymbol{v}\cdot\boldsymbol{d}^*) = 0, \tag{1.4.4}$$

or equivalently

$$\boldsymbol{v}^{\mathrm{T}}\boldsymbol{\alpha}\boldsymbol{u} = 0. \tag{1.4.5}$$

This condition is necessary in order that $\boldsymbol{u}$ and $\boldsymbol{v}$ are along conjugate directions. It is also sufficient. Indeed if $\boldsymbol{u}$ and $\boldsymbol{v}$ are such that (1.4.5) is valid, then taking oblique axes Ox', Oy' with origin O along these vectors, the equation (1.3.9) takes the form (1.2.2). This expresses the fact that Ox' and Oy', and thus $\boldsymbol{u}$ and $\boldsymbol{v}$, are along conjugate directions of the ellipse.

Hence, (1.4.5) is a necessary and sufficient condition that $\boldsymbol{u}$ and $\boldsymbol{v}$ be along conjugate radii of the ellipse with parametric equation (1.3.3) and cartesian equation (1.3.9).

1.5 Principal axes

The values of ϕ corresponding to the principal axes of the ellipse

$$\boldsymbol{r} = \boldsymbol{c}\cos\phi + \boldsymbol{d}\sin\phi, \tag{1.5.1}$$

are now determined. Since $|\boldsymbol{r}|$ has its largest and least values along the principal axes, it follows that for these $\mathrm{d}(\boldsymbol{r}\cdot\boldsymbol{r})/\mathrm{d}\phi = 0$. Now

$$\begin{aligned}\boldsymbol{r}\cdot\boldsymbol{r} &= \boldsymbol{c}\cdot\boldsymbol{c}\cos^2\phi + \boldsymbol{d}\cdot\boldsymbol{d}\sin^2\phi + \boldsymbol{c}\cdot\boldsymbol{d}\sin 2\phi,\\ \frac{\mathrm{d}(\boldsymbol{r}\cdot\boldsymbol{r})}{\mathrm{d}\phi} &= (-\boldsymbol{c}\cdot\boldsymbol{c} + \boldsymbol{d}\cdot\boldsymbol{d})\sin 2\phi + 2\boldsymbol{c}\cdot\boldsymbol{d}\cos 2\phi.\end{aligned} \tag{1.5.2}$$

Thus, the values of $\phi = \phi^*$ (say), corresponding to the principal axes are given by

$$\tan 2\phi^* = \frac{2\boldsymbol{c}\cdot\boldsymbol{d}}{\boldsymbol{c}\cdot\boldsymbol{c} - \boldsymbol{d}\cdot\boldsymbol{d}}. \tag{1.5.3}$$

Let the conjugate radii corresponding to $\hat{\phi}$ and $\hat{\phi} + \frac{1}{2}\pi$ be equal in length. These are called the 'equi-conjugate' radii. Then

$$|\boldsymbol{r}(\hat{\phi})| = |\boldsymbol{r}(\hat{\phi} + \tfrac{1}{2}\pi)|, \tag{1.5.4}$$

or

$$|\boldsymbol{c}\cos\hat{\phi} + \boldsymbol{d}\sin\hat{\phi}| = |-\boldsymbol{c}\sin\hat{\phi} + \boldsymbol{d}\cos\hat{\phi}|. \tag{1.5.5}$$

Thus

$$\tan 2\hat{\phi} = -\frac{\boldsymbol{c}\cdot\boldsymbol{c} - \boldsymbol{d}\cdot\boldsymbol{d}}{2\boldsymbol{c}\cdot\boldsymbol{d}}, \tag{1.5.6}$$

and hence from (1.5.3),

$$\tan 2\phi^* \tan 2\hat{\phi} = -1. \tag{1.5.7}$$

Hence the values ϕ^* corresponding to the principal axes and the values $\hat{\phi}$ corresponding to equi-conjugate radii are related through

$$\hat{\phi} = \phi^* \pm \frac{\pi}{4}. \tag{1.5.8}$$

Exercises 1.2

1. (a) An ellipse has major and minor semi-axes a and b respectively. Determine the lengths of the equi-conjugate radii and the angle between them.
   (b) Determine the equation of the ellipse referred to axes along the equi-conjugate radii.
2. Using the auxiliary circle, interpret the angle ϕ in (1.3.3).
3. Show how to determine the point with position vector (1.3.3) of an ellipse with given conjugate radii $\boldsymbol{c}$ and $\boldsymbol{d}$.
4. Show that the sectional area of the ellipse delimited by the radii corresponding to ϕ_1 and ϕ_2 is $\frac{1}{2}|(\phi_1 - \phi_2)|ab$, where a and b are the principal semi-axes.
5. An ellipse has conjugate radii $\boldsymbol{c} = 2\mathbf{i} - 3\mathbf{j}$ and $\boldsymbol{d} = 4\mathbf{i} + \mathbf{j}$. Determine both its parametric and cartesian equations.
6. Prove

$$|\boldsymbol{r}(\phi)|^2 + |\boldsymbol{r}(\phi + \tfrac{1}{2}\pi)|^2 = \boldsymbol{c}\cdot\boldsymbol{c} + \boldsymbol{d}\cdot\boldsymbol{d},$$
$$|\boldsymbol{r}(\phi) \times \boldsymbol{r}(\phi + \tfrac{1}{2}\pi)| = |\boldsymbol{c} \times \boldsymbol{d}|.$$

1.6 Reciprocal ellipses

Here we consider the ellipse

$$\boldsymbol{r} = \boldsymbol{c}\cos\phi + \boldsymbol{d}\sin\phi, \qquad (1.6.1)$$

which has conjugate radii $(\boldsymbol{c}, \boldsymbol{d})$, and the 'reciprocal ellipse'

$$\boldsymbol{r}^* = \boldsymbol{c}^*\cos\phi + \boldsymbol{d}^*\sin\phi, \qquad (1.6.2)$$

which has conjugate radii $(\boldsymbol{c}^*, \boldsymbol{d}^*)$ reciprocal to the pair $(\boldsymbol{c}, \boldsymbol{d})$. For each pair of conjugate radii of (1.6.1) there is a corresponding reciprocal pair of vectors which are conjugate radii of (1.6.2), and conversely. This may be seen as follows.

For any ϕ, the pair

$$\boldsymbol{r}(\phi) = \boldsymbol{c}\cos\phi + \boldsymbol{d}\sin\phi, \quad \boldsymbol{r}(\phi + \tfrac{1}{2}\pi) = -\boldsymbol{c}\sin\phi + \boldsymbol{d}\cos\phi,$$

are conjugate radii of (1.6.1). Similarly, for any ϕ, the pair of vectors

$$\boldsymbol{r}^*(\phi) = \boldsymbol{c}^*\cos\phi + \boldsymbol{d}^*\sin\phi, \quad \boldsymbol{r}^*(\phi + \tfrac{1}{2}\pi) = -\boldsymbol{c}^*\sin\phi + \boldsymbol{d}^*\cos\phi,$$

are conjugate radii of (1.6.2). Note that

$$\begin{aligned} \boldsymbol{r}(\phi)\cdot\boldsymbol{r}^*(\phi) = 1, \quad \boldsymbol{r}(\phi)\cdot\boldsymbol{r}^*(\phi + \tfrac{1}{2}\pi) = 0, \\ \boldsymbol{r}(\phi + \tfrac{1}{2}\pi)\cdot\boldsymbol{r}^*(\phi) = 0, \quad \boldsymbol{r}(\phi + \tfrac{1}{2}\pi)\cdot\boldsymbol{r}^*(\phi + \tfrac{1}{2}\pi) = 1, \end{aligned} \qquad (1.6.3)$$

which expresses the fact that the pairs $(\boldsymbol{r}(\phi), \boldsymbol{r}(\phi + \tfrac{1}{2}\pi))$ and $(\boldsymbol{r}^*(\phi), \boldsymbol{r}^*(\phi + \tfrac{1}{2}\pi))$ are reciprocal.

Now by (1.3.6), $\boldsymbol{c}^*\cdot\boldsymbol{d}^* = -\kappa(\boldsymbol{c}\cdot\boldsymbol{d})$ where κ is constant for the ellipse (1.6.1). It follows that if $\boldsymbol{c}$ is perpendicular to $\boldsymbol{d}$, so also is $\boldsymbol{c}^*$ perpendicular to $\boldsymbol{d}^*$. Thus the principal axes of (1.6.1) are also principal axes of (1.6.2). Also, by (1.3.6), it follows that if $\boldsymbol{c}\cdot\boldsymbol{c} = \boldsymbol{d}\cdot\boldsymbol{d}$, then $\boldsymbol{c}^*\cdot\boldsymbol{c}^* = \boldsymbol{d}^*\cdot\boldsymbol{d}^*$. Thus the equi-conjugate radii of (1.6.2) are reciprocal to the equi-conjugate radii of (1.6.1).

Now it is shown that the reciprocal ellipse (1.6.2) is the polar reciprocal of the ellipse (1.6.1) with respect to the unit circle.

The polar of the point, with position vector $\boldsymbol{c}\cos\phi + \boldsymbol{d}\sin\phi$ on (1.6.1), with respect to the unit circle, $\boldsymbol{r}^{\mathrm{T}}\boldsymbol{r} = 1$, is

$$\boldsymbol{r}^{\mathrm{T}}\boldsymbol{\alpha}(\boldsymbol{c}\cos\phi + \boldsymbol{d}\sin\phi) = 1.$$

The desired polar reciprocal is the envelope of these lines. This is given by

$$\boldsymbol{r}^{\mathrm{T}}\boldsymbol{\alpha}(\boldsymbol{c}\cos\phi + \boldsymbol{d}\sin\phi) = 1, \quad \boldsymbol{r}^{\mathrm{T}}\boldsymbol{\alpha}(-\boldsymbol{c}\sin\phi + \boldsymbol{d}\cos\phi) = 0. \qquad (1.6.4)$$

The solution of these equations for $\boldsymbol{r}$ is

$$\boldsymbol{r} = \boldsymbol{c}^* \cos\phi + \boldsymbol{d}^* \sin\phi, \tag{1.6.5}$$

the reciprocal ellipse.

Finally, we note that the cartesian forms of (1.6.1) and (1.6.2) are

$$\boldsymbol{r}^{\mathrm{T}}\boldsymbol{\alpha}\boldsymbol{r} = 1, \quad \boldsymbol{r}^{\mathrm{T}}\boldsymbol{\alpha}^*\boldsymbol{r} = 1, \tag{1.6.6}$$

where

$$\boldsymbol{\alpha} = \boldsymbol{c}^* \otimes \boldsymbol{c}^* + \boldsymbol{d}^* \otimes \boldsymbol{d}^*, \quad \boldsymbol{\alpha}^* = \boldsymbol{c} \otimes \boldsymbol{c} + \boldsymbol{d} \otimes \boldsymbol{d}. \tag{1.6.7}$$

Here

$$\boldsymbol{\alpha}\boldsymbol{\alpha}^* = \boldsymbol{c}^* \otimes \boldsymbol{c} + \boldsymbol{d}^* \otimes \boldsymbol{d} = \mathbf{1}, \tag{1.6.8}$$

so that $\boldsymbol{\alpha}^* = \boldsymbol{\alpha}^{-1}$ the matrix inverse to $\boldsymbol{\alpha}$.

Exercise 1.3

Express the two basic invariants $\boldsymbol{c}\cdot\boldsymbol{c} + \boldsymbol{d}\cdot\boldsymbol{d}$ and $|\boldsymbol{c} \times \boldsymbol{d}|$ in terms of the matrix $\boldsymbol{\alpha}^{-1}$.

1.7 Common conjugate directions of a pair of ellipses

It is now shown that any two concentric coplanar ellipses have a pair of conjugate directions in common. Thus, there is a pair of conjugate directions of the first ellipse which are also conjugate for the second ellipse. In general there is just one such pair of common conjugate directions.

Let the equations of the ellipses be

$$\boldsymbol{r}^{\mathrm{T}}\boldsymbol{\alpha}\boldsymbol{r} = 1, \quad x_i\alpha_{ij}x_j = 1, \tag{1.7.1}$$

and

$$\boldsymbol{r}^{\mathrm{T}}\boldsymbol{\beta}\boldsymbol{r} = 1, \quad x_i\beta_{ij}x_j = 1, \tag{1.7.2}$$

where $\boldsymbol{\alpha}$ and $\boldsymbol{\beta}$ are positive definite symmetric (2×2) matrices.

Let $\hat{\lambda}$ be a root of the equation

$$\det(\boldsymbol{\alpha} - \lambda\boldsymbol{\beta}) = 0, \tag{1.7.3}$$

and let $\boldsymbol{u}$ be the corresponding eigenvector of $\boldsymbol{\alpha}$ with respect to $\boldsymbol{\beta}$. Then

$$\boldsymbol{\alpha}\boldsymbol{u} = \hat{\lambda}\boldsymbol{\beta}\boldsymbol{u}. \tag{1.7.4}$$

If $\boldsymbol{v}$ is along a direction conjugate to $\boldsymbol{u}$ for the ellipse (1.7.1), then

$$\boldsymbol{v}^{\mathrm{T}}\boldsymbol{\alpha}\boldsymbol{u} = 0, \tag{1.7.5}$$

and then from (1.7.4)

$$\boldsymbol{v}^{\mathrm{T}}\boldsymbol{\beta}\boldsymbol{u} = 0, \tag{1.7.6}$$

so that $\boldsymbol{v}$ and $\boldsymbol{u}$ are also along conjugate directions for the ellipse (1.7.2).

We note by (1.7.5) and (1.7.6) that the vectors $\boldsymbol{\alpha v}$ and $\boldsymbol{\beta v}$, both in the plane of the ellipse, are orthogonal to $\boldsymbol{u}$. Thus $\boldsymbol{\alpha v}$ is parallel to $\boldsymbol{\beta v}$, and for some λ^*

$$\boldsymbol{\alpha v} = \lambda^* \boldsymbol{\beta v}. \tag{1.7.7}$$

Thus, if the equation (1.7.3) does not have a double root, then $\boldsymbol{u}$ and $\boldsymbol{v}$ must be taken as the eigenvectors of $\boldsymbol{\alpha}$ with respect to $\boldsymbol{\beta}$, corresponding to the two different roots $\hat{\lambda}$, λ^* of (1.7.3) respectively. For this case which is the general case, there is thus one and only one pair of common conjugate directions for the ellipses (1.7.1) and (1.7.2).

If the equation (1.7.3) does have a double root $\tilde{\lambda}$ (say), then (1.7.4) holds for any vector $\boldsymbol{u}$ in the plane of the ellipses and hence $\boldsymbol{\alpha} = \tilde{\lambda}\boldsymbol{\beta}$. Thus any pair of conjugate directions of the first ellipse is also a pair of conjugate directions of the second.

1.7.1 Cartesian form of condition for common conjugate directions

Assuming that (1.7.3) has two different roots $\hat{\lambda}$ and λ^*, it is now shown that the common conjugate directions of the ellipses (1.7.1) and (1.7.2) have cartesian equation

$$\boldsymbol{\alpha x} \times \boldsymbol{\beta x} = \boldsymbol{0}. \tag{1.7.8}$$

This is because $\boldsymbol{x} = \varepsilon \boldsymbol{u}$ (ε arbitrary) satisfies (1.7.8) since

$$(\boldsymbol{\alpha}\varepsilon\boldsymbol{u}) \times (\boldsymbol{\beta}\varepsilon\boldsymbol{u}) = \varepsilon^2 \hat{\lambda}\boldsymbol{\beta u} \times \boldsymbol{\beta u} = \boldsymbol{0},$$

and $\boldsymbol{x} = \eta\boldsymbol{v}$ (η arbitrary) also satisfies (1.7.8) similarly. Further, let us check that no other vector $\boldsymbol{x}$ satisfies (1.7.8). If

$$\boldsymbol{x} = \varepsilon\boldsymbol{u} + \eta\boldsymbol{v}, \quad (\varepsilon, \eta \text{ arbitrary}),$$

so that $\boldsymbol{x}$ is any vector in the plane of $\boldsymbol{u}$ and $\boldsymbol{v}$, then

$$\begin{aligned}\boldsymbol{\alpha x} \times \boldsymbol{\beta x} &= (\varepsilon\hat{\lambda}\boldsymbol{\beta u} + \eta\lambda^*\boldsymbol{\beta v}) \times (\varepsilon\boldsymbol{\beta u} + \eta\boldsymbol{\beta v}) \\ &= \varepsilon\eta(\hat{\lambda} - \lambda^*)\boldsymbol{\beta u} \times \boldsymbol{\beta v}.\end{aligned} \tag{1.7.9}$$

This is not zero if $\hat{\lambda} \neq \lambda^*$ and $\varepsilon, \eta \neq 0$, unless $\boldsymbol{\beta u}$ and $\boldsymbol{\beta v}$ are parallel. Then $\boldsymbol{u}$ and $\boldsymbol{v}$ are parallel, contrary to hypothesis. Indeed, from (1.7.5), $\boldsymbol{u}$ and $\boldsymbol{v}$ may not be parallel since $\boldsymbol{\alpha}$ is positive definite. Thus, $\boldsymbol{x} = \varepsilon \boldsymbol{u}$ and $\boldsymbol{x} = \eta \boldsymbol{v}$ are the only two solutions of (1.7.8) provided $\hat{\lambda} \neq \lambda^*$.

Suppose that the ellipses have cartesian equations

$$ax^2 + 2hxy + by^2 = 1,$$
$$a'x^2 + 2h'xy + b'y^2 = 1. \qquad (1.7.10)$$

Then

$$\boldsymbol{\alpha} = \begin{bmatrix} a & h \\ h & b \end{bmatrix}, \quad \boldsymbol{\beta} = \begin{bmatrix} a' & h' \\ h' & b' \end{bmatrix}, \qquad (1.7.11)$$

and (1.7.8) gives

$$(ax + hy)(h'x + b'y) - (hx + by)(a'x + h'y) = 0,$$

or equivalently

$$(ah' - a'h)x^2 + (ab' - a'b)xy + (hb' - h'b)y^2 = 0. \qquad (1.7.12)$$

Exercise 1.4

Prove that the principal axes of the ellipse $\boldsymbol{x}^{\mathrm{T}}\boldsymbol{\alpha x} = 1$ have cartesian equation

$$\boldsymbol{x} \times \boldsymbol{\alpha x} = \boldsymbol{0}.$$

1.8 Similar and similarly situated ellipses

Two coplanar concentric ellipses are said to be 'similar' if they have the same aspect ratio, that is the ratio of the major semi-axis to the minor semi-axis. The two ellipses are said to be 'similarly situated' if the major axis of one is parallel to the major axis of the other.

The ellipses (1.7.1) and (1.7.2) are similar and similarly situated if the matrices $\boldsymbol{\alpha}$ and $\boldsymbol{\beta}$ are such that $\boldsymbol{\alpha} = \tilde{\lambda}\boldsymbol{\beta}$ for some $\tilde{\lambda}$ (which must be positive because $\boldsymbol{\alpha}$ and $\boldsymbol{\beta}$ are positive definite). Let $x_i = \tilde{\lambda}^{1/2} x_i'$ in equation (1.7.2) which then becomes $x_i'\alpha_{ij}x_j' = 1$, identical in form with (1.7.1).

Suppose now that the ellipses (1.7.1) and (1.7.2) have common conjugate directions along the unit vectors $\hat{\boldsymbol{u}}$ and $\hat{\boldsymbol{v}}$. If the ratio of the lengths of the radii along $\hat{\boldsymbol{u}}$ and $\hat{\boldsymbol{v}}$ for the ellipse (1.7.1) is equal to the ratio of the lengths of the radii along $\hat{\boldsymbol{u}}$ and $\hat{\boldsymbol{v}}$ for the ellipse

(1.7.2), then (1.7.1) and (1.7.2) are similar and similarly situated ellipses.

Let the lengths of the radii along $\hat{\boldsymbol{u}}$ and $\hat{\boldsymbol{v}}$ for (1.7.1) be c and d respectively and for (1.7.2) be p and q respectively. Then $cd^{-1} = pq^{-1}$. Points on (1.7.1) are given by

$$\boldsymbol{r}_1 = c\hat{\boldsymbol{u}} \cos\phi + d\hat{\boldsymbol{v}} \sin\phi, \quad 0 \leqslant \phi \leqslant 2\pi, \tag{1.8.1}$$

and points on (1.7.2) are given by

$$\begin{aligned} \boldsymbol{r}_2 &= p\hat{\boldsymbol{u}} \cos\phi + q\hat{\boldsymbol{v}} \sin\phi \\ &= \frac{q}{d}\boldsymbol{r}_1. \end{aligned} \tag{1.8.2}$$

Hence the radius vector to the first ellipse is a constant scalar multiple of the radius vector to the second. Thus the two ellipses are similar and similarly situated.

If $\boldsymbol{u}^*$ and $\boldsymbol{v}^*$ is the pair reciprocal to $\hat{\boldsymbol{u}}$ and $\hat{\boldsymbol{v}}$, then $\boldsymbol{u}^* \cdot \hat{\boldsymbol{u}} = 1$ etc., and it follows that

$$\boldsymbol{\alpha} = \frac{1}{c^2}\boldsymbol{u}^* \otimes \boldsymbol{u}^* + \frac{1}{d^2}\boldsymbol{v}^* \otimes \boldsymbol{v}^*,$$

$$\boldsymbol{\beta} = \frac{1}{p^2}\boldsymbol{u}^* \otimes \boldsymbol{u}^* + \frac{1}{q^2}\boldsymbol{v}^* \otimes \boldsymbol{v}^*, \tag{1.8.3}$$

and hence

$$\boldsymbol{\alpha} = \frac{q^2}{d^2}\boldsymbol{\beta}.$$

In this case, (1.7.3) has the double root $\bar{\lambda} = q^2 d^{-2}$. All pairs of conjugate directions for the first ellipse are also pairs of conjugate directions for the second ellipse.

2

Bivectors

Bivectors are introduced and their link with ellipses is established. Most of the properties of bivectors which are needed in subsequent chapters are derived here.

2.1 Definitions and operations

If $\boldsymbol{c}$ and $\boldsymbol{d}$ are any vectors of a real three-dimensional euclidean vector space, then $\boldsymbol{A}$ given by

$$\boldsymbol{A} = \boldsymbol{c} + \mathrm{i}\boldsymbol{d}, \tag{2.1.1}$$

is a bivector where $\mathrm{i} = (-1)^{1/2}$. Two bivectors $\boldsymbol{A} = \boldsymbol{c} + \mathrm{i}\boldsymbol{d}$ and $\boldsymbol{A}' = \boldsymbol{c}' + \mathrm{i}\boldsymbol{d}'$ are equal if and only if $\boldsymbol{c} = \boldsymbol{c}'$ and $\boldsymbol{d} = \boldsymbol{d}'$.

In this chapter, upper case italic bold face letters $\boldsymbol{A}, \boldsymbol{B}, \ldots$ represent bivectors whilst lower case italic bold face letters $\boldsymbol{a}, \boldsymbol{b}, \ldots$ represent real vectors. We use the superscripts $+$ and $-$ to denote the real and imaginary parts of a complex quantity whether it is a scalar, vector or tensor. Thus, for example,

$$\boldsymbol{A} = \boldsymbol{A}^+ + \mathrm{i}\boldsymbol{A}^-, \tag{2.1.2}$$

where $\boldsymbol{A}^+ = \boldsymbol{c}, \boldsymbol{A}^- = \boldsymbol{d}$. An overbar over a quantity denotes its complex conjugate. Thus $\bar{\boldsymbol{A}}$ is the complex conjugate of $\boldsymbol{A}$ and is given by

$$\bar{\boldsymbol{A}} = \boldsymbol{A}^+ - \mathrm{i}\boldsymbol{A}^-. \tag{2.1.3}$$

We define the addition of two bivectors $\boldsymbol{A}$ and $\boldsymbol{B}$ by

$$\begin{aligned}\boldsymbol{A} + \boldsymbol{B} &= (\boldsymbol{A}^+ + \mathrm{i}\boldsymbol{A}^-) + (\boldsymbol{B}^+ + \mathrm{i}\boldsymbol{B}^-)\\ &= (\boldsymbol{A}^+ + \boldsymbol{B}^+) + \mathrm{i}(\boldsymbol{A}^- + \boldsymbol{B}^-).\end{aligned} \tag{2.1.4}$$

The multiplication of a bivector $\boldsymbol{A}$ by a scalar λ is given by

$$\begin{aligned}\lambda\boldsymbol{A} &= (\lambda^+ + \mathrm{i}\lambda^-)(\boldsymbol{A}^+ + \mathrm{i}\boldsymbol{A}^-)\\ &= \lambda^+\boldsymbol{A}^+ - \lambda^-\boldsymbol{A}^- + \mathrm{i}(\lambda^-\boldsymbol{A}^+ + \lambda^+\boldsymbol{A}^-).\end{aligned} \tag{2.1.5}$$

With these operations the set of all bivectors is a complex three-dimensional vector space.

The dot product (or scalar product) of two bivectors $\boldsymbol{A}$ and $\boldsymbol{B}$ is denoted by $\boldsymbol{A}\cdot\boldsymbol{B}$ and given by

$$\boldsymbol{A}\cdot\boldsymbol{B} = \boldsymbol{A}^+\cdot\boldsymbol{B}^+ - \boldsymbol{A}^-\cdot\boldsymbol{B}^- + \mathrm{i}(\boldsymbol{A}^+\cdot\boldsymbol{B}^- + \boldsymbol{A}^-\cdot\boldsymbol{B}^+). \qquad (2.1.6)$$

The cross product of two bivectors $\boldsymbol{A}$ and $\boldsymbol{B}$ is denoted by $\boldsymbol{A}\times\boldsymbol{B}$ and given by

$$\boldsymbol{A}\times\boldsymbol{B} = \boldsymbol{A}^+\times\boldsymbol{B}^+ - \boldsymbol{A}^-\times\boldsymbol{B}^- + \mathrm{i}(\boldsymbol{A}^+\times\boldsymbol{B}^- + \boldsymbol{A}^-\times\boldsymbol{B}^+). \qquad (2.1.7)$$

The dot product is not a scalar product of the Hilbert type, because as Synge (1964) pointed out, 'the complex conjugate plays no part at all in it'. The dot product is commutative and linear in each factor. The cross product is anti-commutative and linear in each factor. We note that the dot product of a bivector $\boldsymbol{A}$ with itself may be zero without $\boldsymbol{A}$ being zero because

$$\boldsymbol{A}\cdot\boldsymbol{A} = \boldsymbol{A}^+\cdot\boldsymbol{A}^+ - \boldsymbol{A}^-\cdot\boldsymbol{A}^- + 2\mathrm{i}\boldsymbol{A}^+\cdot\boldsymbol{A}^-. \qquad (2.1.8)$$

Then $\boldsymbol{A}\cdot\boldsymbol{A} = 0$ when $\boldsymbol{A}^+$ and $\boldsymbol{A}^-$ are orthogonal and of equal magnitude. In this case $\boldsymbol{A}$ is said to be **isotropic**.

The cross product of a bivector $\boldsymbol{A}$ with its complex conjugate $\bar{\boldsymbol{A}}$ is given by

$$\boldsymbol{A}\times\bar{\boldsymbol{A}} = -2\mathrm{i}\boldsymbol{A}^+\times\boldsymbol{A}^-. \qquad (2.1.9)$$

Thus $\boldsymbol{A}\times\bar{\boldsymbol{A}} = \boldsymbol{0}$ when $\boldsymbol{A}^+$ and $\boldsymbol{A}^-$ are parallel. In this case we say that 'the bivector $\boldsymbol{A}$ has real direction'. If $\hat{\mathbf{n}}$ is a unit vector in the common direction of $\boldsymbol{A}^+$ and $\boldsymbol{A}^-$, then $\boldsymbol{A}^+ = \lambda^+\hat{\mathbf{n}}$, $\boldsymbol{A}^- = \lambda^-\hat{\mathbf{n}}$ for some scalars λ^+ and λ^-. Thus a bivector $\boldsymbol{A}$ with a real direction may be written

$$\boldsymbol{A} = \lambda\hat{\mathbf{n}}, \qquad (2.1.10)$$

where $\lambda = \lambda^+ + \mathrm{i}\lambda^-$ is some complex scalar. If $\boldsymbol{A}$ does **not** have a real direction, then the bivector $\boldsymbol{A}\times\bar{\boldsymbol{A}}$ does have a real direction which is along the normal to the plane spanned by $\boldsymbol{A}^+$ and $\boldsymbol{A}^-$.

An orthonormal triad $\mathbf{i}, \mathbf{j}, \mathbf{k}$ of real unit vectors is of course a basis for the bivectors. A bivector $\boldsymbol{A} = \boldsymbol{c} + \mathrm{i}\boldsymbol{d}$ may be written in this basis as

$$\begin{aligned}\boldsymbol{A} &= A_1\mathbf{i} + A_2\mathbf{j} + A_3\mathbf{k}\\ &= (c_1 + \mathrm{i}d_1)\mathbf{i} + (c_2 + \mathrm{i}d_2)\mathbf{j} + (c_3 + \mathrm{i}d_3)\mathbf{k},\end{aligned} \qquad (2.1.11)$$

where (c_1, c_2, c_3) and (d_1, d_2, d_3) are the components of the vectors

$\boldsymbol{c}$ and $\boldsymbol{d}$ in this basis. Expanding the dot product (2.1.6) and the cross product (2.1.7) in components in the basis **i**, **j**, **k**, we have, as for real vectors,

$$\boldsymbol{A}\cdot\boldsymbol{B} = A_1B_1 + A_2B_2 + A_3B_3, \tag{2.1.12}$$

$$\boldsymbol{A}\times\boldsymbol{B} = (A_2B_3 - A_3B_2)\mathbf{i} + (A_3B_1 - A_1B_3)\mathbf{j} + (A_1B_2 - A_2B_1)\mathbf{k}. \tag{2.1.13}$$

Now we introduce the ellipse associated with the bivector $\boldsymbol{A}$.

2.2 The directional ellipse

Definition The (directional) ellipse of the bivector $\boldsymbol{A} = \boldsymbol{c} + \mathrm{i}\boldsymbol{d}$ is the ellipse centred at the common origin of $\boldsymbol{c}$ and $\boldsymbol{d}$, and having $\boldsymbol{c}$ and $\boldsymbol{d}$ as conjugate radii. It is oriented from $\boldsymbol{A}^+ (= \boldsymbol{c})$ to $\boldsymbol{A}^- (= \boldsymbol{d})$ (Figure 2.1).

This definition is due to Gibbs. The vector form of the equation of the directional ellipse of $\boldsymbol{A} = \boldsymbol{c} + \mathrm{i}\boldsymbol{d}$ is (1.3.3):

$$\boldsymbol{r} = \boldsymbol{c}\cos\phi + \boldsymbol{d}\sin\phi = \{\boldsymbol{A}\mathrm{e}^{-\mathrm{i}\phi}\}^+. \tag{2.2.1}$$

The orientation is characterized by the sense of increasing ϕ. Note

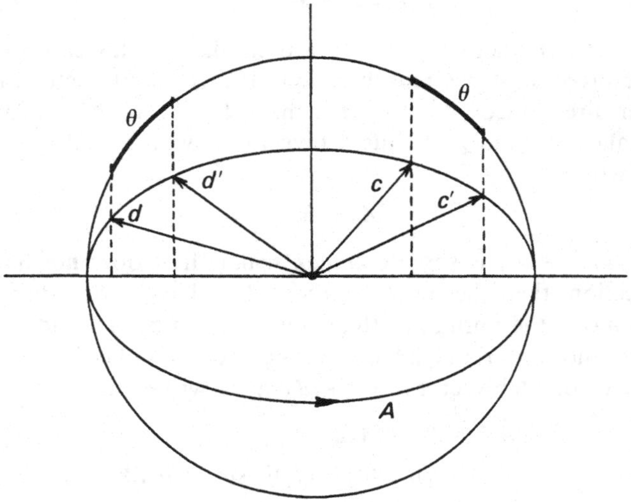

Figure 2.1 *The directional ellipse of a bivector A. MacCullagh's theorem.*

that the ellipse of $\bar{A} = c - \mathrm{i}d$ is the same as the ellipse of A but with the opposite orientation. The plane of the directional ellipse of $A = c + \mathrm{i}d$, that is the plane of c and d, is called **the plane of the bivector A**.

Theorem 2.1 MacCullagh's theorem

Let the bivectors $A = c + \mathrm{i}d$ and $A' = c' + \mathrm{i}d'$ be related by

$$A' = c' + \mathrm{i}d' = \mathrm{e}^{\mathrm{i}\theta}A = \mathrm{e}^{\mathrm{i}\theta}(c + \mathrm{i}d), \tag{2.2.2}$$

where θ is given. Then the directional ellipse of A' is the same as the directional ellipse of A: (c', d') and (c, d) are two pairs of conjugate radii of the same ellipse and the orientation from c' to d' is the same as the orientation from c to d.

Proof The vector equation (2.2.1) of the ellipse of A may be written

$$r = \{A\mathrm{e}^{-\mathrm{i}\phi}\}^{+} = \{A'\mathrm{e}^{-\mathrm{i}\theta}\mathrm{e}^{-\mathrm{i}\phi}\}^{+} = \{A'\mathrm{e}^{-\mathrm{i}(\theta+\phi)}\}^{+}, \tag{2.2.3}$$

that is

$$r = c'\cos(\theta + \phi) + d'\sin(\theta + \phi). \tag{2.2.4}$$

But this is the vector form (with parameter $\theta + \phi$) of the equation of an ellipse having c' and d' as conjugate radii. Thus (c', d') is a pair of conjugate radii of the directional ellipse of A. Moreover, for a given θ, $\theta + \phi$ is increasing together with ϕ, so that the orientation from c' to d' is the same as the orientation from c to d.

From (2.2.2), we note that

$$\begin{aligned} c' &= \{\mathrm{e}^{\mathrm{i}\theta}A\}^{+} = c\cos\theta - d\sin\theta, \\ d' &= \{\mathrm{e}^{\mathrm{i}\theta}A\}^{-} = c\sin\theta + d\cos\theta, \end{aligned} \tag{2.2.5}$$

which shows explicitly how c' and d' are determined from c, d and θ. The angle θ may be easily interpreted using the auxiliary circle of the ellipse of A (see Exercise 1.2.2). As seen in Figure 2.1, the pair of conjugate radii (c, d) is 'rotated' (nonrigidly) clockwise into the pair of conjugate radii (c', d').

2.2.1 *Principal axes of the ellipse of the bivector $A = c + \mathrm{i}d$*

Let $A' = \mathrm{e}^{\mathrm{i}\theta}A$. An appropriate choice of the angle θ will make the real and imaginary parts A'^{+}, A'^{-} of A' be along any selected pair of conjugate radii of the ellipse of $\boldsymbol{A}$. In particular there is a choice of θ such that $\boldsymbol{A}'^{+}$ and $\boldsymbol{A}'^{-}$ are along the principal axes, i.e. such that

$$\boldsymbol{A}' = \mathrm{e}^{\mathrm{i}\theta}\boldsymbol{A} = \boldsymbol{a} + \mathrm{i}\boldsymbol{b}, \quad \text{with } \boldsymbol{a}\cdot\boldsymbol{b} = 0, \quad a^2 > b^2. \tag{2.2.6}$$

The principal semi-axes $\boldsymbol{a}$, $\boldsymbol{b}$ will be determined once this angle θ is known. In order to determine θ from $\boldsymbol{A}$, note that

$$\boldsymbol{A}\cdot\boldsymbol{A} = \mathrm{e}^{-2\mathrm{i}\theta}\boldsymbol{A}'\cdot\boldsymbol{A}' = \mathrm{e}^{-2\mathrm{i}\theta}(a^2 - b^2), \tag{2.2.7}$$

and thus

$$\begin{aligned}(\boldsymbol{A}\cdot\boldsymbol{A})^{+} &= \boldsymbol{c}\cdot\boldsymbol{c} - \boldsymbol{d}\cdot\boldsymbol{d} = (a^2 - b^2)\cos 2\theta,\\ (\boldsymbol{A}\cdot\boldsymbol{A})^{-} &= 2\boldsymbol{c}\cdot\boldsymbol{d} = -(a^2 - b^2)\sin 2\theta.\end{aligned} \tag{2.2.8}$$

It follows that

$$2\theta = -\arg(\boldsymbol{A}\cdot\boldsymbol{A}), \tag{2.2.9}$$

$$\tan 2\theta = \frac{2\boldsymbol{c}\cdot\boldsymbol{d}}{\boldsymbol{d}\cdot\boldsymbol{d} - \boldsymbol{c}\cdot\boldsymbol{c}}. \tag{2.2.10}$$

The quadrant which is chosen for 2θ is the one for which $\sin 2\theta$ has the opposite sign to $\boldsymbol{c}\cdot\boldsymbol{d}$ and $\cos 2\theta$ has the same sign as $\boldsymbol{c}\cdot\boldsymbol{c} - \boldsymbol{d}\cdot\boldsymbol{d}$.

Remarks

(a) If $\boldsymbol{c}\cdot\boldsymbol{c} = \boldsymbol{d}\cdot\boldsymbol{d}$, then $\boldsymbol{c}$ and $\boldsymbol{d}$ are along the equiconjugate diameters of the ellipse of $\boldsymbol{A}$. Then, $2\theta = \frac{1}{2}\pi$ when $\boldsymbol{c}\cdot\boldsymbol{d}$ is negative, and $2\theta = -\frac{1}{2}\pi$ when $\boldsymbol{c}\cdot\boldsymbol{d}$ is positive. In the case when $\boldsymbol{c}\cdot\boldsymbol{d}$ is negative, it follows from (2.2.6) that the principal axes are given by

$$\boldsymbol{a} + \mathrm{i}\boldsymbol{b} = \exp(\mathrm{i}\tfrac{1}{4}\pi)(\boldsymbol{c} + \mathrm{i}\boldsymbol{d}) = \frac{1}{2^{1/2}}\{(\boldsymbol{c} - \boldsymbol{d}) + \mathrm{i}(\boldsymbol{c} + \boldsymbol{d})\}. \tag{2.2.11}$$

Thus $\boldsymbol{a} = 2^{-1/2}(\boldsymbol{c} - \boldsymbol{d})$ and $\boldsymbol{b} = 2^{-1/2}(\boldsymbol{c} + \boldsymbol{d})$. The case when $\boldsymbol{c}\cdot\boldsymbol{d}$ is positive is similar.

(b) If $\boldsymbol{c}\cdot\boldsymbol{c} = \boldsymbol{d}\cdot\boldsymbol{d}$ and also $\boldsymbol{c}\cdot\boldsymbol{d} = 0$ then $\boldsymbol{A}$ is isotropic: $\boldsymbol{A}\cdot\boldsymbol{A} = 0$, and the directional ellipse of $\boldsymbol{A}$ is a circle (oriented from $\boldsymbol{c}$ to $\boldsymbol{d}$). The angle θ is undetermined.

(c) When $\boldsymbol{d}$ is parallel to $\boldsymbol{c}$, that is, when the bivector $\boldsymbol{A}$ has a real direction, the ellipse degenerates into a segment. Writing $\boldsymbol{A} = (c + \mathrm{i}d)\hat{\mathbf{n}}$ we note from (2.2.9) that $\theta = -\arg(c + \mathrm{i}d)$, so that $c + \mathrm{i}d = (c^2 + d^2)^{1/2}\mathrm{e}^{-\mathrm{i}\theta}$. In this case (2.2.6) yields

$$\boldsymbol{a} + \mathrm{i}\boldsymbol{b} = \mathrm{e}^{\mathrm{i}\theta}\boldsymbol{A} = \mathrm{e}^{\mathrm{i}\theta}(c + \mathrm{i}d)\hat{\mathbf{n}} = (c^2 + d^2)^{1/2}\hat{\mathbf{n}}. \tag{2.2.12}$$

Thus $\boldsymbol{a} = (c^2 + d^2)^{1/2}\hat{\mathbf{n}}$ and $\boldsymbol{b} = \mathbf{0}$.

Exercises 2.1

1. Determine the equation of the (directional) ellipse of the bivector $\boldsymbol{A} = (2 - 6\mathrm{i})\mathbf{i} + (3 + \mathrm{i})\mathbf{j}$.
2. Prove that $(\boldsymbol{c}' \cdot \hat{\mathbf{n}})^2 + (\boldsymbol{d}' \cdot \hat{\mathbf{n}})^2 = (\boldsymbol{c} \cdot \hat{\mathbf{n}})^2 + (\boldsymbol{d} \cdot \hat{\mathbf{n}})^2$, where $\hat{\mathbf{n}}$ is an arbitrary unit vector. Interpret this result.
3. Determine the principal axes $\boldsymbol{a}$, $\boldsymbol{b}$ of the directional ellipse of the bivector $\boldsymbol{A} = (2 - 6\mathrm{i})\mathbf{i} + (3 + \mathrm{i})\mathbf{j}$.
4. Determine the principal axes $\boldsymbol{a}$, $\boldsymbol{b}$ of the directional ellipse of the bivector $\boldsymbol{A} = (2 + 3\mathrm{i})\mathbf{i} - (3 + 2\mathrm{i})\mathbf{j}$.

2.3 Parallelism of bivectors

The bivectors $\boldsymbol{B}$ and $\boldsymbol{A}$ ($\neq \mathbf{0}$) are said to be parallel if there exists a (complex) scalar λ such that

$$\boldsymbol{B} = \lambda \boldsymbol{A}. \tag{2.3.1}$$

Let $\boldsymbol{A} = \boldsymbol{c} + \mathrm{i}\boldsymbol{d}$, and $\lambda = |\lambda| \mathrm{e}^{\mathrm{i}\theta}$. Then

$$\boldsymbol{B} = |\lambda| \mathrm{e}^{\mathrm{i}\theta}(\boldsymbol{c} + \mathrm{i}\boldsymbol{d}) = |\lambda|(\boldsymbol{c}' + \mathrm{i}\boldsymbol{d}'), \tag{2.3.2}$$

and by MacCullagh's theorem, $\boldsymbol{c}'$ and $\boldsymbol{d}'$ are thus two conjugate radii of the directional ellipse of the bivector $\boldsymbol{A}$. Thus (2.3.2) shows that two conjugate radii of the directional ellipse of the bivector $\boldsymbol{B}$ are obtained by subjecting $\boldsymbol{c}'$ and $\boldsymbol{d}'$ to the uniform extension $|\lambda|$. Thus, the ellipse of $\boldsymbol{B}$ is similar to the ellipse of $\boldsymbol{A}$, so that the two ellipses have the same aspect ratio, and the major and minor axes of the ellipse of $\boldsymbol{B}$ are respectively along the major and minor axes of the

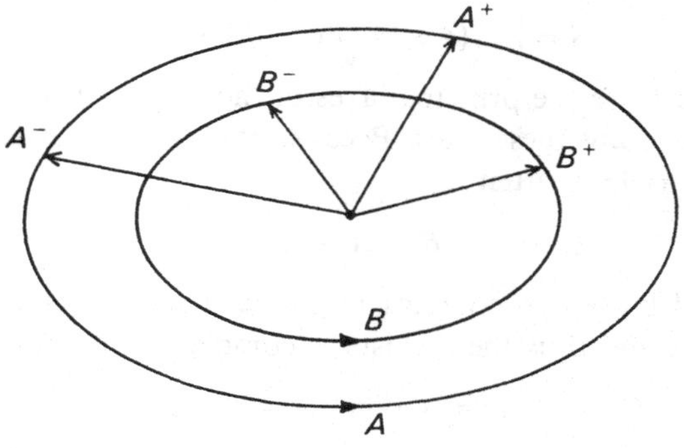

Figure 2.2 *Parallelism of bivectors:* $\boldsymbol{B} \parallel \boldsymbol{A}$.

ellipse of $\boldsymbol{A}$. Furthermore, the ellipses of $\boldsymbol{A}$ and $\boldsymbol{B}$ have the same orientation. Following Gibbs, we say that the ellipse of $\boldsymbol{B}$ is 'similar and similarly situated' to the ellipse of $\boldsymbol{A}$ (Figure 2.2).

2.4 Orthogonality of bivectors

The bivectors $\boldsymbol{A}$ and $\boldsymbol{B}$ are said to be **orthogonal** (perpendicular) when their dot product is zero:

$$\boldsymbol{A}\cdot\boldsymbol{B}=0. \tag{2.4.1}$$

When each of $\boldsymbol{A}$ and $\boldsymbol{B}$ has a real direction, then they are of the form $\boldsymbol{A}=\lambda\hat{\mathbf{n}}$ and $\boldsymbol{B}=\mu\hat{\mathbf{p}}$ and then (2.4.1) simply means that the directions $\hat{\mathbf{n}}$ and $\hat{\mathbf{p}}$ are orthogonal. Now, (2.4.1) will be interpreted when at least one of the two bivectors $\boldsymbol{A}$ and $\boldsymbol{B}$ does not have a real direction.

Case (a) $\boldsymbol{A}$ has a real direction: $\boldsymbol{A}=\lambda\hat{\mathbf{n}}$

In this case, (2.4.1) reduces to

$$\hat{\mathbf{n}}\cdot\boldsymbol{B}=0, \quad \text{i.e. } \hat{\mathbf{n}}\cdot\boldsymbol{B}^{+}=\hat{\mathbf{n}}\cdot\boldsymbol{B}^{-}=0. \tag{2.4.2}$$

This means that the plane of the bivector $\boldsymbol{B}$ is orthogonal to $\hat{\mathbf{n}}$. The ellipse of $\boldsymbol{B}$ lies in the plane $\hat{\mathbf{n}}\cdot\boldsymbol{r}=0$.

Case (b) The ellipses of $\boldsymbol{A}$ and $\boldsymbol{B}$ are coplanar

From (2.2.6), $\boldsymbol{A}$ may be written

$$\boldsymbol{A}=\mathrm{e}^{-\mathrm{i}\theta}(\boldsymbol{a}+\mathrm{i}\boldsymbol{b})=\mathrm{e}^{-\mathrm{i}\theta}(a\mathbf{i}+\mathrm{i}b\mathbf{j}), \tag{2.4.3}$$

where $\boldsymbol{a}$ and $\boldsymbol{b}$ are the principal axes of the ellipse of $\boldsymbol{A}$, and $\mathbf{i}$, $\mathbf{j}$ are unit vectors along these axes. Because the bivector $\boldsymbol{B}$ is coplanar with $\boldsymbol{A}$, it may be written

$$\boldsymbol{B}=\alpha\mathbf{i}+\beta\mathbf{j}, \tag{2.4.4}$$

where α and β are two complex numbers. Then (2.4.1) implies that $\alpha a+\mathrm{i}\beta b=0$, and thus there exists λ (complex) such that

$$\alpha=-\mathrm{i}\lambda b, \quad \beta=\lambda a. \tag{2.4.5}$$

Thus

$$\boldsymbol{B}=\lambda\boldsymbol{B}', \quad \text{with } \boldsymbol{B}'=a\mathbf{j}-\mathrm{i}b\mathbf{i}. \tag{2.4.6}$$

Hence, the major and minor axes of the ellipse of $\boldsymbol{B}'$ are respectively along the minor and major axes of the ellipse of $\boldsymbol{A}$. Also, the ellipses of $\boldsymbol{A}$ and $\boldsymbol{B}'$ have the same aspect ratio (ab^{-1}) and the same orientation. The ellipse of $\boldsymbol{B}'$ is the ellipse of $\boldsymbol{A}$ rotated through a quadrant in its plane. It follows that the ellipse of $\boldsymbol{B}$ is similar and similarly situated to the ellipse of $\boldsymbol{A}$ rotated through a quadrant in its plane.

*Case (c) The ellipses of **A** and **B** are not coplanar*

First note that the planes of $\boldsymbol{A}$ and $\boldsymbol{B}$ may not be orthogonal. Indeed, the normals to the planes of $\boldsymbol{A}$ and of $\boldsymbol{B}$ are along $\boldsymbol{A} \times \bar{\boldsymbol{A}}$ and $\boldsymbol{B} \times \bar{\boldsymbol{B}}$, respectively, and because $\boldsymbol{A} \cdot \boldsymbol{B} = 0$, it follows that

$$(\boldsymbol{A} \times \bar{\boldsymbol{A}}) \cdot (\boldsymbol{B} \times \bar{\boldsymbol{B}}) = -(\boldsymbol{A} \cdot \bar{\boldsymbol{B}})(\overline{\boldsymbol{A} \cdot \bar{\boldsymbol{B}}}) \neq 0, \tag{2.4.7}$$

and thus the two normals are not orthogonal. Now, $\boldsymbol{A}$ may be written in the form (2.4.3), and $\boldsymbol{B}$ in the form

$$\boldsymbol{B} = \alpha \mathbf{i} + \beta \mathbf{j} + \gamma \mathbf{k}. \tag{2.4.8}$$

It then follows from $\boldsymbol{A} \cdot \boldsymbol{B} = 0$ that there exists λ (complex) such that

$$\boldsymbol{B} = \lambda \boldsymbol{B}' + \gamma \mathbf{k}, \tag{2.4.9}$$

where

$$\boldsymbol{B}' = \alpha \mathbf{j} - ib\mathbf{i}. \tag{2.4.10}$$

It is clear that the term $\lambda \boldsymbol{B}'$ represents the projection of $\boldsymbol{B}$ upon the plane of $\boldsymbol{A}$. Hence, $\boldsymbol{B}'$ being the same as in (2.4.6), we conclude with Gibbs (1881) that: 'If two bivectors are perpendicular the directional ellipse of either projected upon the plane of the other and rotated through a quadrant in that plane will be similar and similarly situated to the second.'

2.5 The bivector $\boldsymbol{A}_{\perp}$

Consider a bivector $\boldsymbol{A} = \boldsymbol{c} + i\boldsymbol{d}$, and let $\boldsymbol{c}^*$, $\boldsymbol{d}^*$ be the set reciprocal to $\boldsymbol{c}$, $\boldsymbol{d}$ in the plane of $\boldsymbol{A}$:

$$\boldsymbol{c} \cdot \boldsymbol{d}^* = \boldsymbol{d} \cdot \boldsymbol{c}^* = 0, \quad \boldsymbol{c} \cdot \boldsymbol{c}^* = \boldsymbol{d} \cdot \boldsymbol{d}^* = 1. \tag{2.5.1}$$

The bivector $\boldsymbol{A}_{\perp}$ is defined by

$$\boldsymbol{A}_{\perp} = \boldsymbol{c}^* + i\boldsymbol{d}^*, \tag{2.5.2}$$

and we call this the **reciprocal bivector** of the bivector $\boldsymbol{A}$. It is easily seen that

$$A \cdot A_{\perp} = 0, \tag{2.5.3a}$$

$$A \cdot \bar{A}_{\perp} = 2. \tag{2.5.3b}$$

Writing A as in (2.4.3), $A = \mathrm{e}^{-\mathrm{i}\theta}(a\mathbf{i} + \mathrm{i}b\mathbf{j})$, we note from $A \cdot A_{\perp} = 0$ that $A_{\perp}$ may be written

$$A_{\perp} = \lambda(a\mathbf{j} - \mathrm{i}b\mathbf{i}). \tag{2.5.4}$$

From $A \cdot \bar{A}_{\perp} = 2$, it then follows that $\lambda = (\mathrm{i}\mathrm{e}^{-\mathrm{i}\theta})(ab)^{-1}$, and thus

$$A_{\perp} = \mathrm{e}^{-\mathrm{i}\theta}\left(\frac{1}{a}\mathbf{i} + \frac{i}{b}\mathbf{j}\right). \tag{2.5.5}$$

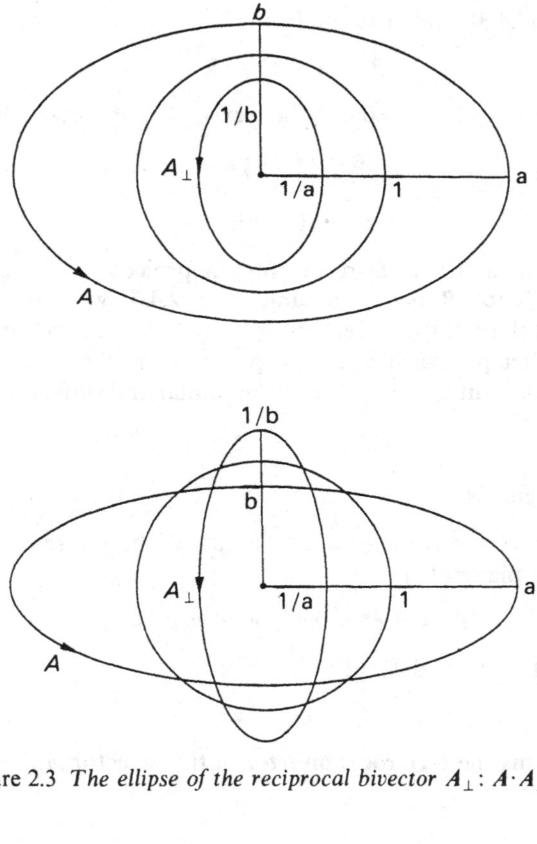

Figure 2.3 *The ellipse of the reciprocal bivector* $A_{\perp}$: $A \cdot A_{\perp} = 0$.

Thus the major and minor axes of the ellipse of $\boldsymbol{A}_\perp$ are respectively along the minor and major axes of the ellipse of $\boldsymbol{A}$, and the lengths of the principal semi-axes of $\boldsymbol{A}_\perp$ are the inverses of the lengths of the principal semi-axes of $\boldsymbol{A}$. It follows that the ellipse of $\boldsymbol{A}_\perp$ is the inverse with respect to the unit circle (the polar reciprocal with respect to the unit circle) of the ellipse of $\boldsymbol{A}$. This is illustrated in Figure 2.3.

Finally, using (2.4.10) and (2.5.5) we note that if $\boldsymbol{A}\cdot\boldsymbol{B}=0$, then $\boldsymbol{B}$ may be written

$$\boldsymbol{B}=\nu\boldsymbol{A}_\perp+\gamma\mathbf{k}, \tag{2.5.6}$$

where $\mathbf{k}$ is a unit vector orthogonal to the plane of the bivector $\boldsymbol{A}$, and ν and γ are constants. It follows that the bivector $\boldsymbol{A}_\perp$ may also be defined as the bivector in the plane of the bivector $\boldsymbol{A}$ such that (2.5.3) is satisfied.

Exercises 2.2

1. Let $\boldsymbol{A}=(1+2\mathrm{i})\mathbf{i}+(2-\mathrm{i})\mathbf{j}-(3-\mathrm{i})\mathbf{k}$. Find $\boldsymbol{A}_\perp$.
2. Let $\boldsymbol{c}$ and $\boldsymbol{d}$ be two conjugate radii of the ellipse of $\boldsymbol{A}$. Show how to construct the conjugate radii $\boldsymbol{c}^*$, $\boldsymbol{d}^*$ of the ellipse of $\boldsymbol{A}_\perp$.

2.6 Isotropic bivectors

When $\boldsymbol{A}\cdot\boldsymbol{A}=0$, the bivector $\boldsymbol{A}$ is said to be isotropic and its directional ellipse is a circle (with an orientation).

Theorem 2.2 (Synge)

If two nonzero isotropic bivectors are orthogonal, they are parallel.

Proof Let $\boldsymbol{A}$ and $\boldsymbol{B}$ be isotropic and orthogonal:

$$\boldsymbol{A}\cdot\boldsymbol{A}=\boldsymbol{B}\cdot\boldsymbol{B}=0, \quad \boldsymbol{A}\cdot\boldsymbol{B}=0. \tag{2.6.1}$$

Let $\mathbf{k}$ be a unit vector orthogonal to the plane of $\boldsymbol{A}$. Since $\boldsymbol{A}$, $\bar{\boldsymbol{A}}$, $\boldsymbol{k}$ are three linearly independent bivectors, any bivector $\boldsymbol{B}$ may be written

$$\boldsymbol{B}=\alpha\boldsymbol{A}+\beta\bar{\boldsymbol{A}}+\gamma\mathbf{k}. \tag{2.6.2}$$

Now, $\boldsymbol{A}\cdot\boldsymbol{B}=0$ implies $\beta=0$ (since $\boldsymbol{A}\cdot\bar{\boldsymbol{A}}\neq 0$) and thus

$$\boldsymbol{B}=\alpha\boldsymbol{A}+\gamma\mathbf{k}. \tag{2.6.3}$$

But then, $\boldsymbol{B}\cdot\boldsymbol{B}=0$ implies $\gamma=0$. Thus, finally,

$$\boldsymbol{B}=\alpha\boldsymbol{A}, \tag{2.6.4}$$

which expresses the fact that $\boldsymbol{B}$ and $\boldsymbol{A}$ are parallel.

Theorem 2.3

If $\boldsymbol{A}$, $\boldsymbol{B}$, $\boldsymbol{C}$ are mutually orthogonal and if $\boldsymbol{A}$ is isotropic, then either $\boldsymbol{B}=\alpha\boldsymbol{A}$ (in which case $\boldsymbol{B}$ is isotropic), or $\boldsymbol{C}=\lambda\boldsymbol{A}$ (in which case $\boldsymbol{C}$ is isotropic).

Proof Again using the basis $\boldsymbol{A}$, $\bar{\boldsymbol{A}}$, $\mathbf{k}$, we write, using $\boldsymbol{A}\cdot\boldsymbol{B}=\boldsymbol{A}\cdot\boldsymbol{C}=0$,

$$\boldsymbol{B}=\alpha\boldsymbol{A}+\gamma\mathbf{k}, \quad \boldsymbol{C}=\lambda\boldsymbol{A}+\nu\mathbf{k}. \tag{2.6.5}$$

Thus $\boldsymbol{B}\cdot\boldsymbol{C}=\gamma\nu=0$, and either ν or γ is zero. Hence, either

$$\boldsymbol{B}=\alpha\boldsymbol{A}, \quad \text{or } \boldsymbol{C}=\lambda\boldsymbol{A}. \tag{2.6.6}$$

From this theorem we immediately deduce the following corollary.

Corollary 2.4 (Synge)

If three linearly independent bivectors are mutually orthogonal, none of them is an isotropic bivector.

Finally, if $\boldsymbol{A}=\boldsymbol{c}+\mathrm{i}\boldsymbol{d}$ is isotropic, i.e. if

$$\boldsymbol{c}\cdot\boldsymbol{c}=\boldsymbol{d}\cdot\boldsymbol{d}, \quad \boldsymbol{c}\cdot\boldsymbol{d}=0, \tag{2.6.7}$$

it is easily seen that $\boldsymbol{c}^*=[\boldsymbol{c}\cdot\boldsymbol{c}]^{-1}\boldsymbol{c}$ and $\boldsymbol{d}^*=[\boldsymbol{d}\cdot\boldsymbol{d}]^{-1}\boldsymbol{d}$. Hence the reciprocal bivector $\boldsymbol{A}_\perp$ of the isotropic bivector $\boldsymbol{A}$ is

$$\boldsymbol{A}_\perp=\boldsymbol{c}^*+\mathrm{i}\boldsymbol{d}^*=2[\boldsymbol{A}\cdot\bar{\boldsymbol{A}}]^{-1}\boldsymbol{A}. \tag{2.6.8}$$

Exercises 2.3

1. Show that the bivector $\boldsymbol{A}$ is isotropic if and only if it is parallel to its reciprocal $\boldsymbol{A}_\perp$.
2. Show that if $\boldsymbol{A}$ is isotropic, and $\boldsymbol{B}$ is in the plane of $\boldsymbol{A}$ with $\boldsymbol{A}\cdot\boldsymbol{B}=0$, then $\boldsymbol{B}$ is parallel to $\boldsymbol{A}$ (and thus also isotropic).

3. Let $\boldsymbol{A} = \hat{\mathbf{m}} - e^{\pm i\theta}\hat{\mathbf{n}}$, where $\hat{\mathbf{m}}$ and $\hat{\mathbf{n}}$ are two unit vectors. Find θ such that $\boldsymbol{A}$ is isotropic.

2.7 Properties of the cross product

Theorem 2.5 (Gibbs)

If $\boldsymbol{A} \times \boldsymbol{B} = \mathbf{0}$, $(\boldsymbol{A} \neq \mathbf{0})$, then $\boldsymbol{B} = \lambda \boldsymbol{A}$.

Proof Without loss of generality, the bivectors $\boldsymbol{A}$ and $\boldsymbol{B}$ may be written

$$\boldsymbol{A} = e^{-i\theta}(a\mathbf{i} + ib\mathbf{j}),$$
$$\boldsymbol{B} = \alpha\mathbf{i} + \beta\mathbf{j} + \gamma\mathbf{k}, \tag{2.7.1}$$

where $\mathbf{i}$, $\mathbf{j}$ are unit vectors along the principal axes of the ellipse of $\boldsymbol{A}$ and where $\mathbf{k} = \mathbf{i} \times \mathbf{j}$. Then $\boldsymbol{A} \times \boldsymbol{B} = \mathbf{0}$ reads

$$\boldsymbol{A} \times \boldsymbol{B} = ib\gamma\mathbf{i} - a\gamma\mathbf{j} + (a\beta - ib\alpha)\mathbf{k} = \mathbf{0}, \tag{2.7.2}$$

which implies $\gamma = 0$ and $\alpha = \mu a$, $\beta = i\mu b$ for some μ (complex). Thus

$$\boldsymbol{B} = \mu(a\mathbf{i} + ib\mathbf{j}) = \lambda\boldsymbol{A}, \quad \text{with} \quad \lambda = e^{i\theta}\mu. \tag{2.7.3}$$

Also, by direct expansion, for any two bivectors $\boldsymbol{A}$ and $\boldsymbol{B}$,

$$\boldsymbol{A}\cdot\boldsymbol{A} \times \boldsymbol{B} = 0, \quad \boldsymbol{B}\cdot\boldsymbol{A} \times \boldsymbol{B} = 0. \tag{2.7.4}$$

Thus the projection of the ellipse of $\boldsymbol{A}$ upon the plane of $\boldsymbol{A} \times \boldsymbol{B}$ is similar and similarly situated to the projection of the ellipse of $\boldsymbol{B}$ upon the plane of $\boldsymbol{A} \times \boldsymbol{B}$. Both projected ellipses, which are similar and similarly situated to each other, are also similar to the ellipse of $\boldsymbol{A} \times \boldsymbol{B}$ when it is rotated through a quadrant in its plane.

When $\boldsymbol{A} \times \boldsymbol{B}$ is isotropic: $(\boldsymbol{A} \times \boldsymbol{B})\cdot(\boldsymbol{A} \times \boldsymbol{B}) = 0$, and its ellipse is a circle. In this case the projections of the ellipses of $\boldsymbol{A}$ and $\boldsymbol{B}$ upon the plane of $\boldsymbol{A} \times \boldsymbol{B}$ are both circles – with the same orientation. Thus $\boldsymbol{A} \times \boldsymbol{B}$ being isotropic means that the ellipses of $\boldsymbol{A}$ and $\boldsymbol{B}$ are similar and similarly situated to two plane sections of the same circular cylinder.

Exercises 2.4

1. Let $\boldsymbol{A} = 2\mathbf{i} + i\mathbf{j}$, $\boldsymbol{B} = \mathbf{k} + 3i\mathbf{i}$. Compute $\boldsymbol{A} \times \boldsymbol{B}$. Determine the projections $\boldsymbol{A}'$, $\boldsymbol{B}'$ of $\boldsymbol{A}$ and $\boldsymbol{B}$ upon the plane of $\boldsymbol{A} \times \boldsymbol{B}$. Check

that $\boldsymbol{B}'$ and $\boldsymbol{A}'$ are parallel. Write equations for the ellipses of $\boldsymbol{A}$, $\boldsymbol{B}$, $\boldsymbol{A} \times \boldsymbol{B}$, $\boldsymbol{A}'$, $\boldsymbol{B}'$.

2. Let $\hat{\mathbf{n}}$ be a unit vector. Interpret $(\hat{\mathbf{n}} \times \boldsymbol{A})\cdot(\hat{\mathbf{n}} \times \boldsymbol{A}) = 0$.

2.8 Bivector identities

Since the expressions (2.1.12) and (2.1.13) of the dot product and cross product of bivectors in components are the same as for real vectors, we have as for the real vectors the identities

$$\boldsymbol{A}\cdot\boldsymbol{B} \times \boldsymbol{C} = \boldsymbol{B}\cdot\boldsymbol{C} \times \boldsymbol{A} = \boldsymbol{C}\cdot\boldsymbol{A} \times \boldsymbol{B}, \qquad (2.8.1)$$

and

$$(\boldsymbol{A} \times \boldsymbol{B}) \times \boldsymbol{C} = (\boldsymbol{A}\cdot\boldsymbol{C})\boldsymbol{B} - (\boldsymbol{B}\cdot\boldsymbol{C})\boldsymbol{A}. \qquad (2.8.2)$$

Using these, the following identities may be proved:

$$(\boldsymbol{A} \times \boldsymbol{B})\cdot(\boldsymbol{C} \times \boldsymbol{D}) = (\boldsymbol{A}\cdot\boldsymbol{C})(\boldsymbol{B}\cdot\boldsymbol{D}) - (\boldsymbol{A}\cdot\boldsymbol{D})(\boldsymbol{B}\cdot\boldsymbol{C}), \qquad (2.8.3)$$

$$(\boldsymbol{A} \times \boldsymbol{B}\cdot\boldsymbol{C})\boldsymbol{D} = (\boldsymbol{B} \times \boldsymbol{C}\cdot\boldsymbol{D})\boldsymbol{A} + (\boldsymbol{C} \times \boldsymbol{A}\cdot\boldsymbol{D})\boldsymbol{B} + (\boldsymbol{A} \times \boldsymbol{B}\cdot\boldsymbol{D})\boldsymbol{C}, \qquad (2.8.4)$$

$$(\boldsymbol{A} \times \boldsymbol{B}\cdot\boldsymbol{C})\boldsymbol{D} = (\boldsymbol{A}\cdot\boldsymbol{D})\boldsymbol{B} \times \boldsymbol{C} + (\boldsymbol{B}\cdot\boldsymbol{D})\boldsymbol{C} \times \boldsymbol{A} + (\boldsymbol{C}\cdot\boldsymbol{D})\boldsymbol{A} \times \boldsymbol{B}. \qquad (2.8.5)$$

Then, the two following theorems due to Gibbs may be proved.

Theorem 2.6

If $\boldsymbol{A}$ and $\boldsymbol{B}$ are not parallel, and if $\boldsymbol{C}$ is orthogonal to $\boldsymbol{A}$ and $\boldsymbol{B}$: $\boldsymbol{A}\cdot\boldsymbol{C} = \boldsymbol{B}\cdot\boldsymbol{C} = 0$, then $\boldsymbol{C} = \lambda\boldsymbol{A} \times \boldsymbol{B}$ for some scalar λ.

Proof Because $\boldsymbol{A}\cdot\boldsymbol{C} = \boldsymbol{B}\cdot\boldsymbol{C} = 0$, it follows from (2.8.2) that $(\boldsymbol{A} \times \boldsymbol{B}) \times \boldsymbol{C} = \mathbf{0}$. Then, as $\boldsymbol{A} \times \boldsymbol{B} \neq \mathbf{0}$ it follows that $\boldsymbol{C} = \lambda\boldsymbol{A} \times \boldsymbol{B}$.

Corollary 2.7

The reciprocal bivector $\boldsymbol{A}_\perp$ of $\boldsymbol{A}$ is given by

$$\boldsymbol{A}_\perp = \{2(\boldsymbol{A} \times \bar{\boldsymbol{A}}) \times \boldsymbol{A}\}\{(\boldsymbol{A} \times \bar{\boldsymbol{A}})\cdot(\boldsymbol{A} \times \bar{\boldsymbol{A}})\}^{-1}. \qquad (2.8.6)$$

Proof Because $\boldsymbol{A}_\perp$ is in the plane of $\boldsymbol{A}$, it is orthogonal to $\boldsymbol{A} \times \bar{\boldsymbol{A}}$. Also $\boldsymbol{A}_\perp$ is orthogonal to $\boldsymbol{A}$. Thus, applying the theorem, we have, for some λ,

$$\boldsymbol{A}_\perp = \lambda(\boldsymbol{A} \times \bar{\boldsymbol{A}}) \times \boldsymbol{A}. \qquad (2.8.7)$$

But, from (2.5.3b), we have, using (2.8.1)

$$\boldsymbol{A}\cdot\bar{\boldsymbol{A}}_{\perp}=\bar{\lambda}\boldsymbol{A}\cdot\{(\bar{\boldsymbol{A}}\times\boldsymbol{A})\times\bar{\boldsymbol{A}}\}=\bar{\lambda}(\bar{\boldsymbol{A}}\times\boldsymbol{A})\cdot(\bar{\boldsymbol{A}}\times\boldsymbol{A})=2, \quad (2.8.8)$$

which gives the value of λ. Inserting this value in (2.8.7) yields the result (2.8.6).

Theorem 2.8

The bivectors $\boldsymbol{A}$, $\boldsymbol{B}$, $\boldsymbol{C}$ are linearly dependent if and only if

$$\boldsymbol{A}\times\boldsymbol{B}\cdot\boldsymbol{C}=0. \quad (2.8.9)$$

Proof If $\boldsymbol{A}\times\boldsymbol{B}\cdot\boldsymbol{C}=0$, then, for any bivector $\boldsymbol{D}$, (2.8.4) reduces to

$$(\boldsymbol{B}\times\boldsymbol{C}\cdot\boldsymbol{D})\boldsymbol{A}+(\boldsymbol{C}\times\boldsymbol{A}\cdot\boldsymbol{D})\boldsymbol{B}+(\boldsymbol{A}\times\boldsymbol{B}\cdot\boldsymbol{D})\boldsymbol{C}=\boldsymbol{0}. \quad (2.8.10)$$

This shows that $\boldsymbol{A}$, $\boldsymbol{B}$, $\boldsymbol{C}$ are linearly dependent since $\boldsymbol{D}$ may be chosen, for instance, to be not orthogonal to $\boldsymbol{A}\times\boldsymbol{B}$. Conversely, if $\boldsymbol{C}=\lambda\boldsymbol{A}+\mu\boldsymbol{B}$, it is easily seen, using (2.7.4) that $\boldsymbol{A}\times\boldsymbol{B}\cdot\boldsymbol{C}=0$.

Exercises 2.5

1. Derive the identities (2.8.3), (2.8.4) and (2.8.5).
2. Prove that $\boldsymbol{A}$, $\boldsymbol{B}$, $\boldsymbol{A}\times\boldsymbol{B}$ (with $\boldsymbol{B}$ not parallel to $\boldsymbol{A}$) are linearly independent if and only if $\boldsymbol{A}\times\boldsymbol{B}$ is not isotropic.
3. Evaluate $\boldsymbol{A}_{\perp}\cdot\boldsymbol{A}_{\perp}$ in terms of $\boldsymbol{A}$ and $\bar{\boldsymbol{A}}$.
4. Evaluate $\boldsymbol{A}\times\boldsymbol{A}_{\perp}$ in terms of $\boldsymbol{A}$ and $\bar{\boldsymbol{A}}$.
5. Prove that if $\boldsymbol{A}\cdot\boldsymbol{A}=\boldsymbol{A}\cdot\boldsymbol{B}=0$ and $\boldsymbol{B}\cdot\boldsymbol{B}\neq 0$, then $\boldsymbol{B}\times\boldsymbol{A}=\lambda\boldsymbol{A}$. Find λ.

2.9 Bivector equations

We consider equations for an unknown bivector $\boldsymbol{X}$, written in terms of given bivectors $\boldsymbol{A}, \boldsymbol{B}, \ldots$. These equations may in general be solved in the same way as the corresponding equations with vectors (e.g. Milne, 1948), except that now $\boldsymbol{A}\cdot\boldsymbol{A}$, $\boldsymbol{B}\cdot\boldsymbol{B}, \ldots$ are no longer necessarily positive quantities. In particular the possibility of $\boldsymbol{A}\cdot\boldsymbol{A}$, $\boldsymbol{B}\cdot\boldsymbol{B}, \ldots$ being zero, without $\boldsymbol{A}$ being zero, must be considered.

We present two examples.

Example 2.1

Solve

$$X \times A = B. \tag{2.9.1}$$

It is clear that A and B must satisfy the compatibility equation

$$A \cdot B = 0. \tag{2.9.2}$$

First we comment that if $X = \hat{X}$ is a solution, then

$$X = \hat{X} + \lambda A \tag{2.9.3}$$

where λ is arbitrary, is also a solution. From (2.9.1), we have

$$\begin{aligned} \bar{A} \times B &= \bar{A} \times (X \times A) \\ &= X(\bar{A} \cdot A) - A(\bar{A} \cdot X). \end{aligned} \tag{2.9.4}$$

Thus

$$(\bar{A} \cdot A)X = \bar{A} \times B + A(\bar{A} \cdot X). \tag{2.9.5}$$

Hence, noting our comment, it follows that

$$X = \frac{\bar{A} \times B}{\bar{A} \cdot A} + \lambda A, \tag{2.9.6}$$

where λ is arbitrary.

Example 2.2

Solve

$$X \times A + \alpha X = 0, \quad (\alpha \neq 0). \tag{2.9.7}$$

We note first that X must satisfy $X \cdot X = 0$: X is isotropic. There are three cases.

Case (a) $A \cdot A + \alpha^2 \neq 0$ Taking the dot product and the cross product with A, and using the identity (2.8.2), yields

$$A \cdot X = 0, \tag{2.9.8a}$$

$$(A \cdot A)X + \alpha A \times X = 0, \tag{2.9.8b}$$

and, eliminating $A \times X$ from (2.9.7) and (2.9.8b), we obtain

$$(A \cdot A + \alpha^2)X = 0. \tag{2.9.9}$$

Hence, $\boldsymbol{X}=\mathbf{0}$. Clearly, when $\boldsymbol{A}$ and α are real, no other case may occur. However, when they are complex, one has to consider the case when $\boldsymbol{A}\cdot\boldsymbol{A}+\alpha^2=0$.

Case (b) $\boldsymbol{A}\cdot\boldsymbol{A}+\alpha^2=0$, $\boldsymbol{A}\times\bar{\boldsymbol{A}}\neq\mathbf{0}$ (*$\boldsymbol{A}$ does not have a real direction*) Since $\boldsymbol{A}$, $\bar{\boldsymbol{A}}$, $\boldsymbol{A}\times\bar{\boldsymbol{A}}$ are linearly independent bivectors, we may write

$$\boldsymbol{X}=\lambda\boldsymbol{A}+\mu\bar{\boldsymbol{A}}+\nu\boldsymbol{A}\times\bar{\boldsymbol{A}}, \tag{2.9.10}$$

for some scalars λ, μ, ν. Introducing this in the equation yields

$$\mu=\alpha\nu, \quad \alpha\mu=-(\boldsymbol{A}\cdot\boldsymbol{A})\nu, \quad \alpha\lambda=(\boldsymbol{A}\cdot\bar{\boldsymbol{A}})\nu. \tag{2.9.11}$$

Because $\boldsymbol{A}\cdot\boldsymbol{A}+\alpha^2=0$, the two first equations are equivalent, and we thus obtain the solutions

$$\boldsymbol{X}=\nu\left\{\frac{1}{\alpha}(\boldsymbol{A}\cdot\bar{\boldsymbol{A}})\boldsymbol{A}+\alpha\bar{\boldsymbol{A}}+\boldsymbol{A}\times\bar{\boldsymbol{A}}\right\}, \tag{2.9.12}$$

which may also be written as

$$\boldsymbol{X}=\nu\boldsymbol{A}\times\left\{\bar{\boldsymbol{A}}+\frac{1}{\alpha}\boldsymbol{A}\times\bar{\boldsymbol{A}}\right\}, \tag{2.9.13}$$

where ν is arbitrary. It is easily checked that the solutions (2.9.13) are isotropic:

$$\boldsymbol{X}\cdot\boldsymbol{X}=0.$$

Case (c) $\boldsymbol{A}\cdot\boldsymbol{A}+\alpha^2=0$, $\boldsymbol{A}\times\bar{\boldsymbol{A}}=\mathbf{0}$ (*$\boldsymbol{A}$ has a real direction*) In this case $\boldsymbol{A}=\mathrm{i}\alpha\hat{\mathbf{n}}$, where $\hat{\mathbf{n}}$ is a real unit vector in the direction of $\boldsymbol{A}$. Choosing $\hat{\mathbf{l}}$, $\hat{\mathbf{m}}$, $\hat{\mathbf{n}}$ to form a real orthonormal basis, we may write

$$\boldsymbol{X}=\lambda\hat{\mathbf{l}}+\mu\hat{\mathbf{m}}+\nu\hat{\mathbf{n}}, \tag{2.9.14}$$

for some scalars λ, μ, ν. Introducing this in the equation with $\boldsymbol{A}=\mathrm{i}\alpha\hat{\mathbf{n}}$, yields

$$\mu=\mathrm{i}\lambda, \quad \nu=0, \tag{2.9.15}$$

and thus the solutions are

$$\boldsymbol{X}=\lambda(\hat{\mathbf{l}}+\mathrm{i}\hat{\mathbf{m}}), \tag{2.9.16}$$

where λ is arbitrary. It is clear that the solutions (2.9.16) are isotropic in the plane orthogonal to $\hat{\mathbf{n}}$.

Exercises 2.6

1. Solve $\boldsymbol{X} \times \boldsymbol{A} = \boldsymbol{A}$ with $\boldsymbol{A}$ isotropic. Show that for all solutions $\boldsymbol{X} \cdot \boldsymbol{X} = -1$.
2. Solve $\boldsymbol{X} \times \boldsymbol{A} + \alpha \boldsymbol{X} = \boldsymbol{B}$ $(\alpha \neq 0)$.
   (a) When $\boldsymbol{A} \cdot \boldsymbol{A} + \alpha^2 \neq 0$, show that the solution is unique.
   (b) When $\boldsymbol{A} \cdot \boldsymbol{A} + \alpha^2 = 0$, show that there are no solutions unless the compatibility condition $\boldsymbol{A} \times [\alpha \boldsymbol{B} + \boldsymbol{A} \times \boldsymbol{B}] = \boldsymbol{0}$ is satisfied.
   (c) When $\boldsymbol{A} \cdot \boldsymbol{A} + \alpha^2 = 0$ and when $\boldsymbol{B}$ is such that the compatibility condition is satisfied, show that one solution may be obtained as a linear combination of $\boldsymbol{A}$ and $\boldsymbol{B}$. Then find all the solutions.
3. Solve the pair of equations $\boldsymbol{X} \times \boldsymbol{A} = \boldsymbol{B}$, $\boldsymbol{X} \cdot \boldsymbol{C} = \alpha$ (with $\boldsymbol{A} \cdot \boldsymbol{B} = 0$).

2.10 Orthonormal bases. Canonical form

Let $\boldsymbol{A}$, $\boldsymbol{B}$, $\boldsymbol{C}$ be an orthonormal basis of bivectors:

$$\boldsymbol{A} \cdot \boldsymbol{B} = \boldsymbol{B} \cdot \boldsymbol{C} = \boldsymbol{C} \cdot \boldsymbol{A} = 0, \tag{2.10.1}$$

$$\boldsymbol{A} \cdot \boldsymbol{A} = \boldsymbol{B} \cdot \boldsymbol{B} = \boldsymbol{C} \cdot \boldsymbol{C} = 1. \tag{2.10.2}$$

Using an orthonormal triad **i**, **j**, **k** of real unit vectors, the bivectors $\boldsymbol{A}$, $\boldsymbol{B}$, $\boldsymbol{C}$ may be written as

$$\begin{aligned} \boldsymbol{A} &= A_1 \mathbf{i} + A_2 \mathbf{j} + A_3 \mathbf{k}, \\ \boldsymbol{B} &= B_1 \mathbf{i} + B_2 \mathbf{j} + B_3 \mathbf{k}, \\ \boldsymbol{C} &= C_1 \mathbf{i} + C_2 \mathbf{j} + C_3 \mathbf{k}. \end{aligned} \tag{2.10.3}$$

Synge (1964) showed how to choose the real orthonormal triad **i**, **j**, **k** so that the expressions (2.10.3) are as simple as possible. Because $\boldsymbol{A} \cdot \boldsymbol{A} = 1$, we have

$$\boldsymbol{A}^+ \cdot \boldsymbol{A}^+ - \boldsymbol{A}^- \cdot \boldsymbol{A}^- = 1, \quad \boldsymbol{A}^+ \cdot \boldsymbol{A}^- = 0. \tag{2.10.4}$$

Then, choosing **i** and **j** along $\boldsymbol{A}^+$ and $\boldsymbol{A}^-$, respectively, we have

$$\boldsymbol{A} = \cosh \lambda \mathbf{i} + \mathrm{i} \sinh \lambda \mathbf{j}, \tag{2.10.5}$$

for some real λ. Now let **k** be defined by $\mathbf{k} = \mathbf{i} \times \mathbf{j}$. Because $\boldsymbol{A} \cdot \boldsymbol{B} = \boldsymbol{A} \cdot \boldsymbol{C} = 0$, we have

$$B_1 \cosh \lambda + \mathrm{i} B_2 \sinh \lambda = 0, \quad C_1 \cosh \lambda + \mathrm{i} C_2 \sinh \lambda = 0, \tag{2.10.6}$$

and hence,

$$B_1 = \mathrm{i} B_0 \sinh \lambda, \quad B_2 = -B_0 \cosh \lambda, \tag{2.10.7}$$

$$C_1 = \mathrm{i}C_0 \sinh\lambda, \quad C_2 = -C_0 \cosh\lambda, \tag{2.10.8}$$

for some complex B_0, C_0. Now, $\boldsymbol{B}\cdot\boldsymbol{B} = \boldsymbol{C}\cdot\boldsymbol{C} = 1$ gives

$$B_0^2 + B_3^2 = 1, \quad C_0^2 + C_3^2 = 1. \tag{2.10.9}$$

Hence, there exist complex numbers Ψ, Ψ' such that

$$B_0 = \sin\Psi, \quad B_3 = \cos\Psi, \tag{2.10.10}$$

$$C_0 = \sin\Psi', \quad C_3 = \cos\Psi'. \tag{2.10.11}$$

But, $\boldsymbol{B}\cdot\boldsymbol{C} = B_0C_0 + B_3C_3 = 0$ yields

$$\cos(\Psi' - \Psi) = 0, \tag{2.10.12}$$

and thus

$$\Psi' - \Psi = \pm\frac{\pi}{2}. \tag{2.10.13}$$

We thus finally obtain the following form, called the 'canonical form', of the orthonormal basis $\boldsymbol{A}$, $\boldsymbol{B}$, $\boldsymbol{C}$:

$$\boldsymbol{A} = \cosh\lambda\mathbf{i} + \mathrm{i}\sinh\lambda\mathbf{j}, \tag{2.10.14a}$$

$$\boldsymbol{B} = \mathrm{i}\sin\Psi(\sinh\lambda\mathbf{i} + \mathrm{i}\cosh\lambda\mathbf{j}) + \cos\Psi\mathbf{k}, \tag{2.10.14b}$$

$$\boldsymbol{C} = \pm\{-\mathrm{i}\cos\Psi(\sinh\lambda\mathbf{i} + \mathrm{i}\cosh\lambda\mathbf{j}) + \sin\Psi\mathbf{k}\}. \tag{2.10.14c}$$

When the '+' sign is chosen in (2.10.13), and thus in (2.10.14c), we have $(\boldsymbol{A}\times\boldsymbol{B})\cdot\boldsymbol{C} = 1$, and when the '−' sign is chosen, we have $(\boldsymbol{A}\times\boldsymbol{B})\cdot\boldsymbol{C} = -1$.

2.11 Reciprocal triads

Let $\boldsymbol{A}$, $\boldsymbol{B}$, $\boldsymbol{C}$ be a basis (in general not orthonormal) for the bivectors. The triad $\boldsymbol{A}^*$, $\boldsymbol{B}^*$, $\boldsymbol{C}^*$ reciprocal to the triad $\boldsymbol{A}$, $\boldsymbol{B}$, $\boldsymbol{C}$ is defined by

$$\boldsymbol{A}\cdot\boldsymbol{B}^* = \boldsymbol{A}\cdot\boldsymbol{C}^* = \boldsymbol{B}\cdot\boldsymbol{A}^* = \boldsymbol{B}\cdot\boldsymbol{C}^* = \boldsymbol{C}\cdot\boldsymbol{A}^* = \boldsymbol{C}\cdot\boldsymbol{B}^* = 0, \tag{2.11.1}$$

$$\boldsymbol{A}\cdot\boldsymbol{A}^* = \boldsymbol{B}\cdot\boldsymbol{B}^* = \boldsymbol{C}\cdot\boldsymbol{C}^* = 1. \tag{2.11.2}$$

Clearly, since the dot product is commutative, the reciprocal triad of the triad $\boldsymbol{A}^*$, $\boldsymbol{B}^*$, $\boldsymbol{C}^*$ is the triad $\boldsymbol{A}$, $\boldsymbol{B}$, $\boldsymbol{C}$ itself. Also, it is easily seen from (2.11.1) and (2.11.2) that

$$\boldsymbol{A}^* = \frac{\boldsymbol{B}\times\boldsymbol{C}}{\boldsymbol{A}\cdot\boldsymbol{B}\times\boldsymbol{C}}, \quad \boldsymbol{B}^* = \frac{\boldsymbol{C}\times\boldsymbol{A}}{\boldsymbol{A}\cdot\boldsymbol{B}\times\boldsymbol{C}}, \quad \boldsymbol{C}^* = \frac{\boldsymbol{A}\times\boldsymbol{B}}{\boldsymbol{A}\cdot\boldsymbol{B}\times\boldsymbol{C}}. \tag{2.11.3}$$

Theorem 2.9 (Brady, 1989)

If $\boldsymbol{A}$ and $\boldsymbol{B}$ are isotropic, then $\boldsymbol{C}^*$ is not isotropic.

Proof Indeed, assuming $\boldsymbol{C}^*$, or equivalently $\boldsymbol{A} \times \boldsymbol{B}$ isotropic, we would have

$$(\boldsymbol{A} \times \boldsymbol{B})\cdot(\boldsymbol{A} \times \boldsymbol{B}) = -(\boldsymbol{A}\cdot\boldsymbol{B})^2 = 0, \tag{2.11.4}$$

and by Theorem 2.2 (Synge's theorem), $\boldsymbol{A}$ and $\boldsymbol{B}$ would be parallel, which is contradictory to the assumption that $\boldsymbol{A}$, $\boldsymbol{B}$, $\boldsymbol{C}$ is a basis.

Theorem 2.10 (Brady, 1989)

If $\boldsymbol{C}$ is isotropic, then $\boldsymbol{A}^*$ and $\boldsymbol{B}^*$ may not both be isotropic.

Proof If $\boldsymbol{A}^*$ and $\boldsymbol{B}^*$ are isotropic, we conclude from Theorem 2.9 (interchanging the roles of the triad $\boldsymbol{A}$, $\boldsymbol{B}$, $\boldsymbol{C}$ and of its reciprocal triad $\boldsymbol{A}^*$, $\boldsymbol{B}^*$, $\boldsymbol{C}^*$), that $\boldsymbol{C}$ may not be isotropic, which is contradictory to the assumption. From Theorem 2.9, we immediately obtain the following corollary.

Corollary 2.11

If $\boldsymbol{A}$, $\boldsymbol{B}$, $\boldsymbol{C}$ are all three isotropic, then none of the bivectors $\boldsymbol{A}^*$, $\boldsymbol{B}^*$, $\boldsymbol{C}^*$ is isotropic.

Remarks

(a) If $\boldsymbol{A}$ and $\boldsymbol{B}$ are isotropic and $\boldsymbol{C}$ is not isotropic, then $\boldsymbol{C}^*$ is not isotropic. However, $\boldsymbol{A}^*$ or $\boldsymbol{B}^*$, or both $\boldsymbol{A}^*$ and $\boldsymbol{B}^*$, may be isotropic.

The bivector $\boldsymbol{A}^*$ is isotropic if

$$(\boldsymbol{B} \times \boldsymbol{C})\cdot(\boldsymbol{B} \times \boldsymbol{C}) = -(\boldsymbol{B}\cdot\boldsymbol{C})^2 = 0, \tag{2.11.5}$$

and the bivector $\boldsymbol{B}^*$ is isotropic if

$$(\boldsymbol{C} \times \boldsymbol{A})\cdot(\boldsymbol{C} \times \boldsymbol{A}) = -(\boldsymbol{C}\cdot\boldsymbol{A})^2 = 0. \tag{2.11.6}$$

Both possibilities (2.11.5) and (2.11.6) are feasible.

(b) If $\boldsymbol{C}$ is isotropic and $\boldsymbol{A}$, $\boldsymbol{B}$ are not isotropic, then $\boldsymbol{C}^*$ may be isotropic, and $\boldsymbol{A}^*$ or $\boldsymbol{B}^*$ (but not both) may also be isotropic.

Indeed, the bivector $\boldsymbol{C}^*$ is isotropic when

$$(\boldsymbol{A}\times\boldsymbol{B})\cdot(\boldsymbol{A}\times\boldsymbol{B})=(\boldsymbol{A}\cdot\boldsymbol{A})(\boldsymbol{B}\cdot\boldsymbol{B})-(\boldsymbol{A}\cdot\boldsymbol{B})^2=0, \qquad (2.11.7)$$

and the bivector $\boldsymbol{A}^*$ ($\boldsymbol{B}^*$) is isotropic when $\boldsymbol{B}\cdot\boldsymbol{C}=0$ ($\boldsymbol{C}\cdot\boldsymbol{A}=0$).

Exercises 2.7

1. Let $\boldsymbol{A}=\mathbf{i}+i\mathbf{j}$, $\boldsymbol{B}=\mathbf{j}+i\mathbf{k}$. Choose $\boldsymbol{C}$ such that $\boldsymbol{A}^*$ and $\boldsymbol{B}^*$ are isotropic. Determine $\boldsymbol{A}^*$, $\boldsymbol{B}^*$, $\boldsymbol{C}^*$.
2. Let $\boldsymbol{A}=\mathbf{i}-i\mathbf{j}-\mathbf{k}$, $\boldsymbol{B}=\mathbf{i}+i\mathbf{j}+\mathbf{k}$, $\boldsymbol{C}=\mathbf{i}+i\mathbf{j}$. Find $\boldsymbol{A}^*$, $\boldsymbol{B}^*$, $\boldsymbol{C}^*$. How many isotropic bivectors are contained in the triad $\boldsymbol{A}$, $\boldsymbol{B}$, $\boldsymbol{C}$ and $\boldsymbol{A}^*$, $\boldsymbol{B}^*$, $\boldsymbol{C}^*$?

2.12 Bivector decompositions

Given an arbitrary bivector $\boldsymbol{C}$ (not with a real direction), it may always be written as a linear combination of two linearly independent bivectors in its plane. For different choices of the two linearly independent bivectors, we obtain different decompositions of the bivector $\boldsymbol{C}$ as a sum of two bivectors. Here we consider the general case and three special cases.

General case

Let $\boldsymbol{A}$ and $\boldsymbol{B}$ be two linearly independent bivectors in the plane of $\boldsymbol{C}$. We may write

$$\boldsymbol{C}=\alpha\boldsymbol{A}+\beta\boldsymbol{B}, \qquad (2.12.1)$$

for some α and β. Introducing the reciprocal bivectors $\boldsymbol{A}_\perp$, $\boldsymbol{B}_\perp$ of $\boldsymbol{A}$ and $\boldsymbol{B}$, respectively, we note that

$$\boldsymbol{C}\cdot\boldsymbol{A}_\perp=\beta\boldsymbol{B}\cdot\boldsymbol{A}_\perp, \quad \boldsymbol{C}\cdot\boldsymbol{B}_\perp=\alpha\boldsymbol{A}\cdot\boldsymbol{B}_\perp, \qquad (2.12.2)$$

and hence

$$\boldsymbol{C}=\frac{\boldsymbol{C}\cdot\boldsymbol{B}_\perp}{\boldsymbol{A}\cdot\boldsymbol{B}_\perp}\boldsymbol{A}+\frac{\boldsymbol{C}\cdot\boldsymbol{A}_\perp}{\boldsymbol{B}\cdot\boldsymbol{A}_\perp}\boldsymbol{B}. \qquad (2.12.3)$$

Remark Because $\boldsymbol{A}$ and $\boldsymbol{B}$ are not parallel, it follows that $\boldsymbol{A}\cdot\boldsymbol{B}_\perp\neq0$ and $\boldsymbol{B}\cdot\boldsymbol{A}_\perp\neq0$. Indeed, we know (see section 2.5) that if two bivectors are orthogonal and in the same plane, each is parallel to the reciprocal of the other. Hence $\boldsymbol{A}\cdot\boldsymbol{B}_{\perp}=0$ (or $\boldsymbol{B}\cdot\boldsymbol{A}_{\perp}=0$) is possible only with $\boldsymbol{B}$ parallel to $\boldsymbol{A}$ and then also $\boldsymbol{B}\cdot\boldsymbol{A}_{\perp}=0$ (or $\boldsymbol{A}\cdot\boldsymbol{B}_{\perp}=0$).

Special case (a)

Let $\boldsymbol{A}$ be an isotropic bivector in the plane of $\boldsymbol{C}$:

$$\boldsymbol{A}\cdot\boldsymbol{A}=0, \quad \boldsymbol{A}\cdot(\boldsymbol{C}\times\bar{\boldsymbol{C}})=0. \tag{2.12.4}$$

Then $\boldsymbol{A}$ and $\bar{\boldsymbol{A}}$ are two linearly independent bivectors in the plane of $\boldsymbol{C}$.

Thus, we may write

$$\boldsymbol{C}=\alpha\boldsymbol{A}+\beta\bar{\boldsymbol{A}}, \tag{2.12.5}$$

for some α and β. Since both $\boldsymbol{A}$ and $\bar{\boldsymbol{A}}$ are isotropic, α and β are given by

$$\alpha=\frac{\bar{\boldsymbol{A}}\cdot\boldsymbol{C}}{\boldsymbol{A}\cdot\bar{\boldsymbol{A}}}, \quad \beta=\frac{\boldsymbol{A}\cdot\boldsymbol{C}}{\boldsymbol{A}\cdot\bar{\boldsymbol{A}}}, \tag{2.12.6}$$

and thus

$$\boldsymbol{C}=\boldsymbol{C}_1+\boldsymbol{C}_2, \quad \text{with} \quad \boldsymbol{C}_1=\frac{\bar{\boldsymbol{A}}\cdot\boldsymbol{C}}{\boldsymbol{A}\cdot\bar{\boldsymbol{A}}}\boldsymbol{A}, \quad \boldsymbol{C}_2=\frac{\boldsymbol{A}\cdot\boldsymbol{C}}{\boldsymbol{A}\cdot\bar{\boldsymbol{A}}}\bar{\boldsymbol{A}}, \tag{2.12.7}$$

which is a decomposition of $\boldsymbol{C}$ into the sum of two isotropic bivectors $\boldsymbol{C}_1$, $\boldsymbol{C}_2$ in its plane. Associated with $\boldsymbol{C}_1$ and $\boldsymbol{C}_2$ are two circles with opposite orientations.

Remarks

(a) This decomposition is unique. Indeed, consider a real orthonormal basis $\mathbf{i}, \mathbf{j}, \mathbf{k}$ with $\mathbf{i}$ and $\mathbf{j}$ in the plane of $\boldsymbol{C}$. Then, either $\boldsymbol{A}=\lambda(\mathbf{i}+i\mathbf{j})$, or $\boldsymbol{A}=\lambda(\mathbf{i}-i\mathbf{j})$ for some λ. Clearly, in each case, $\boldsymbol{C}_1$ and $\boldsymbol{C}_2$ given by equations (2.12.7) do not depend on λ. Moreover, the two possibilities for $\boldsymbol{A}$ correspond to the two orderings $\boldsymbol{C}_1+\boldsymbol{C}_2$ and $\boldsymbol{C}_2+\boldsymbol{C}_1$ of the terms of the decomposition.

(b) When $\boldsymbol{C}$ has a real direction, $\boldsymbol{C}=\lambda\hat{\mathbf{n}}$, then $\boldsymbol{A}$ may be chosen to be isotropic in any plane containing $\hat{\mathbf{n}}$, and the same decomposition (2.12.7) is valid (however, it ceases to be unique). In this case, the circles associated with $\boldsymbol{C}_1$ and $\boldsymbol{C}_2$ are the same ($\boldsymbol{C}_1\cdot\bar{\boldsymbol{C}}_1=\boldsymbol{C}_2\cdot\bar{\boldsymbol{C}}_2$), but, of course, with opposite orientations.

Special case (b)

Let $\boldsymbol{A}$ be any nonisotropic bivector (not with a real direction) in the plane of $\boldsymbol{C}$:

$$\boldsymbol{A}\cdot\boldsymbol{A}\neq 0, \quad \boldsymbol{A}\cdot(\boldsymbol{C}\times\bar{\boldsymbol{C}})=0. \tag{2.12.8}$$

Then the bivector $\boldsymbol{A}$ and its reciprocal bivector $\boldsymbol{A}_\perp$ are two linearly independent bivectors in the plane of $\boldsymbol{C}$. Thus we may write

$$\boldsymbol{C}=\alpha\boldsymbol{A}+\beta\boldsymbol{A}_\perp. \tag{2.12.9}$$

Because $\boldsymbol{A}\cdot\boldsymbol{A}_\perp=0$, α and β are now given by

$$\alpha=\frac{\boldsymbol{A}\cdot\boldsymbol{C}}{\boldsymbol{A}\cdot\boldsymbol{A}}, \quad \beta=\frac{\boldsymbol{A}_\perp\cdot\boldsymbol{C}}{\boldsymbol{A}_\perp\cdot\boldsymbol{A}_\perp}, \tag{2.12.10}$$

and thus

$$\boldsymbol{C}=\boldsymbol{C}_1+\boldsymbol{C}_2, \quad \text{with} \quad \boldsymbol{C}_1=\frac{\boldsymbol{A}\cdot\boldsymbol{C}}{\boldsymbol{A}\cdot\boldsymbol{A}}\boldsymbol{A}, \quad \boldsymbol{C}_2=\frac{\boldsymbol{A}_\perp\cdot\boldsymbol{C}}{\boldsymbol{A}_\perp\cdot\boldsymbol{A}_\perp}\boldsymbol{A}_\perp, \tag{2.12.11}$$

which is a decomposition of $\boldsymbol{C}$ into the sum of two orthogonal bivectors $\boldsymbol{C}_1$, $\boldsymbol{C}_2$ in its plane. The ellipse of $\boldsymbol{C}_2$ is similar and similarly situated to the ellipse of $\boldsymbol{C}_1$ rotated through a quadrant and both ellipses have the same orientation.

Remarks

(a) Clearly, this decomposition is not unique because $\boldsymbol{A}$ may be any nonisotropic bivector in the plane of $\boldsymbol{C}$.

(b) When $\boldsymbol{C}$ is isotropic, $\boldsymbol{C}\cdot\boldsymbol{C}=0$, we have

$$\boldsymbol{C}_2\cdot\boldsymbol{C}_2=-\boldsymbol{C}_1\cdot\boldsymbol{C}_1. \tag{2.12.12}$$

If $\boldsymbol{C}_1$ is written

$$\boldsymbol{C}_1=\mathrm{e}^{-\mathrm{i}\theta}(a\mathbf{i}+\mathrm{i}b\mathbf{j}), \tag{2.12.13}$$

then, because $\boldsymbol{C}_2$ is in the same plane as $\boldsymbol{C}_1$ with $\boldsymbol{C}_1\cdot\boldsymbol{C}_2=0$ and $\boldsymbol{C}_2\cdot\boldsymbol{C}_2=-\boldsymbol{C}_1\cdot\boldsymbol{C}_1$, we have

$$\boldsymbol{C}_2=\pm\mathrm{e}^{-\mathrm{i}\theta}(b\mathbf{i}+\mathrm{i}a\mathbf{j}). \tag{2.12.14}$$

Thus, in particular, the ellipse of $\boldsymbol{C}_2$ is exactly the ellipse of $\boldsymbol{C}_1$ rotated through a quadrant. We also note that, in this case, $\boldsymbol{C}_2^+$ is orthogonal to $\boldsymbol{C}_1^-$, and $\boldsymbol{C}_2^-$ is orthogonal to $\boldsymbol{C}_1^+$ (see Figure 2.4). This is the case considered by Airy for the decomposition of a circularly polarized wave into two elliptically polarized waves.

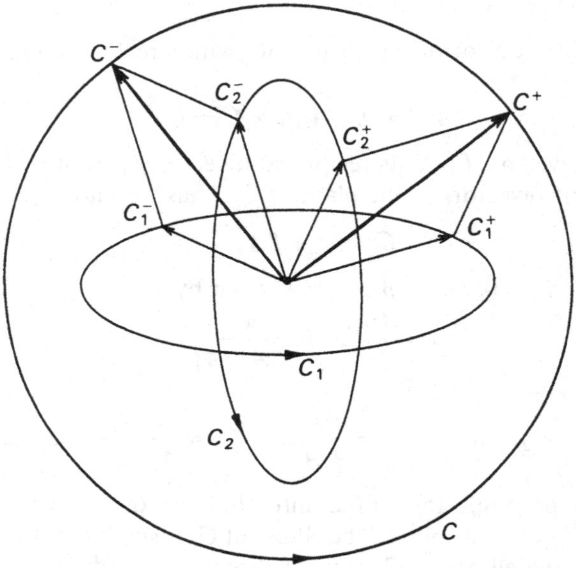

Figure 2.4 *Airy's decomposition of an isotropic bivector.*

Special case (c)

Let $\boldsymbol{A}$ be any bivector (not with a real direction) in the plane of $\boldsymbol{C}$:

$$\boldsymbol{A}\cdot(\boldsymbol{C}\times\bar{\boldsymbol{C}})=0. \tag{2.12.15}$$

Then the bivectors $\boldsymbol{A}$ and $\bar{\boldsymbol{A}}_\perp$ are two linearly independent bivectors in the plane of $\boldsymbol{C}$. We thus have

$$\boldsymbol{C}=\alpha\boldsymbol{A}+\beta\bar{\boldsymbol{A}}_\perp. \tag{2.12.16}$$

Because $\boldsymbol{A}\cdot\boldsymbol{A}_\perp=0$, α and β are here given by

$$\alpha=\frac{\bar{\boldsymbol{A}}\cdot\boldsymbol{C}}{\boldsymbol{A}\cdot\bar{\boldsymbol{A}}},\quad \beta=\frac{\boldsymbol{A}_\perp\cdot\boldsymbol{C}}{\boldsymbol{A}_\perp\cdot\bar{\boldsymbol{A}}_\perp}, \tag{2.12.17}$$

leading to the decomposition

$$\boldsymbol{C}=\boldsymbol{C}_1+\boldsymbol{C}_2,\quad \text{with}\quad \boldsymbol{C}_1=\frac{\bar{\boldsymbol{A}}\cdot\boldsymbol{C}}{\boldsymbol{A}\cdot\bar{\boldsymbol{A}}}\boldsymbol{A},\quad \boldsymbol{C}_2=\frac{\boldsymbol{A}_\perp\cdot\boldsymbol{C}}{\boldsymbol{A}_\perp\cdot\bar{\boldsymbol{A}}_\perp}\bar{\boldsymbol{A}}_\perp, \tag{2.12.18}$$

where $\boldsymbol{C}_1$ and $\boldsymbol{C}_2$ are such that $\boldsymbol{C}_1\cdot\bar{\boldsymbol{C}}_2=0$. Here the ellipse of $\boldsymbol{C}_2$ is similar and similarly situated to the ellipse of $\boldsymbol{C}_1$ rotated through a quadrant, but has opposite orientation.

Exercises 2.8

1. Let $\boldsymbol{C} = (5 + 3\mathrm{i})\mathbf{i} + (-1 + 2\mathrm{i})\mathbf{j}$. Decompose $\boldsymbol{C}$ into the sum of two isotropic bivectors $\boldsymbol{C}_1$ and $\boldsymbol{C}_2$ in its plane.
2. Consider the decomposition (2.12.18) with $\boldsymbol{C}$ isotropic. The bivector $\boldsymbol{C}_1$ being given by (2.12.13), write $\boldsymbol{C}_2$. Represent $\boldsymbol{C}_1$, $\boldsymbol{C}_2$ with their real and imaginary parts and their ellipses on a figure.
3. For real a, interpret the decomposition (Airy)

$$\boldsymbol{A} = a\mathbf{i} = \{(a - k)\mathbf{i} + \mathrm{i}b\mathbf{j}\} + (k\mathbf{i} - \mathrm{i}b\mathbf{j}),$$

where k and b are real. Consider the cases $k = b$ and $a - k = b$.
4. For real k, interpret the decomposition (Airy)

$$\boldsymbol{A} = (1 + k^2)\mathbf{i} = (\mathbf{i} + \mathrm{i}k\mathbf{j}) + (k^2\mathbf{i} - \mathrm{i}k\mathbf{j}).$$

2.13 Tensor product of bivectors

The tensor product $\boldsymbol{A} \otimes \boldsymbol{B}$ of two bivectors $\boldsymbol{A} = \boldsymbol{A}^+ + \mathrm{i}\boldsymbol{A}^-$ and $\boldsymbol{B} = \boldsymbol{B}^+ + \mathrm{i}\boldsymbol{B}^-$ is defined by

$$\boldsymbol{A} \otimes \boldsymbol{B} = \boldsymbol{A}^+ \otimes \boldsymbol{B}^+ - \boldsymbol{A}^- \otimes \boldsymbol{B}^- + \mathrm{i}(\boldsymbol{A}^+ \otimes \boldsymbol{B}^- + \boldsymbol{A}^- \otimes \boldsymbol{B}^+). \quad (2.13.1)$$

It is a complex second order tensor. For any two bivectors $\boldsymbol{X}$, $\boldsymbol{Y}$, the bilinear form $\boldsymbol{X}^{\mathrm{T}}(\boldsymbol{A} \otimes \boldsymbol{B})\boldsymbol{Y}$, where the superscript T denotes the transpose, is given by

$$\boldsymbol{X}^{\mathrm{T}}(\boldsymbol{A} \otimes \boldsymbol{B})\boldsymbol{Y} = (\boldsymbol{X} \cdot \boldsymbol{A})(\boldsymbol{B} \cdot \boldsymbol{Y}). \quad (2.13.2)$$

Theorem 2.12

The equation of the ellipse of the bivector $\boldsymbol{A} = \boldsymbol{c} + \mathrm{i}\boldsymbol{d}$ is

$$\boldsymbol{r}^{\mathrm{T}}\boldsymbol{\alpha}\boldsymbol{r} = 1 \quad \text{with} \quad \boldsymbol{\alpha} = \tfrac{1}{2}(\boldsymbol{A}_\perp \otimes \bar{\boldsymbol{A}}_\perp + \bar{\boldsymbol{A}}_\perp \otimes \boldsymbol{A}_\perp), \quad (2.13.3)$$

where $\boldsymbol{A}_\perp = \boldsymbol{c}^* + \mathrm{i}\boldsymbol{d}^*$ is the reciprocal of the bivector $\boldsymbol{A}$.

Proof From (1.3.9), the equation of the ellipse of the bivector $\boldsymbol{A}$ is

$$\boldsymbol{r}^{\mathrm{T}}\boldsymbol{\alpha}\boldsymbol{r} = 1, \quad \text{with} \quad \boldsymbol{\alpha} = \boldsymbol{c}^* \otimes \boldsymbol{c}^* + \boldsymbol{d}^* \otimes \boldsymbol{d}^*. \quad (2.13.4)$$

But

$$\begin{aligned} \boldsymbol{A}_\perp \otimes \bar{\boldsymbol{A}}_\perp &= (\boldsymbol{c}^* + \mathrm{i}\boldsymbol{d}^*) \otimes (\boldsymbol{c}^* - \mathrm{i}\boldsymbol{d}^*) \\ &= \boldsymbol{c}^* \otimes \boldsymbol{c}^* + \boldsymbol{d}^* \otimes \boldsymbol{d}^* + \mathrm{i}(\boldsymbol{d}^* \otimes \boldsymbol{c}^* - \boldsymbol{c}^* \otimes \boldsymbol{d}^*), \end{aligned} \quad (2.13.5)$$

thus

$$2(\boldsymbol{c}^* \otimes \boldsymbol{c}^* + \boldsymbol{d}^* \otimes \boldsymbol{d}^*) = \boldsymbol{A}_\perp \otimes \bar{\boldsymbol{A}}_\perp + \bar{\boldsymbol{A}}_\perp \otimes \boldsymbol{A}_\perp. \quad (2.13.6)$$

From this theorem it follows that the equation of the ellipse of the bivector $\boldsymbol{A}$ is

$$(\boldsymbol{A}_{\perp}\cdot\boldsymbol{r})(\bar{\boldsymbol{A}}_{\perp}\cdot\boldsymbol{r}) = 1. \tag{2.13.7}$$

Theorem 2.13

If $\boldsymbol{A}$, $\boldsymbol{B}$, $\boldsymbol{C}$ is any triad of linearly independent bivectors and $\boldsymbol{A}^*$, $\boldsymbol{B}^*$, $\boldsymbol{C}^*$ its reciprocal triad, then

$$\boldsymbol{A}\otimes\boldsymbol{A}^* + \boldsymbol{B}\otimes\boldsymbol{B}^* + \boldsymbol{C}\otimes\boldsymbol{C}^* = \mathbf{1}, \tag{2.13.8}$$

$$\boldsymbol{A}^*\otimes\boldsymbol{A} + \boldsymbol{B}^*\otimes\boldsymbol{B} + \boldsymbol{C}^*\otimes\boldsymbol{C} = \mathbf{1}. \tag{2.13.9}$$

Proof For any two bivectors $\boldsymbol{X}$, $\boldsymbol{Y}$ we have

$$\begin{aligned} &\boldsymbol{X}^{\mathrm{T}}(\boldsymbol{A}\otimes\boldsymbol{A}^* + \boldsymbol{B}\otimes\boldsymbol{B}^* + \boldsymbol{C}\otimes\boldsymbol{C}^*)\boldsymbol{Y} \\ &\quad = (\boldsymbol{X}\cdot\boldsymbol{A})(\boldsymbol{A}^*\cdot\boldsymbol{Y}) + (\boldsymbol{X}\cdot\boldsymbol{B})(\boldsymbol{B}^*\cdot\boldsymbol{Y}) + (\boldsymbol{X}\cdot\boldsymbol{C})(\boldsymbol{C}^*\cdot\boldsymbol{Y}) \\ &\quad = \boldsymbol{X}\cdot\{\boldsymbol{A}(\boldsymbol{A}^*\cdot\boldsymbol{Y}) + \boldsymbol{B}(\boldsymbol{B}^*\cdot\boldsymbol{Y}) + \boldsymbol{C}(\boldsymbol{C}^*\cdot\boldsymbol{Y})\}. \end{aligned} \tag{2.13.10}$$

But using (2.11.3) and the identity (2.8.4), we have

$$\boldsymbol{A}(\boldsymbol{A}^*\cdot\boldsymbol{Y}) + \boldsymbol{B}(\boldsymbol{B}^*\cdot\boldsymbol{Y}) + \boldsymbol{C}(\boldsymbol{C}^*\cdot\boldsymbol{Y}) = \boldsymbol{Y}. \tag{2.13.11}$$

Thus

$$\boldsymbol{X}^{\mathrm{T}}(\boldsymbol{A}\otimes\boldsymbol{A}^* + \boldsymbol{B}\otimes\boldsymbol{B}^* + \boldsymbol{C}\otimes\boldsymbol{C}^*)\boldsymbol{Y} = \boldsymbol{X}\cdot\boldsymbol{Y}. \tag{2.13.12}$$

Because this is valid for arbitrary $\boldsymbol{X}$ and $\boldsymbol{Y}$, we obtain (2.13.8). Identity (2.13.9) is the transpose of the identity (2.13.8).

Remark In particular, if $\boldsymbol{A}$, $\boldsymbol{B}$, $\boldsymbol{C}$ is an orthonormal triad of vectors or bivectors, then

$$\boldsymbol{A}\otimes\boldsymbol{A} + \boldsymbol{B}\otimes\boldsymbol{B} + \boldsymbol{C}\otimes\boldsymbol{C} = \mathbf{1}. \tag{2.13.13}$$

Exercises 2.9

1. Show that the equation (2.13.7) may also be written as

$$4\{(\boldsymbol{A}\times\boldsymbol{r})\cdot(\boldsymbol{A}\times\bar{\boldsymbol{A}})\}\{(\bar{\boldsymbol{A}}\times\boldsymbol{r})\cdot(\bar{\boldsymbol{A}}\times\boldsymbol{A})\} = \{(\boldsymbol{A}\times\bar{\boldsymbol{A}})\cdot(\boldsymbol{A}\times\bar{\boldsymbol{A}})\}^2.$$

2. Show that $\frac{1}{2}(\boldsymbol{A}\otimes\bar{\boldsymbol{A}}_{\perp} + \bar{\boldsymbol{A}}\otimes\boldsymbol{A}_{\perp}) = \mathbf{1} - \mathbf{k}\otimes\mathbf{k}$, where $\mathbf{k}$ is the unit vector orthogonal to the plane of $\boldsymbol{A}$.
3. Show that

$$\begin{aligned} (\boldsymbol{A}\times\boldsymbol{B})\otimes(\boldsymbol{A}\times\boldsymbol{B}) = {} & (\boldsymbol{A}\times\boldsymbol{B})\cdot(\boldsymbol{A}\times\boldsymbol{B})\mathbf{1} + (\boldsymbol{A}\cdot\boldsymbol{B})(\boldsymbol{A}\otimes\boldsymbol{B} + \boldsymbol{B}\otimes\boldsymbol{A}) \\ & - (\boldsymbol{B}\cdot\boldsymbol{B})(\boldsymbol{A}\otimes\boldsymbol{A}) - (\boldsymbol{A}\cdot\boldsymbol{A})(\boldsymbol{B}\otimes\boldsymbol{B}). \end{aligned}$$

3

Complex symmetric matrices

Complex symmetric matrices arise naturally in many areas of mechanics and electromagnetism. For example, the study of acoustic wave propagation in elastic materials and the study of electromagnetic waves in crystals are described in terms of such matrices. For that reason, we present in detail here the properties of the eigenvalues and eigenbivectors of such matrices.

3.1 Eigenbivectors and eigenvalues

We consider complex 3×3 matrices

$$Q = Q^+ + \mathrm{i}Q^-, \tag{3.1.1}$$

where Q^+ and Q^- are two real 3×3 symmetric matrices. Thus

$$Q^{\mathrm{T}} = Q. \tag{3.1.2}$$

Definition The bivector $X(\neq 0)$ is said to be an eigenbivector of the complex matrix Q when

$$QX = \lambda X, \tag{3.1.3}$$

where λ is a complex number. Thus number λ is called the eigenvalue corresponding to the eigenbivector X.

Writing $X = X^+ + \mathrm{i}X^-$ and $\lambda = \lambda^+ + \mathrm{i}\lambda^-$, we note that (3.1.3) is equivalent to the system

$$\begin{aligned} Q^+X^+ - Q^-X^- &= \lambda^+X^+ - \lambda^-X^-, \\ Q^-X^+ + Q^+X^- &= \lambda^-X^+ + \lambda^+X^-. \end{aligned} \tag{3.1.4}$$

As for real matrices, the eigenvalues are the solutions of the algebraic equation

$$\det(Q - \lambda \mathbf{1}) = 0. \tag{3.1.5}$$

For a given eigenvalue λ the eigenbivectors are then obtained by solving the algebraic homogeneous system

$$(\boldsymbol{Q}-\lambda\mathbf{1})\boldsymbol{X}=\mathbf{0}. \tag{3.1.6}$$

Theorem 3.1 (Synge)

If $\boldsymbol{A}$ and $\boldsymbol{B}$ are eigenbivectors of the complex symmetric matrix $\boldsymbol{Q}$, corresponding to different eigenvalues, then $\boldsymbol{A}$ and $\boldsymbol{B}$ are linearly independent ($\boldsymbol{B}\neq\alpha\boldsymbol{A}$), and orthogonal: $\boldsymbol{A}\cdot\boldsymbol{B}=0$.

Proof Let λ and μ be the eigenvalues corresponding to the eigenbivectors $\boldsymbol{A}$ and $\boldsymbol{B}$, respectively:

$$\boldsymbol{Q}\boldsymbol{A}=\lambda\boldsymbol{A}, \quad \boldsymbol{Q}\boldsymbol{B}=\mu\boldsymbol{B}, \quad \lambda\neq\mu. \tag{3.1.7}$$

Assuming $\boldsymbol{B}=\alpha\boldsymbol{A}$, we obtain $\boldsymbol{Q}\boldsymbol{B}=\alpha\boldsymbol{Q}\boldsymbol{A}=\alpha\lambda\boldsymbol{A}=\lambda\boldsymbol{B}$, and thus $\mu=\lambda$ which is contrary to the assumption $\mu\neq\lambda$. Thus $\boldsymbol{A}$ and $\boldsymbol{B}$ are linearly independent. Also from (3.1.7) we have

$$\boldsymbol{B}^{\mathrm{T}}\boldsymbol{Q}\boldsymbol{A}=\lambda\boldsymbol{A}\cdot\boldsymbol{B}, \quad \boldsymbol{A}^{\mathrm{T}}\boldsymbol{Q}\boldsymbol{B}=\mu\boldsymbol{A}\cdot\boldsymbol{B}, \tag{3.1.8}$$

and because $\boldsymbol{B}^{\mathrm{T}}\boldsymbol{Q}\boldsymbol{A}=\boldsymbol{A}^{\mathrm{T}}\boldsymbol{Q}\boldsymbol{B}$ ($\boldsymbol{Q}$ is symmetric),

$$(\lambda-\mu)\boldsymbol{A}\cdot\boldsymbol{B}=0, \tag{3.1.9}$$

and thus $\boldsymbol{A}\cdot\boldsymbol{B}=0$ because $\mu\neq\lambda$.

Corollary 3.2 (Synge)

If $\boldsymbol{A}$ and $\boldsymbol{B}$ are eigenbivectors of $\boldsymbol{Q}$ corresponding to different eigenvalues, then $\boldsymbol{A}$ and $\boldsymbol{B}$ cannot both be isotropic.

Proof Indeed, $\boldsymbol{A}$ and $\boldsymbol{B}$ are orthogonal and linearly independent. But, if $\boldsymbol{A}$ and $\boldsymbol{B}$ are both isotropic and orthogonal, then $\boldsymbol{A}$ is parallel to $\boldsymbol{B}$ (Theorem 2.2) which contradicts the statement that $\boldsymbol{A}$ and $\boldsymbol{B}$ are linearly independent.

3.2 Isotropic eigenbivectors

Here we consider the possibility that the complex symmetric matrix $\boldsymbol{Q}$ has isotropic eigenbivectors. The main result is that a necessary and sufficient condition that a complex symmetric 3×3 matrix have an isotropic eigenbivector is that the matrix have a double or triple eigenvalue.

Theorem 3.3

If the complex symmetric matrix $\boldsymbol{Q}$ has an isotropic eigenbivector, then the corresponding eigenvalue must be at least double.

Proof Let $\boldsymbol{A}$ be an isotropic eigenbivector of $\boldsymbol{Q}$:

$$\boldsymbol{Q}\boldsymbol{A} = \lambda \boldsymbol{A}, \quad \boldsymbol{A}\cdot\boldsymbol{A} = 0. \tag{3.2.1}$$

Suppose that $\mathbf{i}$ and $\mathbf{j}$ are orthogonal unit vectors in the plane of $\boldsymbol{A}$. Then, because an eigenbivector is defined up to a factor, we may write, without loss of generality,

$$\boldsymbol{A} = \mathbf{i} + \mathrm{i}\mathbf{j}, \quad \mathbf{i}\cdot\mathbf{i} = \mathbf{j}\cdot\mathbf{j} = 1, \quad \mathbf{i}\cdot\mathbf{j} = 0. \tag{3.2.2}$$

Then let $\mathbf{k} = \mathbf{i} \times \mathbf{j}$. If the elements of $\boldsymbol{Q}$ are now denoted by $Q_{\alpha\beta}$, so that, e.g. $Q_{12} = \mathbf{i}^{\mathrm{T}}\boldsymbol{Q}\mathbf{j}$, it then follows from (3.2.1) that $Q_{31} + \mathrm{i}Q_{32} = 0$. Let $\boldsymbol{T}$ be the matrix whose columns are $\boldsymbol{A}$, $\bar{\boldsymbol{A}}$, $\mathbf{k}$:

$$\boldsymbol{T} = (\boldsymbol{A}\,|\,\bar{\boldsymbol{A}}\,|\,\mathbf{k}). \tag{3.2.3}$$

Because $\boldsymbol{A}$, $\bar{\boldsymbol{A}}$, $\mathbf{k}$ are linearly independent, $\det \boldsymbol{T} \neq 0$. Then it is easily seen that

$$\boldsymbol{T}^{\mathrm{T}}\boldsymbol{T} = \begin{bmatrix} 0 & 2 & 0 \\ 2 & 0 & 0 \\ 0 & 0 & 1 \end{bmatrix}, \quad \boldsymbol{T}^{\mathrm{T}}\boldsymbol{Q}\boldsymbol{T} = \begin{bmatrix} 0 & 2\lambda & 0 \\ 2\lambda & \eta & \varepsilon \\ 0 & \varepsilon & \mu \end{bmatrix}, \tag{3.2.4}$$

where η, ε, μ are the complex numbers

$$\eta = \bar{\boldsymbol{A}}^{\mathrm{T}}\boldsymbol{Q}\bar{\boldsymbol{A}}, \quad \varepsilon = \mathbf{k}^{\mathrm{T}}\boldsymbol{Q}\bar{\boldsymbol{A}}, \quad \mu = \mathbf{k}^{\mathrm{T}}\boldsymbol{Q}\mathbf{k}. \tag{3.2.5}$$

The equation $\det(\boldsymbol{Q} - x\mathbf{1}) = 0$ for the determination of the eigenvalues x is equivalent to the equation $\det(\boldsymbol{T}^{\mathrm{T}}\boldsymbol{Q}\boldsymbol{T} - x\boldsymbol{T}^{\mathrm{T}}\boldsymbol{T}) = 0$ and thus reads

$$4(\lambda - x)^2(\mu - x) = 0. \tag{3.2.6}$$

Thus the eigenvalue λ is double (when $\mu \neq \lambda$) or triple (when $\mu = \lambda$).

Theorem 3.4

If $\boldsymbol{Q}$ has a double eigenvalue λ and a simple eigenvalue $\mu (\mu \neq \lambda)$, then corresponding to λ, **either** (a) There is a double infinity of eigenbivectors, and, among them, there are two linearly independent isotropic eigenbivectors **or** (b) There is a simple infinity of eigenbivectors, all isotropic.

Proof Let $\boldsymbol{C}$ be an eigenbivector corresponding to the simple eigenvalue μ:

$$\boldsymbol{QC} = \mu \boldsymbol{C}. \tag{3.2.7}$$

From Theorem 3.3, it follows that $\boldsymbol{C}$ must not be isotropic, so that, without loss of generality, we take $\boldsymbol{C}\cdot\boldsymbol{C} = 1$. Now, either $\boldsymbol{Q}$ possesses a nonisotropic eigenbivector corresponding to λ, or all eigenbivectors corresponding to λ are isotropic.

(a) Assume that $\boldsymbol{Q}$ possesses a nonisotropic eigenbivector corresponding to λ:

$$\boldsymbol{QB} = \lambda \boldsymbol{B}, \quad \boldsymbol{B}\cdot\boldsymbol{B} \neq 0. \tag{3.2.8}$$

From Theorem 3.1, we know that $\boldsymbol{B}$ is orthogonal to $\boldsymbol{C}$: $\boldsymbol{B}\cdot\boldsymbol{C} = 0$. Without loss of generality, we take $\boldsymbol{B}\cdot\boldsymbol{B} = 1$, and let $\boldsymbol{A} = \boldsymbol{B} \times \boldsymbol{C}$. The bivectors $\boldsymbol{A}$, $\boldsymbol{B}$, $\boldsymbol{C}$ form an orthonormal triad. Let $\boldsymbol{T}$ be the matrix whose columns are $\boldsymbol{A}$, $\boldsymbol{B}$, $\boldsymbol{C}$:

$$\boldsymbol{T} = (\boldsymbol{A} \,|\, \boldsymbol{B} \,|\, \boldsymbol{C}). \tag{3.2.9}$$

Then it is easily seen that

$$\boldsymbol{T}^{\mathrm{T}}\boldsymbol{T} = \mathbf{1}, \quad \boldsymbol{T}^{\mathrm{T}}\boldsymbol{QT} = \begin{bmatrix} \boldsymbol{A}^{\mathrm{T}}\boldsymbol{QA} & 0 & 0 \\ 0 & \lambda & 0 \\ 0 & 0 & \mu \end{bmatrix}. \tag{3.2.10}$$

But λ is a double eigenvalue. Therefore $\lambda = \boldsymbol{A}^{\mathrm{T}}\boldsymbol{QA}$, and thus $\boldsymbol{A}$ is also an eigenbivector of $\boldsymbol{Q}$ corresponding to the eigenvalue λ:

$$\boldsymbol{QA} = \lambda \boldsymbol{A}. \tag{3.2.11}$$

Hence, any linear combination of $\boldsymbol{A}$ and $\boldsymbol{B}$, that is any bivector orthogonal to $\boldsymbol{C}$, is an eigenbivector of $\boldsymbol{Q}$ corresponding to the double eigenvalue λ. Among this double infinity of eigenbivectors, the bivectors $\boldsymbol{A} + \mathrm{i}\boldsymbol{B}$ and $\boldsymbol{A} - \mathrm{i}\boldsymbol{B}$ are isotropic and linearly independent.

(b) Assume now that all eigenbivectors corresponding to λ are isotropic. Let $\boldsymbol{A}$ and $\boldsymbol{A}'$ be any two of them:

$$\boldsymbol{QA} = \lambda \boldsymbol{A}, \quad \boldsymbol{QA}' = \lambda \boldsymbol{A}', \quad \boldsymbol{A}\cdot\boldsymbol{A} = \boldsymbol{A}'\cdot\boldsymbol{A}' = 0. \tag{3.2.12}$$

Then $\boldsymbol{A} + \boldsymbol{A}'$ is also an eigenbivector of $\boldsymbol{Q}$ corresponding to λ. Thus it must be isotropic: $(\boldsymbol{A} + \boldsymbol{A}')\cdot(\boldsymbol{A} + \boldsymbol{A}') = 0$ and hence

$$\boldsymbol{A}\cdot\boldsymbol{A}' = 0. \tag{3.2.13}$$

Thus, from Synge's theorem (Theorem 2.2, section 2.6), it follows that $\boldsymbol{A}'$ is parallel to $\boldsymbol{A}$. This means that all eigenbivectors of $\boldsymbol{Q}$ corresponding to λ are parallel and isotropic.

Theorem 3.5

If $\boldsymbol{Q}$ has a triple eigenvalue λ, then corresponding to λ, **either**

(a) There is a triple infinity of eigenbivectors, and, among them, there are three linearly independent isotropic eigenbivectors, **or**
(b) There is a double infinity of eigenbivectors, and, among them, there is one isotropic eigenbivector (defined up to a factor), **or**
(c) There is a simple infinity of eigenbivectors, all isotropic.

Proof Either $\boldsymbol{Q}$ possesses a nonisotropic eigenbivector corresponding to λ, or all eigenbivectors corresponding to λ are isotropic. ((a) and (b)) Assume first that $\boldsymbol{Q}$ possesses a nonisotropic bivector $\boldsymbol{C}$ corresponding to λ:

$$\boldsymbol{QC} = \lambda\boldsymbol{C}, \quad \boldsymbol{C}\cdot\boldsymbol{C} \neq 0. \tag{3.2.14}$$

Without loss of generality, we take $\boldsymbol{C}\cdot\boldsymbol{C} = 1$. Then choose $\boldsymbol{A}$, $\boldsymbol{B}$ to complete an orthogonal triad $\boldsymbol{A}$, $\boldsymbol{B}$, $\boldsymbol{C}$, and let $\boldsymbol{T}$ be the matrix whose columns are $\boldsymbol{A}$, $\boldsymbol{B}$, $\boldsymbol{C}$:

$$\boldsymbol{T} = (\boldsymbol{A}\,|\,\boldsymbol{B}\,|\,\boldsymbol{C}). \tag{3.2.15}$$

Then it is easily shown that

$$\boldsymbol{T}^{\mathrm{T}}\boldsymbol{T} = \mathbf{1}, \quad \boldsymbol{T}^{\mathrm{T}}\boldsymbol{QT} = \begin{bmatrix} \alpha & \gamma & 0 \\ \gamma & \beta & 0 \\ 0 & 0 & \lambda \end{bmatrix}, \tag{3.2.16}$$

where α, β, γ are the complex numbers

$$\alpha = \boldsymbol{A}^{\mathrm{T}}\boldsymbol{QA}, \quad \beta = \boldsymbol{B}^{\mathrm{T}}\boldsymbol{QB}, \quad \gamma = \boldsymbol{A}^{\mathrm{T}}\boldsymbol{QB} = \boldsymbol{B}^{\mathrm{T}}\boldsymbol{QA}. \tag{3.2.17}$$

But λ is a triple eigenvalue. Therefore $\alpha + \beta = 2\lambda$, $\alpha\beta - \gamma^2 = \lambda^2$, and hence

$$\alpha = \lambda - \mathrm{i}\gamma, \quad \beta = \lambda + \mathrm{i}\gamma. \tag{3.2.18}$$

Let us now distinguish the cases when $\gamma = 0$ and $\gamma \neq 0$.

(a) If $\gamma = 0$, then $\alpha = \beta = \lambda$ and $\boldsymbol{T}^{\mathrm{T}}\boldsymbol{Q}\boldsymbol{T} = \lambda\mathbf{1}$ and thus $\boldsymbol{Q} = \lambda\mathbf{1}$. Every bivector is then an eigenbivector of $\boldsymbol{Q}$ corresponding to λ. Among this triple infinity of eigenbivectors, consider, for instance, the bivectors $\boldsymbol{A} + \mathrm{i}\boldsymbol{B}$, $\boldsymbol{B} + \mathrm{i}\boldsymbol{C}$, $\boldsymbol{C} + \mathrm{i}\boldsymbol{A}$. These are isotropic and linearly independent.

(b) If $\gamma \neq 0$, $\boldsymbol{A} + \mathrm{i}\boldsymbol{B}$ is an isotropic eigenbivector of $\boldsymbol{Q}$ corresponding to λ. Hence any linear combination of $\boldsymbol{A} + \mathrm{i}\boldsymbol{B}$ and $\boldsymbol{C}$ is an eigenbivector of $\boldsymbol{Q}$ corresponding to the double eigenvalue λ. Among this double infinity of eigenvectors, only $\boldsymbol{A} + \mathrm{i}\boldsymbol{B}$ and its scalar multiples are isotropic.

(c) Assume now that all eigenbivectors corresponding to λ are isotropic. Then, using the same argument as in the proof of Theorem 3.3 (case (a)), we note that all these eigenbivectors are parallel. There is thus a simple infinity of eigenbivectors, all isotropic.

From Theorems 3.3–3.5 we immediately obtain the following corollary.

Corollary 3.6

A necessary and sufficient condition that $\boldsymbol{Q}$ have an isotropic eigenbivector corresponding to the eigenvalue λ is that λ be at least double.

Exercises 3.1

1. Let

$$\boldsymbol{Q} = \begin{bmatrix} 2+\mathrm{i} & 1 & 0 \\ 1 & 2-\mathrm{i} & 0 \\ 0 & 0 & \mathrm{i} \end{bmatrix}.$$

Compute the eigenvalues and eigenbivectors.
Find the isotropic eigenbivectors.

2. Let

$$\boldsymbol{Q} = \begin{bmatrix} 1 & 1+\mathrm{i} & 1-\mathrm{i} \\ 1+\mathrm{i} & -1+2\mathrm{i} & 1+\mathrm{i} \\ 1-\mathrm{i} & 1+\mathrm{i} & 1 \end{bmatrix}.$$

Same question.

3. Let

$$Q = \begin{bmatrix} 1+a & ia & b \\ ia & 1-a & ib \\ b & ib & 1 \end{bmatrix},$$

with a and b real. Find the eigenvalues and eigenbivectors. Discuss the different cases that may occur for different choices of a and b.

3.3 Canonical form. Spectral decomposition

Let $\boldsymbol{T}$ be a complex orthogonal 3×3 matrix, that is a complex 3×3 matrix such that

$$\boldsymbol{T}^{\mathrm{T}}\boldsymbol{T} = \mathbf{1} = \boldsymbol{T}\boldsymbol{T}^{\mathrm{T}}. \tag{3.3.1}$$

The transform $\tilde{\boldsymbol{Q}}$ of the complex symmetric 3×3 matrix $\boldsymbol{Q}$ is then defined by

$$\tilde{\boldsymbol{Q}} = \boldsymbol{T}^{\mathrm{T}}\boldsymbol{Q}\boldsymbol{T}. \tag{3.3.2}$$

It is also a symmetric matrix. We show that with an appropriate orthogonal matrix $\boldsymbol{T}$, the matrix $\tilde{\boldsymbol{Q}}$ takes a particularly simple form called the 'canonical form'. Corresponding to the canonical form an expression of $\boldsymbol{Q}$ in terms of its eigenvalues and eigenvectors may be obtained. Such an expression is called the 'spectral decomposition' of $\boldsymbol{Q}$. Three cases have to be distinguished. Either $\boldsymbol{Q}$ possesses three simple eigenvalues λ, μ, ν ($\lambda \neq \mu, \mu \neq \nu, \nu \neq \lambda$), or $\boldsymbol{Q}$ possesses a double eigenvalue λ and a simple eigenvalue $\mu (\mu \neq \lambda)$, or $\boldsymbol{Q}$ possesses a triple eigenvalue λ.

Case (a) Three simple eigenvalues λ, μ, ν

Let $\boldsymbol{A}, \boldsymbol{B}, \boldsymbol{C}$ be eigenbivectors corresponding to the eigenvalues:

$$\boldsymbol{Q}\boldsymbol{A} = \lambda\boldsymbol{A}, \quad \boldsymbol{Q}\boldsymbol{B} = \mu\boldsymbol{B}, \quad \boldsymbol{Q}\boldsymbol{C} = \nu\boldsymbol{C}. \tag{3.3.3}$$

From Theorem 3.3, $\boldsymbol{A}, \boldsymbol{B}, \boldsymbol{C}$ are not isotropic and thus we may take $\boldsymbol{A}\cdot\boldsymbol{A} = \boldsymbol{B}\cdot\boldsymbol{B} = \boldsymbol{C}\cdot\boldsymbol{C} = 1$. From Theorem 3.1, it follows that $\boldsymbol{A}, \boldsymbol{B}, \boldsymbol{C}$ is then an orthonormal triad. The matrix

$$\boldsymbol{T} = (\boldsymbol{A} \,|\, \boldsymbol{B} \,|\, \boldsymbol{C}), \tag{3.3.4}$$

whose columns are $\boldsymbol{A}, \boldsymbol{B}, \boldsymbol{C}$ is thus orthogonal. With this matrix, we have

$$\tilde{Q} = T^{\mathrm{T}} Q T = \begin{bmatrix} \lambda & 0 & 0 \\ 0 & \mu & 0 \\ 0 & 0 & \nu \end{bmatrix}. \tag{3.3.5}$$

This is the canonical form of Q in case (a). Because (3.3.5) may also be written

$$\tilde{Q} = \lambda \mathbf{i} \otimes \mathbf{i} + \mu \mathbf{j} \otimes \mathbf{j} + \nu \mathbf{k} \otimes \mathbf{k}, \tag{3.3.6}$$

it follows that

$$Q = \lambda A \otimes A + \mu B \otimes B + \nu C \otimes C. \tag{3.3.7}$$

This is the spectral decomposition of Q in case (a).

Case (b) One double eigenvalue λ, and one simple eigenvalue μ

Let C be an eigenbivector corresponding to μ:

$$QC = \mu C. \tag{3.3.8}$$

Because C is nonisotropic, we may take $C \cdot C = 1$.

From Theorem 3.4, we know that two subcases have to be considered.

(i) Corresponding to λ, there is a double infinity of eigenbivectors
Then, all the bivectors orthogonal to C are eigenbivectors of Q corresponding to the double eigenvalue λ. Consider any pair A, B of them such that $A \cdot B = 0$, $A \cdot A = B \cdot B = 1$. Then A, B, C is an orthonormal triad and

$$QA = \lambda A, \quad QB = \lambda B. \tag{3.3.9}$$

Thus, with the orthogonal matrix $T = (A|B|C)$, we have

$$\tilde{Q} = T^{\mathrm{T}} Q T = \begin{bmatrix} \lambda & 0 & 0 \\ 0 & \lambda & 0 \\ 0 & 0 & \mu \end{bmatrix}. \tag{3.3.10}$$

This is the canonical form of Q in case (b)(i). As in case (a), it follows from (3.3.10) that we have the spectral decomposition

$$Q = \lambda(A \otimes A + B \otimes B) + \mu C \otimes C. \tag{3.3.11}$$

(ii) Corresponding to λ, there is a simple infinity of eigenbivectors Then, all these eigenbivectors are isotropic and parallel. Let $\boldsymbol{D}$ be one of them:

$$\boldsymbol{QD} = \lambda \boldsymbol{D}, \tag{3.3.12a}$$

$$\boldsymbol{D} \cdot \boldsymbol{D} = 0. \tag{3.3.12b}$$

From Theorem 3.1, we have $\boldsymbol{C} \cdot \boldsymbol{D} = 0$. Then choose $\boldsymbol{A}$, $\boldsymbol{B}$ to complete an orthonormal triad $\boldsymbol{A}, \boldsymbol{B}, \boldsymbol{C}$. Because $\boldsymbol{C} \cdot \boldsymbol{D} = \boldsymbol{D} \cdot \boldsymbol{D} = 0$, we may take, without loss of generality,

$$\boldsymbol{D} = \boldsymbol{A} + \mathrm{i}\boldsymbol{B}. \tag{3.3.13}$$

Taking the dot product of (3.3.12a) with $\boldsymbol{A}$ and $\boldsymbol{B}$ thus yields

$$\begin{aligned} \boldsymbol{A}^{\mathrm{T}}\boldsymbol{QA} + \mathrm{i}\boldsymbol{A}^{\mathrm{T}}\boldsymbol{QB} &= \lambda, \\ \boldsymbol{B}^{\mathrm{T}}\boldsymbol{QA} + \mathrm{i}\boldsymbol{B}^{\mathrm{T}}\boldsymbol{QB} &= \mathrm{i}\lambda, \end{aligned} \tag{3.3.14}$$

and writing

$$\gamma = \boldsymbol{A}^{\mathrm{T}}\boldsymbol{QB} = \boldsymbol{B}^{\mathrm{T}}\boldsymbol{QA}, \tag{3.3.15}$$

it follows that

$$\boldsymbol{A}^{\mathrm{T}}\boldsymbol{QA} = \lambda - \mathrm{i}\gamma, \quad \boldsymbol{B}^{\mathrm{T}}\boldsymbol{QB} = \lambda + \mathrm{i}\gamma. \tag{3.3.16}$$

Thus with the orthogonal matrix $\boldsymbol{T} = (\boldsymbol{A} \mid \boldsymbol{B} \mid \boldsymbol{C})$, we have

$$\tilde{\boldsymbol{Q}} = \boldsymbol{T}^{\mathrm{T}}\boldsymbol{QT} = \begin{bmatrix} \lambda - \mathrm{i}\gamma & \gamma & 0 \\ \gamma & \lambda + \mathrm{i}\gamma & 0 \\ 0 & 0 & \mu \end{bmatrix}. \tag{3.3.17}$$

Corresponding to (3.3.17) we have the spectral decomposition

$$\begin{aligned} \boldsymbol{Q} = (\lambda - \mathrm{i}\gamma)\boldsymbol{A} \otimes \boldsymbol{A} + (\lambda + \mathrm{i}\gamma)\boldsymbol{B} \otimes \boldsymbol{B} + \mu \boldsymbol{C} \otimes \boldsymbol{C} \\ + \gamma(\boldsymbol{A} \otimes \boldsymbol{B} + \boldsymbol{B} \otimes \boldsymbol{A}), \end{aligned} \tag{3.3.18}$$

that is, using (3.3.13),

$$\boldsymbol{Q} = \lambda(\boldsymbol{A} \otimes \boldsymbol{A} + \boldsymbol{B} \otimes \boldsymbol{B}) + \mu \boldsymbol{C} \otimes \boldsymbol{C} - \mathrm{i}\gamma \boldsymbol{D} \otimes \boldsymbol{D}, \tag{3.3.19}$$

or

$$\boldsymbol{Q} = \lambda(\boldsymbol{1} - \boldsymbol{C} \otimes \boldsymbol{C}) + \mu \boldsymbol{C} \otimes \boldsymbol{C} - \mathrm{i}\gamma \boldsymbol{D} \otimes \boldsymbol{D}. \tag{3.3.20}$$

Further simplification of (3.3.17), (3.3.19) and (3.3.20) is possible. Let $\gamma = \mathrm{e}^{2\mathrm{i}\Phi}$ where Φ is a complex number. Then consider the bivector

$$\boldsymbol{D}' = \mathrm{e}^{\mathrm{i}\Phi}\boldsymbol{D} = \boldsymbol{A}' + \mathrm{i}\boldsymbol{B}', \tag{3.3.21}$$

with

$$\boldsymbol{A}' = \cos\Phi\,\boldsymbol{A} - \sin\Phi\,\boldsymbol{B},$$
$$\boldsymbol{B}' = \sin\Phi\,\boldsymbol{A} + \cos\Phi\boldsymbol{B}. \tag{3.3.22}$$

Note that $\boldsymbol{A}'$, $\boldsymbol{B}'$, $\boldsymbol{C}$ form an orthonormal triad, and that

$$\boldsymbol{A}'^{\mathrm{T}}\boldsymbol{Q}\boldsymbol{B}' = 1. \tag{3.3.23}$$

Clearly, (3.3.19) and (3.3.20) may now be written

$$\boldsymbol{Q} = \lambda(\boldsymbol{A}'\otimes\boldsymbol{A}' + \boldsymbol{B}'\otimes\boldsymbol{B}') + \mu\boldsymbol{C}\otimes\boldsymbol{C} - \mathrm{i}\boldsymbol{D}'\otimes\boldsymbol{D}', \tag{3.3.24}$$

or

$$\boldsymbol{Q} = \lambda(\boldsymbol{1} - \boldsymbol{C}\otimes\boldsymbol{C}) + \mu\boldsymbol{C}\otimes\boldsymbol{C} - \mathrm{i}\boldsymbol{D}'\otimes\boldsymbol{D}'. \tag{3.3.25}$$

Thus, with the orthogonal matrix $\boldsymbol{T}' = (\boldsymbol{A}'\,|\,\boldsymbol{B}'\,|\,\boldsymbol{C}')$, we have

$$\tilde{\boldsymbol{Q}}' = \boldsymbol{T}'^{\mathrm{T}}\boldsymbol{Q}\boldsymbol{T}' = \begin{bmatrix} \lambda - \mathrm{i} & 1 & 0 \\ 1 & \lambda + \mathrm{i} & 0 \\ 0 & 0 & \mu \end{bmatrix}. \tag{3.3.26}$$

This is the canonical form of $\boldsymbol{Q}$ in case (b)(ii), the spectral decomposition being (3.3.25).

Case (c) One triple eigenvalue λ

From Theorem 3.5, we know that three subcases have to be considered.

(i) Corresponding to λ there is a triple infinity of eigenbivectors
Then, all the bivectors are eigenbivectors of $\boldsymbol{Q}$, and (see the proof of Theorem 3.5, case (a)),

$$\boldsymbol{Q} = \lambda\boldsymbol{1}. \tag{3.3.27}$$

Then, if $\boldsymbol{A}$, $\boldsymbol{B}$, $\boldsymbol{C}$ is any orthonormal triad, we may write

$$\boldsymbol{Q} = \lambda(\boldsymbol{A}\otimes\boldsymbol{A} + \boldsymbol{B}\otimes\boldsymbol{B} + \boldsymbol{C}\otimes\boldsymbol{C}). \tag{3.3.28}$$

(ii) Corresponding to λ, there is a double infinity of eigenbivectors
Then (see the proof of Theorem 3.5) there is an orthogonal matrix $\boldsymbol{T} = (\boldsymbol{A}\,|\,\boldsymbol{B}\,|\,\boldsymbol{C})$ such that (3.2.16) and (3.2.18) hold with $\gamma \neq 0$, and thus

$$\tilde{\boldsymbol{Q}} = \boldsymbol{T}^{\mathrm{T}}\boldsymbol{Q}\boldsymbol{T} = \begin{bmatrix} \lambda - \mathrm{i}\gamma & \gamma & 0 \\ \gamma & \lambda + \mathrm{i}\gamma & 0 \\ 0 & 0 & \lambda \end{bmatrix}. \tag{3.3.29}$$

The eigenbivectors of $\boldsymbol{Q}$ are $\boldsymbol{C}$ and $\boldsymbol{D}=\boldsymbol{A}+\mathrm{i}\boldsymbol{B}$ and any linear combination of them. It follows from (3.3.29) that

$$\boldsymbol{Q}=\lambda\mathbf{1}-\mathrm{i}\gamma\boldsymbol{D}\otimes\boldsymbol{D}. \tag{3.3.30}$$

Further simplification of (3.3.23) and (3.3.24) is possible as in case (a)(ii). Considering $\boldsymbol{D}'=\boldsymbol{A}'+\mathrm{i}\boldsymbol{B}'$ defined by (3.3.21), with $\boldsymbol{A}'$, $\boldsymbol{B}'$ defined by (3.3.22), we have

$$\boldsymbol{Q}=\lambda\mathbf{1}-\mathrm{i}\boldsymbol{D}'\otimes\boldsymbol{D}'. \tag{3.3.31}$$

Thus, with the orthogonal matrix $\boldsymbol{T}'=(\boldsymbol{A}'|\boldsymbol{B}'|\boldsymbol{C})$, we have

$$\tilde{\boldsymbol{Q}}'=\boldsymbol{T}'^{\mathrm{T}}\boldsymbol{Q}\boldsymbol{T}'=\begin{bmatrix}\lambda-\mathrm{i} & 1 & 0\\ 1 & \lambda+\mathrm{i} & 0\\ 0 & 0 & \lambda\end{bmatrix}. \tag{3.3.32}$$

This is the canonical form of $\boldsymbol{Q}$ in case (c)(ii). The spectral decomposition is given by (3.3.31).

(iii) Corresponding to λ there is a simple infinity of eigenbivectors Then, all the eigenbivectors are isotropic and parallel. Let $\boldsymbol{D}$ be one of them:

$$\boldsymbol{Q}\boldsymbol{D}=\lambda\boldsymbol{D}, \tag{3.3.33a}$$

$$\boldsymbol{D}\cdot\boldsymbol{D}=0. \tag{3.3.33b}$$

Without loss of generality we may write

$$\boldsymbol{D}=\boldsymbol{A}+\mathrm{i}\boldsymbol{B}, \tag{3.3.34}$$

where $\boldsymbol{A}$ and $\boldsymbol{B}$ are such that $\boldsymbol{A}\cdot\boldsymbol{A}=\boldsymbol{B}\cdot\boldsymbol{B}=1$ and $\boldsymbol{A}\cdot\boldsymbol{B}=0$. Let $\boldsymbol{C}=\boldsymbol{A}\times\boldsymbol{B}$ so that $\boldsymbol{A}$, $\boldsymbol{B}$, $\boldsymbol{C}$ is an orthonormal triad. Taking the dot product of (3.3.33a) with $\boldsymbol{A}$, $\boldsymbol{B}$, $\boldsymbol{C}$ yields

$$\begin{aligned}\boldsymbol{A}^{\mathrm{T}}\boldsymbol{Q}\boldsymbol{A}+\mathrm{i}\boldsymbol{A}^{\mathrm{T}}\boldsymbol{Q}\boldsymbol{B}&=\lambda,\\ \boldsymbol{B}^{\mathrm{T}}\boldsymbol{Q}\boldsymbol{A}+\mathrm{i}\boldsymbol{B}^{\mathrm{T}}\boldsymbol{Q}\boldsymbol{B}&=\mathrm{i}\lambda,\\ \boldsymbol{C}^{\mathrm{T}}\boldsymbol{Q}\boldsymbol{A}+\mathrm{i}\boldsymbol{C}^{\mathrm{T}}\boldsymbol{Q}\boldsymbol{B}&=0.\end{aligned} \tag{3.3.35}$$

Thus, writing

$$\gamma=\boldsymbol{A}^{\mathrm{T}}\boldsymbol{Q}\boldsymbol{B}=\boldsymbol{B}^{\mathrm{T}}\boldsymbol{Q}\boldsymbol{A},\quad \delta=\boldsymbol{C}^{\mathrm{T}}\boldsymbol{Q}\boldsymbol{A}, \tag{3.3.36}$$

it follows that

$$\boldsymbol{A}^{\mathrm{T}}\boldsymbol{Q}\boldsymbol{A}=\lambda-\mathrm{i}\gamma,\quad \boldsymbol{B}^{\mathrm{T}}\boldsymbol{Q}\boldsymbol{B}=\lambda+\mathrm{i}\gamma,\quad \boldsymbol{C}^{\mathrm{T}}\boldsymbol{Q}\boldsymbol{B}=\mathrm{i}\delta. \tag{3.3.37}$$

Thus, with the orthogonal matrix $\boldsymbol{T}=(\boldsymbol{A}\,|\,\boldsymbol{B}\,|\,\boldsymbol{C})$, we have

$$\tilde{\boldsymbol{Q}}=\boldsymbol{T}^{\mathrm{T}}\boldsymbol{Q}\boldsymbol{T}=\begin{bmatrix}\lambda-\mathrm{i}\gamma & \gamma & \delta\\ \gamma & \lambda+\mathrm{i}\gamma & \mathrm{i}\delta\\ \delta & \mathrm{i}\delta & \boldsymbol{C}^{\mathrm{T}}\boldsymbol{Q}\boldsymbol{C}\end{bmatrix}. \tag{3.3.38}$$

But λ is a triple eigenvalue. Hence $\boldsymbol{C}^{\mathrm{T}}\boldsymbol{Q}\boldsymbol{C}=\lambda$ and

$$\tilde{\boldsymbol{Q}}=\boldsymbol{T}^{\mathrm{T}}\boldsymbol{Q}\boldsymbol{T}=\begin{bmatrix}\lambda-\mathrm{i}\gamma & \gamma & \delta\\ \gamma & \lambda+\mathrm{i}\gamma & \mathrm{i}\delta\\ \delta & \mathrm{i}\delta & \lambda\end{bmatrix}. \tag{3.3.39}$$

Corresponding to (3.3.39) we have the spectral decomposition

$$\begin{aligned}\boldsymbol{Q}=&(\lambda-\mathrm{i}\gamma)\boldsymbol{A}\otimes\boldsymbol{A}+(\lambda+\mathrm{i}\gamma)\boldsymbol{B}\otimes\boldsymbol{B}+\lambda\boldsymbol{C}\otimes\boldsymbol{C}\\ &+\gamma(\boldsymbol{A}\otimes\boldsymbol{B}+\boldsymbol{B}\otimes\boldsymbol{A})+\delta(\boldsymbol{A}\otimes\boldsymbol{C}+\boldsymbol{C}\otimes\boldsymbol{A})\\ &+\mathrm{i}\delta(\boldsymbol{B}\otimes\boldsymbol{C}+\boldsymbol{C}\otimes\boldsymbol{A}),\end{aligned} \tag{3.3.40}$$

that is, using (3.3.34),

$$\boldsymbol{Q}=\lambda\boldsymbol{1}-\mathrm{i}\gamma\boldsymbol{D}\otimes\boldsymbol{D}+\delta(\boldsymbol{D}\otimes\boldsymbol{C}+\boldsymbol{C}\otimes\boldsymbol{D}). \tag{3.3.41}$$

Further simplication of (3.3.39) and (3.3.41) is possible. Consider

$$\boldsymbol{D}'=\delta\boldsymbol{D},\quad \boldsymbol{C}'=\boldsymbol{C}-\frac{\mathrm{i}}{2}\gamma\delta^{-1}\boldsymbol{D}. \tag{3.3.42}$$

Then, (3.3.41) may be written as

$$\boldsymbol{Q}=\lambda\boldsymbol{1}+\boldsymbol{D}'\otimes\boldsymbol{C}'+\boldsymbol{C}'\otimes\boldsymbol{D}'. \tag{3.3.43}$$

Clearly $\boldsymbol{C}'\cdot\boldsymbol{C}'=1$, $\boldsymbol{C}'\cdot\boldsymbol{D}'=0$ and $\boldsymbol{D}'\cdot\boldsymbol{D}'=0$. It is then possible to find $\boldsymbol{A}'$, $\boldsymbol{B}'$ such that $\boldsymbol{A}'$, $\boldsymbol{B}'$, $\boldsymbol{C}'$ form an orthonormal triad and

$$\boldsymbol{D}'=\boldsymbol{A}'+\mathrm{i}\boldsymbol{B}'. \tag{3.3.44}$$

Indeed, writing the conditions $\boldsymbol{B}'\cdot\boldsymbol{A}'=\mathrm{i}(\boldsymbol{A}'-\boldsymbol{D}')\cdot\boldsymbol{A}'=0$, $\boldsymbol{C}'\cdot\boldsymbol{A}'=0$, $\boldsymbol{A}'\cdot\boldsymbol{A}'=1$, we find

$$\begin{aligned}\boldsymbol{A}'=&\frac{\delta}{2}\left(1+\delta^{-2}+\gamma^2\frac{\delta^{-4}}{4}\right)\boldsymbol{A}\\ &+\mathrm{i}\frac{\delta}{2}\left(1-\delta^{-2}+\gamma^2\frac{\delta^{-4}}{4}\right)\boldsymbol{B}+\frac{\mathrm{i}}{2}\gamma\delta^{-2}\boldsymbol{C},\end{aligned} \tag{3.3.45}$$

and hence, from (3.3.44),

$$B' = -\mathrm{i}\frac{\delta}{2}\left(1 - \delta^{-2} - \gamma^2\frac{\delta^{-4}}{4}\right)A$$
$$+\frac{\delta}{2}\left(1 + \delta^{-2} - \gamma^2\frac{\delta^{-4}}{4}\right)B - \frac{1}{2}\gamma\delta^{-2}C. \tag{3.3.46}$$

Thus, from (3.3.43) and (3.3.44), we conclude that with the orthogonal matrix $T' = (A' \mid B' \mid C')$, we have

$$\tilde{Q}' = T'^{\mathrm{T}}QT' = \begin{bmatrix} \lambda & 0 & 1 \\ 0 & \lambda & \mathrm{i} \\ 1 & \mathrm{i} & \lambda \end{bmatrix}. \tag{3.3.47}$$

This is the canonical form of Q in case (c)(iii). The corresponding spectral decomposition is given by (3.3.43).

Remarks In case (b)(ii), we note that any matrix of the form (3.3.20), with an arbitrary complex $\gamma \neq 0$, has D (isotropic) and $C(C \cdot C = 1)$, and their scalar multiples as eigenbivectors corresponding to the eigenvalues λ (double) and μ (simple), respectively.

In case (c)(ii), any matrix of the form (3.3.30), with an arbitrary complex $\gamma \neq 0$, has D (isotropic) and every bivector orthogonal to D as eigenbivectors corresponding to the eigenvalue λ (triple).

In case (c)(iii), any matrix of the form (3.3.41), with arbitrary complex γ and $\delta \neq 0$, has D (isotropic) and its scalar multiples as eigenbivectors corresponding to the eigenvalue λ (triple).

Exercises 3.2

1. Let $A = (1, \mathrm{i}, 1)$, $B = \frac{1}{2}(2^{1/2})(\mathrm{i}, -2, \mathrm{i})$, $C = \frac{1}{2}(2^{1/2})(-1, 0, 1)$. Check that this is an orthonormal triad. Construct the symmetric matrix Q such that A, B, C are the eigenbivectors corresponding to the eigenvalues $\lambda = 1$, $\mu = 2$, $\nu = -2\mathrm{i}$, respectively.
2. Let $D = (\frac{1}{2}(2^{1/2}), \mathrm{i}, \frac{1}{2}(2^{1/2}))$ and $C = (\frac{1}{2}(2^{1/2}), 0, -\frac{1}{2}(2^{1/2}))$. Check that $D \cdot D = 0$, $C \cdot C = 1$, $C \cdot D = 0$. Find all the symmetric matrices Q having D and its scalar multiples as eigenbivectors corresponding to the eigenvalue $\lambda = 1$ (double), and having C and its scalar multiples corresponding to the eigenvalue $\mu = -1$ (simple).

3. Let

$$Q = \begin{bmatrix} \lambda & \frac{1}{2}(1+\mathrm{i}) & 0 \\ \frac{1}{2}(1+\mathrm{i}) & \lambda & \frac{1}{2}(1-\mathrm{i}) \\ 0 & \frac{1}{2}(1-\mathrm{i}) & \lambda \end{bmatrix}.$$

Find $\boldsymbol{T}$ orthogonal such that $\boldsymbol{T}^{\mathrm{T}}\boldsymbol{Q}\boldsymbol{T}$ is canonical.

3.4 Hamiltonian cyclic form

Here we show that any complex symmetric 3×3 matrix $\boldsymbol{Q}$ may be written in the form

$$\boldsymbol{Q} = x\mathbf{1} + \tfrac{1}{2}(\boldsymbol{M}_x \otimes \boldsymbol{P}_x + \boldsymbol{P}_x \otimes \boldsymbol{M}_x), \tag{3.4.1}$$

where x is any of the eigenvalues of $\boldsymbol{Q}$ and where $\boldsymbol{P}_x$, $\boldsymbol{M}_x$ are two bivectors. Of course, the bivectors $\boldsymbol{M}_x$, $\boldsymbol{P}_x$ are different for different choices of the eigenvalue x.

Let us consider the different cases for the spectral decomposition of $\boldsymbol{Q}$.

Case (a) Three simple eigenvalues λ, μ, ν

The spectral decomposition is given by (3.3.7). Using (2.12.13), we may write (3.3.7) as

$$\boldsymbol{Q} = \lambda\mathbf{1} + (\nu - \lambda)\boldsymbol{C} \otimes \boldsymbol{C} - (\lambda - \mu)\boldsymbol{B} \otimes \boldsymbol{B}. \tag{3.4.2}$$

Thus, taking

$$\begin{aligned} \boldsymbol{M}_\lambda &= \sqrt{\nu - \lambda}\,\boldsymbol{C} - \sqrt{\lambda - \mu}\,\boldsymbol{B}, \\ \boldsymbol{P}_\lambda &= \sqrt{\nu - \lambda}\,\boldsymbol{C} + \sqrt{\lambda - \mu}\,\boldsymbol{B}, \end{aligned} \tag{3.4.3}$$

the expression (3.4.2) becomes

$$\boldsymbol{Q} = \lambda\mathbf{1} + \tfrac{1}{2}(\boldsymbol{M}_\lambda \otimes \boldsymbol{P}_\lambda + \boldsymbol{P}_\lambda \otimes \boldsymbol{M}_\lambda). \tag{3.4.4}$$

Analogously, we derive

$$\boldsymbol{Q} = \mu\mathbf{1} + \tfrac{1}{2}(\boldsymbol{M}_\mu \otimes \boldsymbol{P}_\mu + \boldsymbol{P}_\mu \otimes \boldsymbol{M}_\mu), \tag{3.4.5}$$

with

$$\begin{aligned} \boldsymbol{M}_\mu &= \sqrt{\lambda - \mu}\,\boldsymbol{A} - \sqrt{\mu - \nu}\,\boldsymbol{C}, \\ \boldsymbol{P}_\mu &= \sqrt{\lambda - \mu}\,\boldsymbol{A} + \sqrt{\mu - \nu}\,\boldsymbol{C}, \end{aligned} \tag{3.4.6}$$

and also

$$Q = \nu \mathbf{1} + \tfrac{1}{2}(M_\nu \otimes P_\nu + P_\nu \otimes M_\nu), \tag{3.4.7}$$

with

$$M_\nu = \sqrt{\mu - \nu}\,B - \sqrt{\nu - \lambda}\,A,$$
$$P_\nu = \sqrt{\mu - \nu}\,B + \sqrt{\nu - \lambda}\,A. \tag{3.4.8}$$

We note that M_λ, P_λ, M_μ, P_μ and M_ν, P_ν are not isotropic because $\mu \neq \nu$, $\nu \neq \lambda$, and $\lambda \neq \mu$, and that $M_\lambda \cdot M_\lambda = P_\lambda \cdot P_\lambda$ and $M_\mu \cdot M_\mu = P_\mu \cdot P_\mu$, $M_\nu \cdot M_\nu = P_\nu \cdot P_\nu$. Also, the eigenbivectors A, B, C of Q are given by

$$2\sqrt{\nu - \lambda}\,C = P_\lambda + M_\lambda, \quad 2\sqrt{\lambda - \mu}\,B = P_\lambda - M_\lambda, \tag{3.4.9}$$

$$2\sqrt{\nu - \lambda}\,\sqrt{\lambda - \mu}\,A = P_\lambda \times M_\lambda. \tag{3.4.10}$$

Analogous formulae with P_μ, M_μ or P_ν, M_ν may be obtained by cyclic permutations of A, B, C and λ, μ, ν.

Remark When Q is real, the eigenvalues λ, μ, ν are real and they may be ordered $\lambda > \mu > \nu$. Also A, B, C may be taken to be real. Then M_μ, P_μ are real whilst M_λ, P_λ and M_ν, P_ν are complex. Thus (3.4.5) gives an expression of Q in terms of two real vectors. This result is due to Hamilton, and is called the 'Hamiltonian cyclic form' of the real matrix Q.

Case (b) One double eigenvalue λ, and one simple eigenvalue μ

(i) Corresponding to λ there is a double infinity of eigenbivectors The spectral decomposition is then given by (3.3.11). Using (2.12.13), we may write (3.3.11) as

$$Q = \lambda \mathbf{1} + (\mu - \lambda)\,C \otimes C, \tag{3.4.11}$$

which is of the form (3.4.4) with

$$M_\lambda = P_\lambda = \sqrt{\mu - \lambda}\,C. \tag{3.4.12}$$

Also, using (2.12.13), we may write (3.3.11) as

$$Q = \mu \mathbf{1} + (\lambda - \mu)(A \otimes A + B \otimes B). \tag{3.4.13}$$

Thus, taking

$$M_\mu = \sqrt{\lambda - \mu}\,(A - \mathrm{i}B),$$
$$P_\mu = \sqrt{\lambda - \mu}\,(A + \mathrm{i}B), \tag{3.4.14}$$

the expression (3.4.13) takes the form (3.4.5). We note that the bivectors $\boldsymbol{M}_\mu$ and $\boldsymbol{P}_\mu$ defined by (3.4.14) are both isotropic, with $\boldsymbol{P}_\mu \cdot \boldsymbol{M}_\mu \neq 0$. Also, the eigenbivectors of $\boldsymbol{Q}$ are

$$\boldsymbol{C} = \boldsymbol{M}_\lambda = \boldsymbol{P}_\lambda = \frac{\mathrm{i}}{2(\lambda - \mu)} \boldsymbol{P}_\mu \times \boldsymbol{M}_\mu, \tag{3.4.15}$$

and all the bivectors orthogonal to $\boldsymbol{C}$, that is all the linear combinations of $\boldsymbol{M}_\mu$ and $\boldsymbol{P}_\mu$.

(ii) Corresponding to λ there is a simple infinity of eigenbivectors
The spectral decomposition is then given by (3.3.24) or equivalently (3.3.25). Thus

$$\boldsymbol{Q} = \lambda \boldsymbol{1} + (\mu - \lambda)\boldsymbol{C} \otimes \boldsymbol{C} - \mathrm{i}\boldsymbol{D}' \otimes \boldsymbol{D}'. \tag{3.4.16}$$

Taking

$$\boldsymbol{M}_\lambda = \sqrt{\mu - \lambda}\,\boldsymbol{C} - \sqrt{\mathrm{i}}\,\boldsymbol{D}',$$
$$\boldsymbol{P}_\lambda = \sqrt{\mu - \lambda}\,\boldsymbol{C} + \sqrt{\mathrm{i}}\,\boldsymbol{D}', \tag{3.4.17}$$

the expression (3.4.16) takes the form (3.4.4). We note that the bivectors $\boldsymbol{M}_\lambda$ and $\boldsymbol{P}_\lambda$ defined by (3.4.17) are **not** isotropic and that $\boldsymbol{M}_\lambda \cdot \boldsymbol{M}_\lambda = \boldsymbol{P}_\lambda \cdot \boldsymbol{P}_\lambda = \boldsymbol{P}_\lambda \cdot \boldsymbol{M}_\lambda$. The eigenbivectors of $\boldsymbol{Q}$ are given by

$$2\sqrt{\mu - \lambda}\,\boldsymbol{C} = \boldsymbol{P}_\lambda + \boldsymbol{M}_\lambda, \tag{3.4.18}$$

$$2\sqrt{\mathrm{i}}\,\boldsymbol{D}' = \boldsymbol{P}_\lambda - \boldsymbol{M}_\lambda = \frac{\mathrm{i}}{(\mu - \lambda)^{1/2}} \boldsymbol{M}_\lambda \times \boldsymbol{P}_\lambda. \tag{3.4.19}$$

Note that here the cross product $\boldsymbol{M}_\lambda \times \boldsymbol{P}_\lambda$ is isotropic, whilst in (3.4.10) it is not isotropic (because $\boldsymbol{A} \cdot \boldsymbol{A} = 1$). Also, because $\boldsymbol{A}'$, $\boldsymbol{B}'$, $\boldsymbol{C}$ is an orthonormal triad, we may write the spectral decomposition (3.3.24) as

$$\boldsymbol{Q} = \mu \boldsymbol{1} + (\lambda - \mu)(\boldsymbol{A}' \otimes \boldsymbol{A}' + \boldsymbol{B}' \otimes \boldsymbol{B}') - \mathrm{i}\boldsymbol{D}' \otimes \boldsymbol{D}'. \tag{3.4.20}$$

But, from (3.3.21) we have

$$\boldsymbol{A}' \otimes \boldsymbol{A}' + \boldsymbol{B}' \otimes \boldsymbol{B}' = \tfrac{1}{2}(\boldsymbol{D}' \otimes \hat{\boldsymbol{D}}' + \hat{\boldsymbol{D}}' \otimes \boldsymbol{D}'), \tag{3.4.21}$$

with

$$\hat{\boldsymbol{D}}' = \boldsymbol{A}' - \mathrm{i}\boldsymbol{B}'. \tag{3.4.22}$$

Hence, (3.4.20) becomes

$$\boldsymbol{Q} = \mu \boldsymbol{1} + \tfrac{1}{2}\boldsymbol{D}' \otimes \{(\lambda - \mu)\hat{\boldsymbol{D}}' - \mathrm{i}\boldsymbol{D}'\}$$
$$+ \tfrac{1}{2}\{(\lambda - \mu)\hat{\boldsymbol{D}}' - \mathrm{i}\boldsymbol{D}'\} \otimes \boldsymbol{D}', \tag{3.4.23}$$

which is of the form (3.4.5) with

$$\boldsymbol{M}_\mu = (\lambda - \mu)\hat{\boldsymbol{D}}' - \mathrm{i}\boldsymbol{D}' = (\lambda - \mu - \mathrm{i})\boldsymbol{A}' - \mathrm{i}(\lambda - \mu + \mathrm{i})\boldsymbol{B}',$$
$$\boldsymbol{P}_\mu = \boldsymbol{D}' = \boldsymbol{A}' + \mathrm{i}\boldsymbol{B}'. \tag{3.4.24}$$

We note that the bivector $\boldsymbol{M}_\mu$ is not isotropic but that $\boldsymbol{P}_\mu$ is isotropic, with $\boldsymbol{P}_\mu \cdot \boldsymbol{M}_\mu \neq 0$, and that $2\boldsymbol{P}_\mu \cdot \boldsymbol{M}_\mu = \mathrm{i}\boldsymbol{M}_\mu \cdot \boldsymbol{M}_\mu$. Also the eigenbivectors of $\boldsymbol{Q}$ are given by

$$2\mathrm{i}(\lambda - \mu)\boldsymbol{C} = \boldsymbol{P}_\mu \times \boldsymbol{M}_\mu, \quad \boldsymbol{D}' = \boldsymbol{P}_\mu. \tag{3.4.25}$$

Case (c) One triple eigenvalue λ

(i) Corresponding to λ there is a triple infinity of eigenbivectors In this case, we know that $\boldsymbol{Q} = \lambda\mathbf{1}$, and thus (3.4.4) is valid with $\boldsymbol{M}_\lambda = \mathbf{0}$ or $\boldsymbol{P}_\lambda = \mathbf{0}$.

(ii) Corresponding to λ there is a double infinity of eigenbivectors The spectral decomposition is then given by (3.3.31), which is of the form (3.4.4) with

$$\boldsymbol{M}_\lambda = \boldsymbol{P}_\lambda = \mathrm{i}\sqrt{2\mathrm{i}}\,\boldsymbol{D}'. \tag{3.4.26}$$

Thus here, $\boldsymbol{M}_\lambda$ and $\boldsymbol{P}_\lambda$ are both isotropic, with $\boldsymbol{P}_\lambda \cdot \boldsymbol{M}_\lambda = 0$. The eigenbivectors of $\boldsymbol{Q}$ are $\boldsymbol{M}_\lambda = \boldsymbol{P}_\lambda$ and any bivector orthogonal to $\boldsymbol{M}_\lambda = \boldsymbol{P}_\lambda$.

(iii) Corresponding to λ there is a simple infinity of eigenbivectors The spectral decomposition is then given by (3.3.43), which is also of the form (3.4.4) with

$$\boldsymbol{M}_\lambda = \boldsymbol{C}', \quad \boldsymbol{P}_\lambda = 2\boldsymbol{D}' = 2(\boldsymbol{A}' + \mathrm{i}\boldsymbol{B}'). \tag{3.4.27}$$

Here $\boldsymbol{M}_\lambda$ is not isotropic (because $\boldsymbol{C}' \cdot \boldsymbol{C}' = 1$), $\boldsymbol{P}_\lambda$ is isotropic, and $\boldsymbol{M}_\lambda \cdot \boldsymbol{P}_\lambda = 0$ (because $\boldsymbol{C}' \cdot \boldsymbol{D}' = 0$). The eigenbivectors are $\boldsymbol{P}_\lambda$ (and its scalar multiples).

Using the different results obtained here we establish the following classification of the matrices $\boldsymbol{Q}$ written in the form (3.4.1): for eight different possible choices of $\boldsymbol{M}_x$ and $\boldsymbol{P}_x$ (covering all the possibilities), we give the corresponding cases for the canonical form and spectral decomposition, and also the eigenvalues and eigenbivectors.

Classification of $\boldsymbol{Q} = x\mathbf{1} + \frac{1}{2}(\boldsymbol{M}_x \otimes \boldsymbol{P}_x + \boldsymbol{P}_x \otimes \boldsymbol{M}_x)$

(*a*) $\boldsymbol{P}_x \cdot \boldsymbol{P}_x \neq 0$, $\boldsymbol{M}_x \cdot \boldsymbol{M}_x \neq 0$, $(\boldsymbol{P}_x \times \boldsymbol{M}_x) \cdot (\boldsymbol{P}_x \times \boldsymbol{M}_x) \neq 0$.

Without loss of generality, we take $\boldsymbol{P}_x \cdot \boldsymbol{P}_x = \boldsymbol{M}_x \cdot \boldsymbol{M}_x = \gamma$ (say).

Case (a) Three simple eigenvalues:

eigenvalues: $x + \frac{1}{2}(\gamma + \boldsymbol{P}_x \cdot \boldsymbol{M}_x)$, $x - \frac{1}{2}(\gamma - \boldsymbol{P}_x \cdot \boldsymbol{M}_x)$, x;
eigenbivectors: $\boldsymbol{P}_x + \boldsymbol{M}_x$, $\boldsymbol{P}_x - \boldsymbol{M}_x$, $\boldsymbol{P}_x \times \boldsymbol{M}_x$.

(*b*) $\boldsymbol{P}_x \cdot \boldsymbol{P}_x \neq 0$, $\boldsymbol{M}_x \cdot \boldsymbol{M}_x \neq 0$, $(\boldsymbol{P}_x \times \boldsymbol{M}_x) \cdot (\boldsymbol{P}_x \times \boldsymbol{M}_x) = 0$ *with* $\boldsymbol{P}_x \times \boldsymbol{M}_x \neq \mathbf{0}$.

Without loss of generality, we take $\boldsymbol{P}_x \cdot \boldsymbol{P}_x = \boldsymbol{M}_x \cdot \boldsymbol{M}_x = \boldsymbol{P}_x \cdot \boldsymbol{M}_x = \gamma$ (say).

Case (b)(ii) One double eigenvalue and one simple eigenvalue:

eigenvalues: x (double), $x + \gamma$ (simple);
eigenbivectors: $\boldsymbol{P}_x \times \boldsymbol{M}_x$, $\boldsymbol{P}_x + \boldsymbol{M}_x$.

(*c*) $\boldsymbol{P}_x \cdot \boldsymbol{P}_x \neq 0$, $\boldsymbol{M}_x \cdot \boldsymbol{M}_x \neq 0$, $\boldsymbol{P}_x \times \boldsymbol{M}_x = \mathbf{0}$.

Without loss of generality, we take $\boldsymbol{P}_x = \boldsymbol{M}_x$. Let $\boldsymbol{P}_x \cdot \boldsymbol{P}_x = \boldsymbol{M}_x \cdot \boldsymbol{M}_x = \boldsymbol{P}_x \cdot \boldsymbol{M}_x = \gamma$ (say).

Case (b)(i) One double eigenvalue and one simple eigenvalue:

eigenvalues: x (double), $x + \gamma$ (simple);
eigenbivectors: all bivectors orthogonal to $\boldsymbol{M}_x = \boldsymbol{P}_x$, $\boldsymbol{M}_x = \boldsymbol{P}_x$.

(*d*) $\boldsymbol{P}_x \cdot \boldsymbol{P}_x = 0$, $\boldsymbol{M}_x \cdot \boldsymbol{M}_x \neq 0$, $\boldsymbol{P}_x \cdot \boldsymbol{M}_x \neq 0$

Case (b)(ii) One double eigenvalue and one simple eigenvalue:

eigenvalues: x (simple), $x + \frac{1}{2}\boldsymbol{M}_x \cdot \boldsymbol{P}_x$ (double);
eigenbivectors: $\boldsymbol{P}_x \times \boldsymbol{M}_x$, $\boldsymbol{P}_x$.

(*e*) $\boldsymbol{P}_x \cdot \boldsymbol{P}_x = 0$, $\boldsymbol{M}_x \cdot \boldsymbol{M}_x \neq 0$, $\boldsymbol{P}_x \cdot \boldsymbol{M}_x = 0$.

Case (c)(iii) One triple eigenvalue:

eigenvalue: x (triple);
eigenbivectors: $\boldsymbol{P}_x$.

(*f*) $\boldsymbol{P}_x \cdot \boldsymbol{P}_x = \boldsymbol{M}_x \cdot \boldsymbol{M}_x = 0$, $\boldsymbol{P}_x \cdot \boldsymbol{M}_x \neq 0$.

Case (b)(i) One double eigenvalue and one simple eigenvalue:

eigenvalues: x (simple), $x + \frac{1}{2}\boldsymbol{M}_x \cdot \boldsymbol{P}_x$ (double);
eigenbivectors: $\boldsymbol{P}_x \times \boldsymbol{M}_x$, all bivectors orthogonal to $\boldsymbol{P}_x \times \boldsymbol{M}_x$.

(*g*) $\boldsymbol{P}_x \cdot \boldsymbol{P}_x = \boldsymbol{M}_x \cdot \boldsymbol{M}_x = \boldsymbol{P}_x \cdot \boldsymbol{M}_x = 0$.
Case (c)(ii) One triple eigenvalue:

eigenvalue: x (triple);
eigenbivectors: $\boldsymbol{M}_\lambda \parallel \boldsymbol{P}_\lambda$ and all bivectors orthogonal to $\boldsymbol{M}_\lambda \parallel \boldsymbol{P}_\lambda$.

(*h*) $\boldsymbol{P}_x = \boldsymbol{0}$ *or* $\boldsymbol{M}_x = 0$.
Case (c)(i) One triple eigenvalue:

eigenvalue: x (triple);
eigenbivectors: all bivectors.

Exercises 3.3

1. Find Hamiltonian cyclic forms of

$$\boldsymbol{Q} = \begin{bmatrix} 3-\mathrm{i} & 0 & 2\sqrt{2}\mathrm{i} \\ 0 & \mathrm{i} & 0 \\ 2\sqrt{2}\mathrm{i} & 0 & -3+\mathrm{i} \end{bmatrix}.$$

2. Find Hamiltonian cyclic forms of

$$\boldsymbol{Q} = \begin{bmatrix} -\mathrm{i} & \sqrt{2} & 1-\mathrm{i} \\ \sqrt{2} & 1+2\mathrm{i} & \sqrt{2} \\ 1-\mathrm{i} & \sqrt{2} & -\mathrm{i} \end{bmatrix}.$$

3. Find Hamiltonian cyclic forms of

$$\boldsymbol{Q} = \begin{bmatrix} \lambda & \frac{1}{2}(1+\mathrm{i}) & 0 \\ \frac{1}{2}(1+\mathrm{i}) & \lambda & \frac{1}{2}(1-\mathrm{i}) \\ 0 & \frac{1}{2}(1-\mathrm{i}) & \lambda \end{bmatrix}.$$

4. Let $\boldsymbol{Q} = \boldsymbol{1} + \frac{1}{2}(\boldsymbol{M} \otimes \boldsymbol{P} + \boldsymbol{P} \otimes \boldsymbol{M})$ with $\boldsymbol{M} = (1, -\mathrm{i}, 1)$, $\boldsymbol{P} = (1, \mathrm{i}, 0)$. Find the eigenvalues and eigenbivectors of $\boldsymbol{Q}$.
5. Same question with $\boldsymbol{M} = \boldsymbol{P} = (1, \mathrm{i}, 0)$.

3.5 Eigenbivectors with a real direction

Here we consider the possibility that the complex symmetric 3×3 matrix $\boldsymbol{Q}$ have eigenbivectors with real directions, or equivalently, because eigenbivectors are defined up to a scalar factor, to have real eigenvectors.

Theorem 3.7

If Q commutes with its complex conjugate, $Q\bar{Q} = \bar{Q}Q$, then Q has three orthogonal real eigenvectors. Moreover, in the case when Q has three simple eigenvalues, all the eigenbivectors of Q have a real direction.

Proof Writing Q in the form (3.1.1), we obtain

$$\begin{aligned} Q\bar{Q} &= Q^{+2} + Q^{-2} - \mathrm{i}(Q^{+}Q^{-} - Q^{-}Q^{+}), \\ \bar{Q}Q &= Q^{+2} + Q^{-2} + \mathrm{i}(Q^{+}Q^{-} - Q^{-}Q^{+}). \end{aligned} \tag{3.5.1}$$

Thus, the assumption $Q\bar{Q} = \bar{Q}Q$ is equivalent to $Q^{+}Q^{-} = Q^{-}Q^{+}$. Because the real symmetric matrices Q^{+} and Q^{-} commute, they have in common three orthogonal (unit) eigenvectors $\mathbf{i}$, $\mathbf{j}$, $\mathbf{k}$ (say). Thus,

$$\begin{aligned} Q^{+}\mathbf{i} &= \lambda^{+}\mathbf{i}, \quad & Q^{-}\mathbf{i} &= \lambda^{-}\mathbf{i}, \\ Q^{+}\mathbf{j} &= \mu^{+}\mathbf{j}, \quad & Q^{-}\mathbf{j} &= \mu^{-}\mathbf{j}, \\ Q^{+}\mathbf{k} &= \nu^{+}\mathbf{k}, \quad & Q^{-}\mathbf{k} &= \nu^{-}\mathbf{k}. \end{aligned} \tag{3.5.2}$$

It follows that

$$Q\mathbf{i} = \lambda\mathbf{i}, \quad Q\mathbf{j} = \mu\mathbf{j}, \quad Q\mathbf{k} = \nu\mathbf{k}, \tag{3.5.3}$$

with

$$\lambda = \lambda^{+} + \mathrm{i}\lambda^{-}, \quad \mu = \mu^{+} + \mathrm{i}\mu^{-}, \quad \nu = \nu^{+} + \mathrm{i}\nu^{-}. \tag{3.5.4}$$

Thus, $\mathbf{i}$, $\mathbf{j}$, $\mathbf{k}$ are real eigenvectors of the complex symmetric matrix Q. Moreover, when the three eigenvalues λ, μ, ν are simple ($\lambda \neq \mu$, $\mu \neq \nu$, $\nu \neq \lambda$), then the eigenbivectors of Q are all scalar multiples (possibly complex) of $\mathbf{i}$, $\mathbf{j}$, $\mathbf{k}$.

Let us now consider the matrix

$$\boldsymbol{\omega} = \frac{\mathrm{i}}{2}(Q\bar{Q} - \bar{Q}Q). \tag{3.5.5}$$

Writing Q in the form (3.1.1), we note that

$$\boldsymbol{\omega} = Q^{+}Q^{-} - Q^{-}Q^{+}. \tag{3.5.6}$$

Thus $\boldsymbol{\omega}$ is a real 3×3 skew-symmetric matrix. Let us denote by $\boldsymbol{w}$ the real vector associated with $\boldsymbol{\omega}$:

$$w_k = \tfrac{1}{2}e_{ikj}\omega_{ij}, \quad \omega_{ij} = e_{ikj}w_k. \tag{3.5.7}$$

For every bivector $\boldsymbol{A}$, we have

$$\boldsymbol{\omega A} = \boldsymbol{w} \times \boldsymbol{A}. \tag{3.5.8}$$

Theorem 3.8

A necessary and sufficient condition that $\boldsymbol{Q}$ have an eigenbivector with a real direction is that $\boldsymbol{Q}$ and $\boldsymbol{\omega}^2$ commute:

$$\boldsymbol{Q\omega}^2 = \boldsymbol{\omega}^2\boldsymbol{Q}, \tag{3.5.9}$$

or equivalently that

$$\boldsymbol{w} \times \boldsymbol{Qw} = \boldsymbol{0}. \tag{3.5.10}$$

Proof Let us first show that (3.5.9) is equivalent to (3.5.10). Indeed, (3.5.9) holds if and only if, for every bivector $\boldsymbol{A}$, we have

$$(\boldsymbol{Q\omega}^2 - \boldsymbol{\omega}^2\boldsymbol{Q})\boldsymbol{A} = \boldsymbol{0}. \tag{3.5.11}$$

But, using (3.5.8), this may be written as

$$\boldsymbol{Q}\{\boldsymbol{w} \times (\boldsymbol{w} \times \boldsymbol{A})\} - \boldsymbol{w} \times (\boldsymbol{w} \times \boldsymbol{QA}) = \boldsymbol{0}, \tag{3.5.12}$$

or

$$(\boldsymbol{A}\cdot\boldsymbol{w})\boldsymbol{Qw} - (\boldsymbol{A}^{\mathrm{T}}\boldsymbol{Qw})\boldsymbol{w} = \boldsymbol{0}, \tag{3.5.13}$$

or again

$$(\boldsymbol{w} \times \boldsymbol{Qw}) \times \boldsymbol{A} = \boldsymbol{0}. \tag{3.5.14}$$

But this is valid for every bivector $\boldsymbol{A}$ if and only if (3.5.10) holds. Thus (3.5.9) and (3.5.10) are equivalent.

Let us now show that (3.5.9) or (3.5.10) is a necessary and sufficient condition that $\boldsymbol{Q}$ have an eigenbivector with a real direction.

(a) The condition is necessary Assume that $\boldsymbol{Q}$ has an eigenbivector with a real direction, or equivalently, a real eigenvector $\boldsymbol{a}(\neq \boldsymbol{0})$:

$$\boldsymbol{Qa} = \lambda \boldsymbol{a}, \tag{3.5.15}$$

or,

$$\boldsymbol{Q}^+\boldsymbol{a} = \lambda^+\boldsymbol{a}, \quad \boldsymbol{Q}^-\boldsymbol{a} = \lambda^-\boldsymbol{a}. \tag{3.5.16}$$

It then follows from (3.5.6) that

$$\boldsymbol{\omega a} = \boldsymbol{w} \times \boldsymbol{a} = \boldsymbol{0}, \tag{3.5.17}$$

and thus either $\boldsymbol{w} = \boldsymbol{0}$ or $\boldsymbol{w}$ is parallel to $\boldsymbol{a}$. Hence (3.5.10) holds because $\boldsymbol{Qw} = \lambda \boldsymbol{w}$.

(b) The condition is sufficient Assume now that (3.5.10) holds. Then, either $\boldsymbol{w} = \mathbf{0}$, or $\boldsymbol{Q}\boldsymbol{w} = \lambda \boldsymbol{w}$, $(\boldsymbol{w} \neq \mathbf{0})$. If $\boldsymbol{w} = \mathbf{0}$, it follows from Theorem 3.7 that $\boldsymbol{Q}$ has three orthogonal real eigenvectors. If $\boldsymbol{Q}\boldsymbol{w} = \lambda \boldsymbol{w}$, the vector $\boldsymbol{w}$ is a real eigenvector of the complex symmetric matrix $\boldsymbol{Q}$.

Remarks The equivalence of (3.5.15) and (3.5.16) means that $\boldsymbol{a}$ is a real eigenbivector of $\boldsymbol{Q} = \boldsymbol{Q}^+ + \mathrm{i}\boldsymbol{Q}^-$ if and only if it is a common eigenvector of the real symmetric matrices $\boldsymbol{Q}^+$ and $\boldsymbol{Q}^-$. Hence we may establish the following classification for the number of eigenbivectors with a real direction.

Case (a) $\boldsymbol{Q}^+$ and $\boldsymbol{Q}^-$ have no common eigenvectors, or equivalently, the quadrics $\boldsymbol{x}^{\mathrm{T}}\boldsymbol{Q}^+\boldsymbol{x} = 1$ and $\boldsymbol{x}^{\mathrm{T}}\boldsymbol{Q}^-\boldsymbol{x} = 1$ have no common principal axes. Then the complex matrix $\boldsymbol{Q}$ has no eigenbivectors with real direction: $\boldsymbol{w} \times \boldsymbol{Q}\boldsymbol{w} \neq \mathbf{0}$.
Case (b) $\boldsymbol{Q}^+$ and $\boldsymbol{Q}^-$ have just one common eigenvector (up to a scalar factor), or equivalently, the quadrics $\boldsymbol{x}^{\mathrm{T}}\boldsymbol{Q}^+\boldsymbol{x} = 1$ and $\boldsymbol{x}^{\mathrm{T}}\boldsymbol{Q}^-\boldsymbol{x} = 1$ have just one common axis. Then the complex matrix $\boldsymbol{Q}$ has just one eigenbivector with a real direction (defined up to a scalar factor): $\boldsymbol{w} \neq \mathbf{0}$ and $\boldsymbol{Q}\boldsymbol{w} \parallel \boldsymbol{w}$.
Case (c) $\boldsymbol{Q}^+$ and $\boldsymbol{Q}^-$ have three common orthogonal eigenvectors, or equivalently, the quadrics $\boldsymbol{x}^{\mathrm{T}}\boldsymbol{Q}^+\boldsymbol{x} = 1$ and $\boldsymbol{x}^{\mathrm{T}}\boldsymbol{Q}^-\boldsymbol{w} = 1$ have the same principal axes. Then the complex matrix $\boldsymbol{Q}$ has three orthogonal eigenbivectors with real direction: $\boldsymbol{w} = \mathbf{0}$.

Exercises 3.4

1. Let

$$\boldsymbol{Q} = \begin{bmatrix} 1 & \mathrm{i} & 1 \\ \mathrm{i} & a+\mathrm{i} & \mathrm{i} \\ 1 & \mathrm{i} & 1 \end{bmatrix},$$

where a is real. Compute $\boldsymbol{w}$. Discuss the possibility of $\boldsymbol{Q}$ having eigenbivectors with real direction. Find the real eigenbivectors.

2. Let

$$\boldsymbol{Q} = \begin{bmatrix} 1+\mathrm{i} & a & b \\ a & 2+\mathrm{i} & 0 \\ b & 0 & 2+2\mathrm{i} \end{bmatrix},$$

where a and b are real. Compute $\boldsymbol{w}$. Discuss the possibility of $\boldsymbol{Q}$ having eigenbivectors with real directions.

3. Let

$$\boldsymbol{Q}^+ = \begin{bmatrix} a_{11} & a_{12} & a_{13} \\ a_{12} & a_{22} & a_{23} \\ a_{13} & a_{23} & a_{33} \end{bmatrix}, \quad \boldsymbol{Q}^- = \begin{bmatrix} b_1 & 0 & 0 \\ 0 & b_2 & 0 \\ 0 & 0 & b_3 \end{bmatrix},$$

with $b_1 \neq b_2$, $b_2 \neq b_3$, $b_3 \neq b_1$. Discuss the possibility of $\boldsymbol{Q}$ having eigenbivectors with real directions.

4

Complex orthogonal matrices and complex skew-symmetric matrices

Here we complete the study of Chapter 3 by considering complex orthogonal 3×3 matrices, and the closely related complex skew-symmetric matrices. These have a strong link with bivectors, because with each complex 3×3 skew-symmetric matrix may be associated a bivector, in a one-to-one correspondence.

4.1 Eigenbivectors and eigenvalues

In this chapter, we consider complex 3×3 matrices $\boldsymbol{R}$ satisfying

$$\boldsymbol{R}^{\mathrm{T}}\boldsymbol{R} = \mathbf{1} = \boldsymbol{R}\boldsymbol{R}^{\mathrm{T}}. \qquad (4.1.1)$$

These are said to be orthogonal. Such matrices define (in a real orthonormal basis) linear transformations of the three-dimensional bivector space which preserve the dot product. Indeed, let $\boldsymbol{A}$, $\boldsymbol{B}$ be any two bivectors and let $\boldsymbol{A}'$, $\boldsymbol{B}'$ be their transforms, defined (in a real orthonormal basis) by

$$\boldsymbol{A}' = \boldsymbol{R}\boldsymbol{A}, \quad \boldsymbol{B}' = \boldsymbol{R}\boldsymbol{B}, \qquad (4.1.2)$$

where the matrix $\boldsymbol{R}$ satisfies (4.1.1). Then,

$$\boldsymbol{A}'\cdot\boldsymbol{B}' = (\boldsymbol{R}\boldsymbol{A})\cdot(\boldsymbol{R}\boldsymbol{B}) = \boldsymbol{A}^{\mathrm{T}}\boldsymbol{R}^{\mathrm{T}}\boldsymbol{R}\boldsymbol{B} = \boldsymbol{A}\cdot\boldsymbol{B}. \qquad (4.1.3)$$

In particular, orthogonal bivectors are transformed into orthogonal bivectors and isotropic bivectors are transformed into isotropic bivectors.

Taking the determinant of (4.1.1), we have $(\det \boldsymbol{R})^2 = 1$, so that

$$\det \boldsymbol{R} = \pm 1 \equiv \varepsilon \,(\text{say}). \qquad (4.1.4)$$

The matrix $\boldsymbol{R}$ is said to be a proper orthogonal matrix when $\det \boldsymbol{R} = +1$ and is said to be improper when $\det \boldsymbol{R} = -1$.

The properties of the eigenvalues and eigenbivectors of complex orthogonal matrices are now studied in detail. We first recall their definition.

Definition The bivector $\boldsymbol{X}(\neq \boldsymbol{0})$ is said to be an eigenbivector of the complex matrix $\boldsymbol{R}$ when

$$\boldsymbol{R}\boldsymbol{X} = \lambda \boldsymbol{X}, \tag{4.1.5}$$

where λ is a number. This number λ is called the eigenvalue corresponding to the eigenbivector $\boldsymbol{X}$.

Because the eigenvalues are the solutions of the algebraic equation

$$\det(\boldsymbol{R} - \lambda \boldsymbol{1}) = 0, \tag{4.1.6}$$

the product of the three eigenvalues is equal to ε, that is to $+1$ for proper orthogonal matrices and to -1 for improper orthogonal matrices. In particular $\boldsymbol{R}$ may not have the eigenvalue zero.

We also note that if $\boldsymbol{X}$ is an eigenbivector of $\boldsymbol{R}$ corresponding to the eigenvalue λ, then $\boldsymbol{X}$ is also an eigenbivector of $\boldsymbol{R}^{\mathrm{T}}$ corresponding to the eigenvalue λ^{-1}. Indeed, multiplying (4.1.5) by $\boldsymbol{R}^{\mathrm{T}}$, and using (4.1.1), we obtain $\boldsymbol{X} = \lambda \boldsymbol{R}^{\mathrm{T}} \boldsymbol{X}$, that is

$$\boldsymbol{R}^{\mathrm{T}} \boldsymbol{X} = \lambda^{-1} \boldsymbol{X}. \tag{4.1.7}$$

Theorem 4.1

The eigenvalues of the complex 3×3 orthogonal matrix $\boldsymbol{R}$ are of the form λ, λ^{-1}, ε, where λ is in general complex, and where $\varepsilon = \pm 1$, according to whether the matrix $\boldsymbol{R}$ is proper or improper ($\varepsilon = \det \boldsymbol{R}$).

Proof First assume that $\boldsymbol{R}$ possesses an eigenvalue λ different from $+\varepsilon$ and $-\varepsilon$:

$$\boldsymbol{R}\boldsymbol{X} = \lambda \boldsymbol{X}, \quad \boldsymbol{R}^{\mathrm{T}} \boldsymbol{X} = \lambda^{-1} \boldsymbol{X}. \tag{4.1.8}$$

But because $\boldsymbol{R}^{\mathrm{T}}$ has the same eigenvalues as $\boldsymbol{R}$ has, it follows that λ^{-1} is another eigenvalue of $\boldsymbol{R}$. Because the product of the three eigenvalues is ε, the third eigenvalue must be ε.

Suppose now that $\boldsymbol{R}$ possesses no eigenvalues other than $+\varepsilon$ or $-\varepsilon$. Then, as the product of the three eigenvalues is ε, the only two possibilities are: a double eigenvalue $-\varepsilon$ and a simple eigenvalue $+\varepsilon(\lambda=\lambda^{-1}=-\varepsilon)$, or, a triple eigenvalue $\varepsilon(\lambda=\lambda^{-1}=\varepsilon)$.

Remark An eigenbivector $\boldsymbol{C}$ of $\boldsymbol{R}$ corresponding to the eigenvalue ε is also an eigenbivector of $\boldsymbol{R}^{\mathrm{T}}$ corresponding to the same eigenvalue ε:

$$\boldsymbol{R}\boldsymbol{C}=\varepsilon\boldsymbol{C}, \quad \boldsymbol{R}^{\mathrm{T}}\boldsymbol{C}=\varepsilon\boldsymbol{C}. \tag{4.1.9}$$

Theorem 4.2

If $\boldsymbol{C}$ is an eigenbivector of the complex orthogonal matrix $\boldsymbol{R}$ corresponding to the eigenvalue ε, and if $\boldsymbol{D}$ is an eigenbivector corresponding to another eigenvalue $\lambda(\neq\varepsilon)$, then $\boldsymbol{C}$ and $\boldsymbol{D}$ are linearly independent ($\boldsymbol{D}\neq\alpha\boldsymbol{C}$), and orthogonal: $\boldsymbol{C}\cdot\boldsymbol{D}=0$.

Proof By hypothesis

$$\boldsymbol{R}\boldsymbol{C}=\varepsilon\boldsymbol{C}, \tag{4.1.10a}$$

$$\boldsymbol{R}^{\mathrm{T}}\boldsymbol{C}=\varepsilon\boldsymbol{C}, \tag{4.1.10b}$$

$$\boldsymbol{R}\boldsymbol{D}=\lambda\boldsymbol{D}, \quad \lambda\neq\varepsilon. \tag{4.1.10c}$$

Assuming $\boldsymbol{D}=\alpha\boldsymbol{C}$, we easily obtain $\boldsymbol{R}\boldsymbol{D}=\varepsilon\boldsymbol{D}$, and thus $\lambda=\varepsilon$, which is contrary to the assumption $\lambda\neq\varepsilon$. Thus $\boldsymbol{C}$ and $\boldsymbol{D}$ are linearly independent. Also from (4.1.10b), (4.1.10c) we have

$$\boldsymbol{C}^{\mathrm{T}}\boldsymbol{R}\boldsymbol{D}=\lambda\boldsymbol{C}\cdot\boldsymbol{D}, \quad \boldsymbol{D}^{\mathrm{T}}\boldsymbol{R}^{\mathrm{T}}\boldsymbol{C}=\varepsilon\boldsymbol{C}\cdot\boldsymbol{D}, \tag{4.1.11}$$

and because $\boldsymbol{C}^{\mathrm{T}}\boldsymbol{R}\boldsymbol{D}=\boldsymbol{D}^{\mathrm{T}}\boldsymbol{R}^{\mathrm{T}}\boldsymbol{C}$,

$$(\lambda-\varepsilon)\boldsymbol{C}\cdot\boldsymbol{D}=0. \tag{4.1.12}$$

Thus $\boldsymbol{C}\cdot\boldsymbol{D}=0$ because $\lambda\neq\varepsilon$.

4.2 Isotropic eigenbivectors

Here special consideration is given to isotropic eigenbivectors. Unlike complex symmetric matrices, complex orthogonal matrices always have isotropic eigenbivectors.

Theorem 4.3

If the complex orthogonal matrix $\boldsymbol{R}$ possesses an isotropic eigenbivector corresponding to the eigenvalue $\varepsilon(\varepsilon=\det\boldsymbol{R})$, then this eigenvalue is triple.

Proof Let $\boldsymbol{A}$ be an isotropic eigenbivector of $\boldsymbol{R}$ corresponding to the eigenvalue ε:

$$\boldsymbol{RA}=\varepsilon\boldsymbol{A},\quad \boldsymbol{R}^{\mathrm{T}}\boldsymbol{A}=\varepsilon\boldsymbol{A},\quad \boldsymbol{A}\cdot\boldsymbol{A}=0. \tag{4.2.1}$$

Without loss of generality, we may write

$$\boldsymbol{A}=\mathbf{i}+\mathrm{i}\mathbf{j},\quad \mathbf{i}\cdot\mathbf{i}=\mathbf{j}\cdot\mathbf{j}=1,\quad \mathbf{i}\cdot\mathbf{j}=0. \tag{4.2.2}$$

Then, let $\mathbf{k}=\mathbf{i}\times\mathbf{j}$, and let $\boldsymbol{T}$ be the matrix whose columns are $\boldsymbol{A}, \bar{\boldsymbol{A}}, \mathbf{k}$:

$$\boldsymbol{T}=(\boldsymbol{A}\,|\,\bar{\boldsymbol{A}}\,|\,\mathbf{k}). \tag{4.2.3}$$

Because $\boldsymbol{A}$, $\bar{\boldsymbol{A}}$, $\mathbf{k}$ are linearly independent, $\det\boldsymbol{T}\neq 0$. In fact, $\det\boldsymbol{T}=-2\mathrm{i}$. Then, it is easily seen that

$$\boldsymbol{T}^{\mathrm{T}}\boldsymbol{T}=\begin{bmatrix}0&2&0\\2&0&0\\0&0&1\end{bmatrix},\quad \boldsymbol{T}^{\mathrm{T}}\boldsymbol{RT}=\begin{bmatrix}0&2\varepsilon&0\\2\varepsilon&\eta&\gamma\\0&\delta&\mu\end{bmatrix}, \tag{4.2.4}$$

where η, μ, γ, δ are the numbers

$$\eta=\bar{\boldsymbol{A}}^{\mathrm{T}}\boldsymbol{R}\bar{\boldsymbol{A}},\quad \mu=\mathbf{k}^{\mathrm{T}}\boldsymbol{R}\mathbf{k},\quad \gamma=\bar{\boldsymbol{A}}^{\mathrm{T}}\boldsymbol{R}\mathbf{k},\quad \delta=\mathbf{k}^{\mathrm{T}}\boldsymbol{R}\bar{\boldsymbol{A}}. \tag{4.2.5}$$

But, because $\det(\boldsymbol{T}^{\mathrm{T}}\boldsymbol{RT})=(\det\boldsymbol{T})^2\det\boldsymbol{R}=-4\varepsilon$, we have $\mu=\varepsilon$. The equation $\det(\boldsymbol{R}-x\mathbf{1})=0$ for the determination of the eigenvalues x, which is equivalent to the equation $\det(\boldsymbol{T}^{\mathrm{T}}\boldsymbol{RT}-x\boldsymbol{T}^{\mathrm{T}}\boldsymbol{T})=0$, thus reads

$$4(\varepsilon-x)^3=0. \tag{4.2.6}$$

Hence the eigenvalue ε is triple.

We immediately note the following corollary.

Corollary 4.4

If the eigenvalues of $\boldsymbol{R}$ are λ, λ^{-1}, ε with $\lambda\neq\varepsilon$, then the eigenbivector corresponding to the eigenvalue ε (defined up to a scalar factor) is not isotropic.

Theorem 4.5 Simple eigenvalues

Assume that the complex orthogonal matrix $\boldsymbol{R}$ has three simple eigenvalues λ, λ^{-1}, ε (with $\lambda\neq\pm\varepsilon$). If $\boldsymbol{D}$ and $\boldsymbol{E}$ are eigenbivectors of $\boldsymbol{R}$ corresponding to the eigenvalues λ, λ^{-1}, respectively, then $\boldsymbol{D}$ and $\boldsymbol{E}$ are isotropic and linearly independent.

Proof By hypothesis,

$$\boldsymbol{RD} = \lambda \boldsymbol{D}, \quad \boldsymbol{R}^{\mathrm{T}}\boldsymbol{D} = \lambda^{-1}\boldsymbol{D}. \tag{4.2.7}$$

Hence,

$$\boldsymbol{D}^{\mathrm{T}}\boldsymbol{RD} = \lambda \boldsymbol{D}\cdot\boldsymbol{D}, \quad \boldsymbol{D}^{\mathrm{T}}\boldsymbol{R}^{\mathrm{T}}\boldsymbol{D} = \lambda^{-1}\boldsymbol{D}\cdot\boldsymbol{D}, \tag{4.2.8}$$

and because $\boldsymbol{D}^{\mathrm{T}}\boldsymbol{RD} = \boldsymbol{D}^{\mathrm{T}}\boldsymbol{R}^{\mathrm{T}}\boldsymbol{D}$,

$$(\lambda - \lambda^{-1})\boldsymbol{D}\cdot\boldsymbol{D} = 0. \tag{4.2.9}$$

Thus, $\boldsymbol{D}\cdot\boldsymbol{D} = 0$ because, by assumption, $\lambda^{-1} \neq \lambda$. Similarly, it is proved that $\boldsymbol{E}\cdot\boldsymbol{E} = 0$. Also, $\boldsymbol{D}$ and $\boldsymbol{E}$ are linearly independent and thus, using Synge's theorem (Theorem 2.2, section 2.6), $\boldsymbol{D}\cdot\boldsymbol{E} \neq 0$. Indeed, assuming $\boldsymbol{D} = \alpha\boldsymbol{E}$ implies $\lambda^{-1} = \lambda$ which is contrary to hypothesis.

Theorem 4.6 Double eigenvalue

If $\boldsymbol{R}$ has the double eigenvalue $\lambda = \lambda^{-1} = -\varepsilon$ and the simple eigenvalue ε ($\varepsilon = \det \boldsymbol{R}$), then corresponding to $-\varepsilon$ there is a double infinity of eigenbivectors, and, among them, there are two linearly independent isotropic eigenbivectors.

Proof Let $\boldsymbol{C}$ be an eigenbivector corresponding to the simple eigenvalue ε:

$$\boldsymbol{RC} = \varepsilon\boldsymbol{C}, \quad \boldsymbol{R}^{\mathrm{T}}\boldsymbol{C} = \varepsilon\boldsymbol{C}. \tag{4.2.10}$$

From Corollary 4.4 of Theorem 4.3, we know that $\boldsymbol{C}$ may not be isotropic, so that, without loss of generality, we take $\boldsymbol{C}\cdot\boldsymbol{C} = 1$. Then, choose $\boldsymbol{A}$, $\boldsymbol{B}$ to complete an orthonormal triad $\boldsymbol{A}$, $\boldsymbol{B}$, $\boldsymbol{C}$. Let $\boldsymbol{T} = (\boldsymbol{A}\,|\,\boldsymbol{B}\,|\,\boldsymbol{C})$ be the matrix whose columns are $\boldsymbol{A}$, $\boldsymbol{B}$, $\boldsymbol{C}$. It is readily seen that

$$\boldsymbol{T}^{\mathrm{T}}\boldsymbol{T} = \mathbf{1}, \quad \boldsymbol{T}^{\mathrm{T}}\boldsymbol{RT} = \begin{bmatrix} \alpha & \gamma & 0 \\ \delta & \beta & 0 \\ 0 & 0 & \varepsilon \end{bmatrix}, \tag{4.2.11}$$

where α, β, γ, δ are the numbers given by

$$\alpha = \boldsymbol{A}^{\mathrm{T}}\boldsymbol{RA}, \quad \beta = \boldsymbol{B}^{\mathrm{T}}\boldsymbol{RB}, \quad \gamma = \boldsymbol{A}^{\mathrm{T}}\boldsymbol{RB}, \quad \delta = \boldsymbol{B}^{\mathrm{T}}\boldsymbol{RA}. \tag{4.2.12}$$

But $\det \boldsymbol{R} = \varepsilon$ and $\lambda = \lambda^{-1} = -\varepsilon$ is a double eigenvalue. Therefore $\alpha\beta - \gamma\delta = 1$, $\alpha + \beta = -2\varepsilon$, and hence

$$\alpha = -\varepsilon + \mathrm{i}\sqrt{\gamma\delta}, \quad \beta = -\varepsilon - \mathrm{i}\sqrt{\gamma\delta}. \tag{4.2.13}$$

But, because $\boldsymbol{R}$ and also $\boldsymbol{T}^{\mathrm{T}}\boldsymbol{R}\boldsymbol{T}$ are orthogonal, $\alpha^2+\delta^2=1$ and $\gamma^2+\beta^2=1$, and so

$$\delta^2-\gamma\delta-2\mathrm{i}\varepsilon\sqrt{\gamma\delta}=0, \quad \gamma^2-\gamma\delta+2\mathrm{i}\varepsilon\sqrt{\gamma\delta}=0. \qquad (4.2.14)$$

Taking the sum and the difference of these equations yields $(\gamma-\delta)^2=0$ and $(\gamma\delta)^{1/2}=0$, and hence, using (4.2.13),

$$\alpha=\beta=-\varepsilon, \quad \gamma=\delta=0. \qquad (4.2.15)$$

Thus $\boldsymbol{T}^{\mathrm{T}}\boldsymbol{R}\boldsymbol{T}$ is diagonal and $\boldsymbol{A}$ and $\boldsymbol{B}$ are eigenbivectors of the matrix $\boldsymbol{R}$ corresponding to the eigenvalue $\lambda=\lambda^{-1}=-\varepsilon$. Hence, any linear combination of $\boldsymbol{A}$ and $\boldsymbol{B}$, that is, any bivector orthogonal to $\boldsymbol{C}$, is an eigenbivector of $\boldsymbol{R}$ corresponding to the double eigenvalue $-\varepsilon$. Among this double infinity of eigenbivectors, the bivectors $\boldsymbol{A}+\mathrm{i}\boldsymbol{B}$ and $\boldsymbol{A}-\mathrm{i}\boldsymbol{B}$ are isotropic and linearly independent.

Theorem 4.7 Triple eigenvalue

If $\boldsymbol{R}$ has the triple eigenvalue ε ($\varepsilon=\det\boldsymbol{R}$), then corresponding to ε, **either**

(a) there is a triple infinity of eigenbivectors, and, among them, there are three linearly independent eigenbivectors, **or**
(b) there is a simple infinity of eigenbivectors, all isotropic.

Proof Either $\boldsymbol{R}$ possesses a nonisotropic eigenbivector corresponding to ε, or all eigenbivectors corresponding to ε are isotropic.

(a) Assume first that $\boldsymbol{R}$ possesses a nonisotropic eigenbivector $\boldsymbol{C}$:

$$\boldsymbol{R}\boldsymbol{C}=\varepsilon\boldsymbol{C}, \quad \boldsymbol{R}^{\mathrm{T}}\boldsymbol{C}=\varepsilon\boldsymbol{C}, \quad \boldsymbol{C}\cdot\boldsymbol{C}\neq 0. \qquad (4.2.16)$$

Without loss of generality, we take $\boldsymbol{C}\cdot\boldsymbol{C}=1$. Then choose $\boldsymbol{A}$, $\boldsymbol{B}$ to complete an orthonormal triad $\boldsymbol{A}$, $\boldsymbol{B}$, $\boldsymbol{C}$ and let $\boldsymbol{T}=(\boldsymbol{A}\,|\,\boldsymbol{B}\,|\,\boldsymbol{C})$ be the matrix whose columns are $\boldsymbol{A}$, $\boldsymbol{B}$, $\boldsymbol{C}$. It is easily seen that

$$\boldsymbol{T}^{\mathrm{T}}\boldsymbol{T}=\boldsymbol{1}, \quad \boldsymbol{T}^{\mathrm{T}}\boldsymbol{R}\boldsymbol{T}=\begin{bmatrix}\alpha & \gamma & 0\\ \delta & \beta & 0\\ 0 & 0 & \varepsilon\end{bmatrix}, \qquad (4.2.17)$$

with α, β, γ, δ given by (4.2.12). But $\det\boldsymbol{R}=\varepsilon$, the eigenvalue ε is triple, and $\boldsymbol{R}$ is orthogonal. Hence, proceeding as in the proof of Theorem 4.6, we obtain

$$\alpha=\beta=\varepsilon, \quad \gamma=\delta=0. \qquad (4.2.18)$$

Thus $\boldsymbol{T}^{\mathrm{T}}\boldsymbol{R}\boldsymbol{T}=\varepsilon\mathbf{1}$ and $\boldsymbol{R}=\varepsilon\mathbf{1}$. Every bivector is an eigenbivector of $\boldsymbol{R}$ corresponding to ε. Among this triple infinity of eigenbivectors consider, for instance, the bivectors $\boldsymbol{A}+\mathrm{i}\boldsymbol{B}$, $\boldsymbol{B}+\mathrm{i}\boldsymbol{C}$, $\boldsymbol{C}+\mathrm{i}\boldsymbol{A}$. These are isotropic and linearly independent.

(b) Assume now that all eigenbivectors corresponding to ε are isotropic. Then, using the same argument as in the proof of Theorem 3.8 (case (b)), we note that all these eigenbivectors are parallel. There is thus a simple infinity of eigenbivectors, all isotropic.

Exercises 4.1

1. Let

$$\boldsymbol{R}=\begin{bmatrix} \cosh a & \mathrm{i}\sinh a & 0 \\ -\mathrm{i}\sinh a & \cosh a & 0 \\ 0 & 0 & -1 \end{bmatrix},$$

with a real. Check that it is orthogonal.

Find the eigenvalues and eigenbivectors. Find, in particular, the isotropic eigenbivectors. Discuss the cases $a=0$ and $a\neq 0$.

2. Let

$$\boldsymbol{R}=\begin{bmatrix} 2 & \sqrt{2}(-1+\mathrm{i}) & 1+2\mathrm{i} \\ \sqrt{2}(1+\mathrm{i}) & -1 & \sqrt{2}(-1+\mathrm{i}) \\ 1-2\mathrm{i} & \sqrt{2}(1+\mathrm{i}) & 2 \end{bmatrix}.$$

Check that it is orthogonal.

Find the eigenvalues and eigenbivectors.

4.3 Canonical form. Spectral decomposition

Let $\boldsymbol{T}$ be a complex orthogonal 3×3 matrix. The transform $\tilde{\boldsymbol{R}}$ of the complex orthogonal 3×3 matrix $\boldsymbol{R}$, by the matrix $\boldsymbol{T}$, is defined by

$$\tilde{\boldsymbol{R}}=\boldsymbol{T}^{\mathrm{T}}\boldsymbol{R}\boldsymbol{T}. \tag{4.3.1}$$

This is also an orthogonal matrix and it is proper or improper according to whether $\boldsymbol{R}$ is proper or improper. We show here that with an appropriate choice of the orthogonal matrix $\boldsymbol{T}$, the matrix $\tilde{\boldsymbol{R}}$ takes a particularly simple form called the 'canonical form'. Corresponding to the canonical form an expression for $\boldsymbol{R}$ in terms of its eigenvalues and its eigenbivectors may be obtained. Such an expression is called the 'spectral decomposition' of $\boldsymbol{R}$.

Let $\varepsilon = \det \boldsymbol{R} = \pm 1$. Three cases have to be distinguished. **Either** (a) $\boldsymbol{R}$ possesses three simple eigenvalues ε, λ, λ^{-1} (with $\lambda \neq \pm\varepsilon$), **or** (b) $\boldsymbol{R}$ possesses the simple value ε and the double eigenvalue $\lambda = \lambda^{-1} = -\varepsilon$, **or** (c) $\boldsymbol{R}$ possesses the triple eigenvalue ε.

Case (a) Three simple eigenvalues ε, λ, λ^{-1} (with $\lambda \neq \pm\varepsilon$)

Let $\boldsymbol{C}$, $\boldsymbol{D}$, $\boldsymbol{E}$ be eigenbivectors corresponding to these eigenvalues:

$$\boldsymbol{RC} = \varepsilon \boldsymbol{C}, \tag{4.3.2a}$$

$$\boldsymbol{R}^{\mathrm{T}}\boldsymbol{C} = \varepsilon \boldsymbol{C}, \tag{4.3.2b}$$

$$\boldsymbol{RD} = \lambda \boldsymbol{D}, \tag{4.3.2c}$$

$$\boldsymbol{RE} = \lambda^{-1} \boldsymbol{E}. \tag{4.3.2d}$$

From the Corollary 4.4 of Theorem 4.3, $\boldsymbol{C}$ is not isotropic and thus we may take $\boldsymbol{C}\cdot\boldsymbol{C} = 1$. Then choose $\boldsymbol{A}$, $\boldsymbol{B}$ to complete an orthonormal triad $\boldsymbol{A}$, $\boldsymbol{B}$, $\boldsymbol{C}$. From Theorem 4.2, $\boldsymbol{D}$ and $\boldsymbol{E}$ are orthogonal to $\boldsymbol{C}$, and, from Theorem 4.5, they are isotropic and linearly independent:

$$\boldsymbol{C}\cdot\boldsymbol{D} = \boldsymbol{C}\cdot\boldsymbol{E} = 0, \quad \boldsymbol{D}\cdot\boldsymbol{D} = \boldsymbol{E}\cdot\boldsymbol{E} = 0, \quad \boldsymbol{D}\cdot\boldsymbol{E} \neq 0.$$

Hence, without loss of generality, we may take

$$\boldsymbol{D} = \boldsymbol{A} + \mathrm{i}\boldsymbol{B}, \quad \boldsymbol{E} = \boldsymbol{A} - \mathrm{i}\boldsymbol{B}. \tag{4.3.3}$$

Writing $\lambda = \mathrm{e}^{-\mathrm{i}\Theta}$, with Θ complex (in general), and taking the dot product of (4.3.2c), (4.3.2d) with $\boldsymbol{A}$ and $\boldsymbol{B}$ yields

$$\begin{aligned} \boldsymbol{A}^{\mathrm{T}}\boldsymbol{RA} + \mathrm{i}\boldsymbol{A}^{\mathrm{T}}\boldsymbol{RB} &= \cos\Theta - \mathrm{i}\sin\Theta, \\ \boldsymbol{B}^{\mathrm{T}}\boldsymbol{RA} + \mathrm{i}\boldsymbol{B}^{\mathrm{T}}\boldsymbol{RB} &= \mathrm{i}\cos\Theta + \sin\Theta. \end{aligned} \tag{4.3.4}$$

$$\begin{aligned} \boldsymbol{A}^{\mathrm{T}}\boldsymbol{RA} - \mathrm{i}\boldsymbol{A}^{\mathrm{T}}\boldsymbol{RB} &= \cos\Theta + \mathrm{i}\sin\Theta, \\ \boldsymbol{B}^{\mathrm{T}}\boldsymbol{RA} - \mathrm{i}\boldsymbol{B}^{\mathrm{T}}\boldsymbol{RB} &= -\mathrm{i}\cos\Theta + \sin\Theta. \end{aligned} \tag{4.3.5}$$

Hence,

$$\begin{aligned} \boldsymbol{A}^{\mathrm{T}}\boldsymbol{RA} = \cos\Theta, &\quad \boldsymbol{A}^{\mathrm{T}}\boldsymbol{RB} = -\sin\Theta, \\ \boldsymbol{B}^{\mathrm{T}}\boldsymbol{RA} = \sin\Theta, &\quad \boldsymbol{B}^{\mathrm{T}}\boldsymbol{RB} = \cos\Theta. \end{aligned} \tag{4.3.6}$$

Thus, with the orthogonal matrix $\boldsymbol{T} = (\boldsymbol{A}\,|\,\boldsymbol{B}\,|\,\boldsymbol{C})$, we have

$$\tilde{\boldsymbol{R}} = \boldsymbol{T}^{\mathrm{T}}\boldsymbol{RT} = \begin{bmatrix} \cos\Theta & -\sin\Theta & 0 \\ \sin\Theta & \cos\Theta & 0 \\ 0 & 0 & \varepsilon \end{bmatrix}. \tag{4.3.7}$$

This is the canonical form of $\boldsymbol{R}$ in case (a). It follows that $\boldsymbol{R}$ may be written as

$$\begin{aligned}\boldsymbol{R} = \cos\Theta(\boldsymbol{A}\otimes\boldsymbol{A} + \boldsymbol{B}\otimes\boldsymbol{B}) \\ + \sin\Theta(\boldsymbol{B}\otimes\boldsymbol{A} - \boldsymbol{A}\otimes\boldsymbol{B}) + \varepsilon\boldsymbol{C}\otimes\boldsymbol{C},\end{aligned} \tag{4.3.8}$$

or

$$\boldsymbol{R} = \tfrac{1}{2}(\lambda\boldsymbol{D}\otimes\boldsymbol{E} + \lambda^{-1}\boldsymbol{E}\otimes\boldsymbol{D}) + \varepsilon\boldsymbol{C}\otimes\boldsymbol{C}. \tag{4.3.9}$$

This is the spectral decomposition of $\boldsymbol{R}$ in case (a).

Case (b) One simple eigenvalue ε and one double eigenvalue $\lambda = \lambda^{-1} = -\varepsilon$

Let $\boldsymbol{C}$ be an eigenbivector corresponding to ε:

$$\boldsymbol{R}\boldsymbol{C} = \varepsilon\boldsymbol{C}, \quad \boldsymbol{R}^{\mathrm{T}}\boldsymbol{C} = \varepsilon\boldsymbol{C}. \tag{4.3.10}$$

Because $\boldsymbol{C}$ is not isotropic (Corollary 4.4), we may take $\boldsymbol{C}\cdot\boldsymbol{C} = 1$. From Theorems 4.2 and 4.6, we know that all the bivectors orthogonal to $\boldsymbol{C}$ are eigenbivectors of $\boldsymbol{R}$ corresponding to the double eigenvalue $-\varepsilon$. Consider any pair $\boldsymbol{A}$, $\boldsymbol{B}$ of these eigenbivectors such that $\boldsymbol{A}\cdot\boldsymbol{B} = 0$, $\boldsymbol{A}\cdot\boldsymbol{A} = \boldsymbol{B}\cdot\boldsymbol{B} = 1$. Then $\boldsymbol{A}$, $\boldsymbol{B}$, $\boldsymbol{C}$ is an orthonormal triad, and

$$\boldsymbol{R}\boldsymbol{A} = -\varepsilon\boldsymbol{A}, \quad \boldsymbol{R}^{\mathrm{T}}\boldsymbol{A} = -\varepsilon\boldsymbol{A}, \quad \boldsymbol{R}\boldsymbol{B} = -\varepsilon\boldsymbol{B}, \quad \boldsymbol{R}^{\mathrm{T}}\boldsymbol{B} = -\varepsilon\boldsymbol{B}. \tag{4.3.11}$$

Thus, with the orthogonal matrix $\boldsymbol{T} = (\boldsymbol{A}\,|\,\boldsymbol{B}\,|\,\boldsymbol{C})$, we have

$$\tilde{\boldsymbol{R}} = \boldsymbol{T}^{\mathrm{T}}\boldsymbol{R}\boldsymbol{T} = \begin{bmatrix} -\varepsilon & 0 & 0 \\ 0 & -\varepsilon & 0 \\ 0 & 0 & \varepsilon \end{bmatrix}. \tag{4.3.12}$$

This is the canonical form of $\boldsymbol{R}$ in case (b). It follows that the spectral decomposition of $\boldsymbol{R}$ is

$$\boldsymbol{R} = \varepsilon(\boldsymbol{C}\otimes\boldsymbol{C} - \boldsymbol{A}\otimes\boldsymbol{A} - \boldsymbol{B}\otimes\boldsymbol{B}), \tag{4.3.13}$$

or

$$\boldsymbol{R} = \varepsilon(2\boldsymbol{C}\otimes\boldsymbol{C} - \mathbf{1}). \tag{4.3.14}$$

We note that (4.3.12) or (4.3.13) is a special case of (4.3.7) or (4.3.8). Indeed (4.3.7) and (4.3.8) reduce to (4.3.12) and (4.3.13) for $\Theta = \pi$ (when $\varepsilon = +1$) or $\Theta = 0$ (when $\varepsilon = -1$).

Case (c) One triple eigenvalue ε

From Theorem 4.7, we know that two subcases have to be considered.

(i) Corresponding to ε, there is a triple infinity of eigenbivectors Then, all the bivectors are eigenbivectors of $\boldsymbol{Q}$, and (see the proof of Theorem 4.7, case (a)),

$$\boldsymbol{R} = \varepsilon \mathbf{1}. \tag{4.3.15}$$

We note that (4.3.15) is a special case of (4.3.8). Indeed (4.3.8) reduces to (4.3.15) for $\Theta = 0$ (when $\varepsilon = +1$) or $\Theta = \pi$ (when $\varepsilon = -1$).

(ii) Corresponding to ε, there is a simple infinity of eigenbivectors Then, all the eigenbivectors are isotropic and parallel. Let $\boldsymbol{D}$ be one of them:

$$\boldsymbol{R}\boldsymbol{D} = \varepsilon \boldsymbol{D}, \tag{4.3.16a}$$

$$\boldsymbol{R}^{\mathrm{T}}\boldsymbol{D} = \varepsilon \boldsymbol{D}, \tag{4.3.16b}$$

$$\boldsymbol{D}\cdot\boldsymbol{D} = 0. \tag{4.3.16c}$$

Without loss of generality we may write

$$\boldsymbol{D} = \boldsymbol{A} + \mathrm{i}\boldsymbol{B}, \tag{4.3.17}$$

where $\boldsymbol{A}$ and $\boldsymbol{B}$ are such that $\boldsymbol{A}\cdot\boldsymbol{A} = \boldsymbol{B}\cdot\boldsymbol{B} = 1$ and $\boldsymbol{A}\cdot\boldsymbol{B} = 0$. Let $\boldsymbol{C} = \boldsymbol{A} \times \boldsymbol{B}$ so that $\boldsymbol{A}$, $\boldsymbol{B}$, $\boldsymbol{C}$ is an orthonormal triad. Taking the dot product of (4.3.16a), (4.3.16b) with $\boldsymbol{A}$, $\boldsymbol{B}$, $\boldsymbol{C}$ yields

$$\begin{aligned} &\boldsymbol{A}^{\mathrm{T}}\boldsymbol{R}\boldsymbol{A} + \mathrm{i}\boldsymbol{A}^{\mathrm{T}}\boldsymbol{R}\boldsymbol{B} = \varepsilon, \quad &&\boldsymbol{A}^{\mathrm{T}}\boldsymbol{R}\boldsymbol{A} + \mathrm{i}\boldsymbol{A}^{\mathrm{T}}\boldsymbol{R}^{\mathrm{T}}\boldsymbol{B} = \varepsilon,\\ &\boldsymbol{B}^{\mathrm{T}}\boldsymbol{R}\boldsymbol{A} + \mathrm{i}\boldsymbol{B}^{\mathrm{T}}\boldsymbol{R}\boldsymbol{B} = \mathrm{i}\varepsilon, \quad &&\boldsymbol{B}^{\mathrm{T}}\boldsymbol{R}^{\mathrm{T}}\boldsymbol{A} + \mathrm{i}\boldsymbol{B}^{\mathrm{T}}\boldsymbol{R}^{\mathrm{T}}\boldsymbol{B} = \mathrm{i}\varepsilon,\\ &\boldsymbol{C}^{\mathrm{T}}\boldsymbol{R}\boldsymbol{A} + \mathrm{i}\boldsymbol{C}^{\mathrm{T}}\boldsymbol{R}\boldsymbol{B} = 0, \quad &&\boldsymbol{C}^{\mathrm{T}}\boldsymbol{R}^{\mathrm{T}}\boldsymbol{A} + \mathrm{i}\boldsymbol{C}^{\mathrm{T}}\boldsymbol{R}^{\mathrm{T}}\boldsymbol{B} = 0. \end{aligned} \tag{4.3.18}$$

Thus writing

$$\delta = \varepsilon \boldsymbol{A}^{\mathrm{T}}\boldsymbol{R}\boldsymbol{C}, \quad \gamma = \varepsilon \boldsymbol{C}^{\mathrm{T}}\boldsymbol{R}\boldsymbol{A}, \quad \beta = \varepsilon \boldsymbol{A}^{\mathrm{T}}\boldsymbol{R}\boldsymbol{B} = \varepsilon \boldsymbol{B}^{\mathrm{T}}\boldsymbol{R}\boldsymbol{A}, \tag{4.3.19}$$

it follows that

$$\begin{aligned} &\boldsymbol{A}^{\mathrm{T}}\boldsymbol{R}\boldsymbol{A} = \varepsilon(1 - \mathrm{i}\beta), \quad &&\boldsymbol{B}^{\mathrm{T}}\boldsymbol{R}\boldsymbol{B} = \varepsilon(1 + \mathrm{i}\beta),\\ &\boldsymbol{B}^{\mathrm{T}}\boldsymbol{R}\boldsymbol{C} = \mathrm{i}\varepsilon\delta, \quad &&\boldsymbol{C}^{\mathrm{T}}\boldsymbol{R}\boldsymbol{B} = \mathrm{i}\varepsilon\gamma. \end{aligned} \tag{4.3.20}$$

Thus, with the orthogonal matrix $\boldsymbol{T}=(\boldsymbol{A}\,|\,\boldsymbol{B}\,|\,\boldsymbol{C})$, we have

$$\tilde{\boldsymbol{R}}=\boldsymbol{T}^{\mathrm{T}}\boldsymbol{R}\boldsymbol{T}=\varepsilon\begin{bmatrix}1-\mathrm{i}\beta & \beta & \delta\\ \beta & 1+\mathrm{i}\beta & \mathrm{i}\delta\\ \gamma & \mathrm{i}\gamma & \varepsilon\boldsymbol{C}^{\mathrm{T}}\boldsymbol{R}\boldsymbol{C}\end{bmatrix}. \tag{4.3.21}$$

But ε is a triple eigenvalue, so that $\boldsymbol{C}^{\mathrm{T}}\boldsymbol{R}\boldsymbol{C}=\varepsilon$, and $\tilde{\boldsymbol{R}}$ must be orthogonal, and hence

$$\gamma=-\delta,\quad \beta=-\mathrm{i}\frac{\delta^2}{2}. \tag{4.3.22}$$

Thus,

$$\tilde{\boldsymbol{R}}=\boldsymbol{T}^{\mathrm{T}}\boldsymbol{R}\boldsymbol{T}=\varepsilon\begin{bmatrix}1-\frac{1}{2}\delta^2 & -\frac{1}{2}\mathrm{i}\delta^2 & \delta\\ -\frac{1}{2}\mathrm{i}\delta^2 & 1+\frac{1}{2}\delta^2 & \mathrm{i}\delta\\ -\delta & -\mathrm{i}\delta & 1\end{bmatrix}. \tag{4.3.23}$$

Corresponding to (4.3.23) we have the spectral decomposition

$$\varepsilon\boldsymbol{R}=\boldsymbol{A}\otimes\boldsymbol{A}+\boldsymbol{B}\otimes\boldsymbol{B}+\boldsymbol{C}\otimes\boldsymbol{C}-\frac{\delta^2}{2}(\boldsymbol{A}+\mathrm{i}\boldsymbol{B})\otimes(\boldsymbol{A}+\mathrm{i}\boldsymbol{B})$$

$$+\,\delta\{(\boldsymbol{A}+\mathrm{i}\boldsymbol{B})\otimes\boldsymbol{C}-\boldsymbol{C}\otimes(\boldsymbol{A}+\mathrm{i}\boldsymbol{B})\}, \tag{4.3.24}$$

or

$$\varepsilon\boldsymbol{R}=\mathbf{1}-\frac{\delta^2}{2}\boldsymbol{D}\otimes\boldsymbol{D}+\delta(\boldsymbol{D}\otimes\boldsymbol{C}-\boldsymbol{C}\otimes\boldsymbol{D}). \tag{4.3.25}$$

Further simplification of (4.3.23), (4.3.25) is possible. Let $\delta=\mathrm{e}^{\mathrm{i}\Phi}$ where Φ is a complex number. Then consider the bivector $\boldsymbol{D}'$ given by

$$\boldsymbol{D}'=\mathrm{e}^{\mathrm{i}\Phi}\boldsymbol{D}=\boldsymbol{A}'+\mathrm{i}\boldsymbol{B}', \tag{4.3.26}$$

with

$$\boldsymbol{A}'=\cos\Phi\,\boldsymbol{A}-\sin\Phi\,\boldsymbol{B},$$
$$\boldsymbol{B}'=\sin\Phi\,\boldsymbol{A}+\cos\Phi\,\boldsymbol{B}. \tag{4.3.27}$$

Note that $\boldsymbol{A}'$, $\boldsymbol{B}'$, $\boldsymbol{C}$ form an orthonormal triad. Clearly, (4.3.25) may now be written

$$\varepsilon\boldsymbol{R}=\mathbf{1}-\tfrac{1}{2}\boldsymbol{D}'\otimes\boldsymbol{D}'+(\boldsymbol{D}'\otimes\boldsymbol{C}-\boldsymbol{C}\otimes\boldsymbol{D}'). \tag{4.3.28}$$

Because $\boldsymbol{D}'=\boldsymbol{A}'+\mathrm{i}\boldsymbol{B}'$, we have now, with the orthogonal matrix $\boldsymbol{T}'=(\boldsymbol{A}'\,|\,\boldsymbol{B}'\,|\,\boldsymbol{C})$,

$$\tilde{\boldsymbol{R}}'=\boldsymbol{T}'^{\mathrm{T}}\boldsymbol{R}\boldsymbol{T}'=\varepsilon\begin{bmatrix}\frac{1}{2} & -\frac{1}{2}\mathrm{i} & 1\\ -\frac{1}{2}\mathrm{i} & \frac{3}{2} & \mathrm{i}\\ -1 & -\mathrm{i} & 1\end{bmatrix}. \tag{4.3.29}$$

This is the canonical form of $\boldsymbol{R}$ in case (c)(ii), the spectral decomposition being (4.3.28).

Remark In case (c)(ii), we note that any matrix of the form (4.3.25), with an arbitrary complex $\delta \neq 0$, has $\boldsymbol{D}$ (isotropic) and its scalar multiples as eigenbivectors corresponding to the eigenvalue ε (triple).

Exercises 4.2

1. Consider the orthonormal triad $\boldsymbol{A} = \frac{1}{2}(2^{1/2})(\mathrm{i}, -2, \mathrm{i})$, $\boldsymbol{B} = \frac{1}{2}(2^{1/2})(-1, 0, 1)$, $\boldsymbol{C} = \boldsymbol{A} \times \boldsymbol{B} = (-1, -\mathrm{i}, -1)$. Construct the orthogonal matrix $\boldsymbol{R}$ such that $\boldsymbol{D} = \boldsymbol{A} + \mathrm{i}\boldsymbol{B}$, $\boldsymbol{E} = \boldsymbol{A} - \mathrm{i}\boldsymbol{B}$, $\boldsymbol{C}$ are the eigenbivectors corresponding to the eigenvalues $\lambda = \frac{1}{2}$, $\lambda^{-1} = 2$, $\varepsilon = +1$, respectively.
2. Let

$$\boldsymbol{R} = \begin{bmatrix} 2 & \sqrt{2}(-1+\mathrm{i}) & 1+2\mathrm{i} \\ \sqrt{2}(1+\mathrm{i}) & -1 & \sqrt{2}(-1+\mathrm{i}) \\ 1-2\mathrm{i} & \sqrt{2}(1+\mathrm{i}) & 2 \end{bmatrix}.$$

   Find $\boldsymbol{T}$ orthogonal such that $\boldsymbol{T}^{\mathrm{T}}\boldsymbol{R}\boldsymbol{T}$ is canonical (use results of Exercise 4.1.2).

4.4 Skew-symmetric matrices and bivectors

Here we consider complex 3×3 skew-symmetric matrices, that is matrices $\boldsymbol{W}$ such that

$$\boldsymbol{W}^{\mathrm{T}} = -\boldsymbol{W}. \tag{4.4.1}$$

The link between these matrices and orthogonal matrices is established in the next section.

With a complex 3×3 skew-symmetric matrix $\boldsymbol{W}$ may be associated a bivector $\boldsymbol{\Omega}$ (in a one-to-one correspondence) through

$$\Omega_i = -\tfrac{1}{2} e_{ijk} W_{jk}, \quad W_{ij} = -e_{ijk}\Omega_k, \tag{4.4.2}$$

where e_{ijk} denotes the alternating symbol. Here $\boldsymbol{W}$ and $\boldsymbol{\Omega}$ are referred to a real orthonormal basis. For any bivector $\boldsymbol{X}$ we have

$$\boldsymbol{W}\boldsymbol{X} = \boldsymbol{\Omega} \times \boldsymbol{X}. \tag{4.4.3}$$

It is thus clear that $\boldsymbol{\Omega}$ is an eigenbivector of $\boldsymbol{W}$ corresponding to the eigenvalue zero:

$$\boldsymbol{W\Omega} = \boldsymbol{0}. \tag{4.4.4}$$

Let us now distinguish between the case when $\boldsymbol{\Omega}$ is not isotropic and the case when $\boldsymbol{\Omega}$ is isotropic.

Case (a) $\boldsymbol{\Omega}$ is not isotropic: $\boldsymbol{\Omega}\cdot\boldsymbol{\Omega} \neq 0$

We may write $\Omega = \Theta \boldsymbol{C}$, where $\boldsymbol{C}$ is such that $\boldsymbol{C}\cdot\boldsymbol{C} = 1$ and Θ is some complex scalar. Complete $\boldsymbol{C}$ with $\boldsymbol{A}$, $\boldsymbol{B}$ to form an orthonormal triad $\boldsymbol{A}$, $\boldsymbol{B}$, $\boldsymbol{C}$ such that $\boldsymbol{A}\cdot\boldsymbol{B}\times\boldsymbol{C} = 1$ (direct orthonormal triad). Let $\boldsymbol{T} = (\boldsymbol{A}|\boldsymbol{B}|\boldsymbol{C})$ be the orthogonal matrix whose columns are $\boldsymbol{A}$, $\boldsymbol{B}$, $\boldsymbol{C}$. Then, because

$$\boldsymbol{WC} = 0, \quad \boldsymbol{B}^{\mathrm{T}}\boldsymbol{WA} = -\boldsymbol{A}^{\mathrm{T}}\boldsymbol{WB} = -\boldsymbol{A}\cdot\boldsymbol{\Omega}\times\boldsymbol{B} = \Theta, \tag{4.4.5}$$

we have

$$\tilde{\boldsymbol{W}} = \boldsymbol{T}^{\mathrm{T}}\boldsymbol{WT} = \begin{bmatrix} 0 & -\Theta & 0 \\ \Theta & 0 & 0 \\ 0 & 0 & 0 \end{bmatrix}. \tag{4.4.6}$$

This is the canonical form of $\boldsymbol{W}$ in case (a).

From this we note that the skew-symmetric matrix $\boldsymbol{W}$ has the eigenbivectors $\boldsymbol{D} = \boldsymbol{A} + \mathrm{i}\boldsymbol{B}$, $\boldsymbol{E} = \boldsymbol{A} - \mathrm{i}\boldsymbol{B}$ and $\boldsymbol{C}$ corresponding to the eigenvalues $-\mathrm{i}\Theta$, $\mathrm{i}\Theta$ and zero, respectively. Corresponding to (4.4.6) we have the spectral decomposition

$$\boldsymbol{W} = \Theta(\boldsymbol{B}\otimes\boldsymbol{A} - \boldsymbol{A}\otimes\boldsymbol{B}) = \frac{\mathrm{i}}{2}\Theta(\boldsymbol{E}\otimes\boldsymbol{D} - \boldsymbol{D}\otimes\boldsymbol{E}). \tag{4.4.7}$$

Case (b) $\boldsymbol{\Omega}$ is isotropic: $\boldsymbol{\Omega}\cdot\boldsymbol{\Omega} = 0$

Any isotropic bivector $\boldsymbol{D}$ parallel to $\boldsymbol{\Omega}$ is an eigenbivector of $\boldsymbol{W}$ corresponding to the eigenvalue zero. Without loss of generality, write

$$\boldsymbol{D} = \boldsymbol{A} + \mathrm{i}\boldsymbol{B}, \quad \text{with} \quad \boldsymbol{A}\cdot\boldsymbol{A} = \boldsymbol{B}\cdot\boldsymbol{B} = 1 \quad \text{and} \quad \boldsymbol{A}\cdot\boldsymbol{B} = 0.$$

Thus

$$\boldsymbol{WD} = \boldsymbol{0}, \quad \boldsymbol{WA} = -\mathrm{i}\boldsymbol{WB}. \tag{4.4.8}$$

Then let $\boldsymbol{C} = \boldsymbol{A}\times\boldsymbol{B}$ so that $\boldsymbol{A}$, $\boldsymbol{B}$, $\boldsymbol{C}$ is an orthonormal triad. Thus, with the orthogonal matrix $\boldsymbol{T} = (\boldsymbol{A}|\boldsymbol{B}|\boldsymbol{C})$ we have

$$\tilde{\boldsymbol{W}} = \boldsymbol{T}^{\mathrm{T}}\boldsymbol{WT} = \begin{bmatrix} 0 & 0 & \delta \\ 0 & 0 & \mathrm{i}\delta \\ -\delta & -\mathrm{i}\delta & 0 \end{bmatrix}, \tag{4.4.9}$$

with

$$\delta = \boldsymbol{A}^{\mathrm{T}}\boldsymbol{W}\boldsymbol{C} = \boldsymbol{A}\cdot(\boldsymbol{\Omega}\times\boldsymbol{C}). \tag{4.4.10}$$

From this we note that the eigenvalue 0 of $\boldsymbol{W}$ is triple and that corresponding to this triple eigenvalue, there is a simple infinity of eigenbivectors, all parallel to $\boldsymbol{D}$ and $\boldsymbol{\Omega}$. Corresponding to (4.4.9) we have the spectral decomposition

$$\boldsymbol{W} = \delta\{(\boldsymbol{A}+\mathrm{i}\boldsymbol{B})\otimes\boldsymbol{C} - \boldsymbol{C}\otimes(\boldsymbol{A}+\mathrm{i}\boldsymbol{B})\} = \delta(\boldsymbol{D}\otimes\boldsymbol{C} - \boldsymbol{C}\otimes\boldsymbol{D}). \tag{4.4.11}$$

Further simplification is possible. Let $\delta = \mathrm{e}^{\mathrm{i}\Phi}$ where Φ is a complex number, and consider the bivector $\boldsymbol{D}' = \mathrm{e}^{\mathrm{i}\Phi}\boldsymbol{D}$. Clearly, (4.4.11) may be written

$$\boldsymbol{W} = \boldsymbol{D}'\otimes\boldsymbol{C} - \boldsymbol{C}\otimes\boldsymbol{D}'. \tag{4.4.12}$$

Defining $\boldsymbol{A}'$, $\boldsymbol{B}'$ as in (4.3.27), we have thus, with the orthogonal matrix $\boldsymbol{T}' = (\boldsymbol{A}'|\boldsymbol{B}'|\boldsymbol{C})$,

$$\tilde{\boldsymbol{W}}' = \boldsymbol{T}'^{\mathrm{T}}\boldsymbol{W}\boldsymbol{T}' = \begin{bmatrix} 0 & 0 & 1 \\ 0 & 0 & \mathrm{i} \\ -1 & -\mathrm{i} & 0 \end{bmatrix}. \tag{4.4.13}$$

This is the canonical form of $\boldsymbol{W}$ in case (b), the spectral decomposition being (4.4.12).

We finally note that the bivector $\boldsymbol{\Omega}$ associated with $\boldsymbol{W}$ given by (4.4.12) is the isotropic bivector

$$\boldsymbol{\Omega} = -\mathrm{i}\boldsymbol{D}' = \boldsymbol{B}' - \mathrm{i}\boldsymbol{A}'. \tag{4.4.14}$$

Indeed, because $\boldsymbol{A}'$, $\boldsymbol{B}'$, $\boldsymbol{C}$ is an orthonormal triad, we have for any bivector $\boldsymbol{X}$ that

$$\begin{aligned}\boldsymbol{\Omega}\times\boldsymbol{X} &= (\boldsymbol{B}'-\mathrm{i}\boldsymbol{A}')\times\{(\boldsymbol{X}\cdot\boldsymbol{A}')\boldsymbol{A}' + (\boldsymbol{X}\cdot\boldsymbol{B}')\boldsymbol{B}' + (\boldsymbol{X}\cdot\boldsymbol{C})\boldsymbol{C}\} \\ &= (\boldsymbol{X}\cdot\boldsymbol{C})\boldsymbol{D}' - (\boldsymbol{X}\cdot\boldsymbol{D}')\boldsymbol{C} = (\boldsymbol{D}'\otimes\boldsymbol{C} - \boldsymbol{C}\otimes\boldsymbol{D}')\boldsymbol{X}.\end{aligned} \tag{4.4.15}$$

Exercises 4.3

1. Let

$$\boldsymbol{W} = \begin{bmatrix} 0 & -2 & 2\mathrm{i} \\ 2 & 0 & -2 \\ -2\mathrm{i} & 2 & 0 \end{bmatrix}.$$

Find the associated bivector $\boldsymbol{\Omega}$. Find the eigenvalues and eigenbivectors of $\boldsymbol{W}$. Find $\boldsymbol{T}$ such that $\boldsymbol{T}^{\mathrm{T}}\boldsymbol{W}\boldsymbol{T}$ is canonical.

2. For

$$W = \begin{bmatrix} 0 & -1 & \mathrm{i}\sqrt{2} \\ 1 & 0 & -1 \\ -\mathrm{i}\sqrt{2} & 1 & 0 \end{bmatrix},$$

find the associated bivector, the eigenvalues and the eigenbivectors. Find $\boldsymbol{T}$ such that $\boldsymbol{T}^{\mathrm{T}}\boldsymbol{W}\boldsymbol{T}$ is canonical.

4.5 Link between skew-symmetric and orthogonal matrices (exponential of a skew-symmetric matrix)

Here we show that a complex orthogonal 3×3 matrix may always be written as plus (if it is proper) or minus (if it is improper) the exponential of a complex skew-symmetric 3×3 matrix. We first note that if $\boldsymbol{R}$ is an improper orthogonal 3×3 matrix ($\det \boldsymbol{R} = -1$), then $-\boldsymbol{R}$ is a proper orthogonal 3×3 matrix ($\det(-\boldsymbol{R}) = +1$). It is thus sufficient to show here that a complex proper orthogonal matrix $\boldsymbol{R}$ ($\det \boldsymbol{R} = +1$) may always be written as the exponential of a complex skew-symmetric matrix.

In section 4.3, it has been shown that a complex proper orthogonal matrix $\boldsymbol{R}$ may be written either as (cases (a), (b) and (c) (i))

$$\begin{aligned} \boldsymbol{R} = {} & \cos\Theta(\boldsymbol{A}\otimes\boldsymbol{A} + \boldsymbol{B}\otimes\boldsymbol{B}) \\ & + \sin\Theta(\boldsymbol{B}\otimes\boldsymbol{A} - \boldsymbol{A}\otimes\boldsymbol{B}) + \boldsymbol{C}\otimes\boldsymbol{C}, \end{aligned} \tag{4.5.1}$$

where $\boldsymbol{A}, \boldsymbol{B}, \boldsymbol{C}$ is an appropriate orthonormal triad, or as (case (c)(ii))

$$\boldsymbol{R} = \mathbf{1} - \tfrac{1}{2}\boldsymbol{D}'\otimes\boldsymbol{D}' + (\boldsymbol{D}'\otimes\boldsymbol{C} - \boldsymbol{C}\otimes\boldsymbol{D}'), \tag{4.5.2}$$

where $\boldsymbol{D}' = \boldsymbol{A}' + \mathrm{i}\boldsymbol{B}'$, and $\boldsymbol{A}'$, $\boldsymbol{B}'$, $\boldsymbol{C}$ is an appropriate orthonormal triad.

(a) We first show that $\boldsymbol{R}$ given by (4.5.1) may be written as the exponential of a skew-symmetric matrix $\boldsymbol{W}$ (associated with a non-isotropic bivector).

Let $\boldsymbol{\Gamma}$ be the skew-symmetric matrix associated with the bivector $\boldsymbol{C}$. From 4.4 (case (a)), we know that

$$\boldsymbol{\Gamma} = \boldsymbol{B}\otimes\boldsymbol{A} - \boldsymbol{A}\otimes\boldsymbol{B}. \tag{4.5.3}$$

Also, it is easily seen that

$$\boldsymbol{\Gamma}^2 = -(\boldsymbol{A}\otimes\boldsymbol{A} + \boldsymbol{B}\otimes\boldsymbol{B}), \quad \mathbf{1} + \boldsymbol{\Gamma}^2 = \boldsymbol{C}\otimes\boldsymbol{C}. \tag{4.5.4}$$

Thus, the spectral decomposition (4.5.1) of $\boldsymbol{R}$ may be written as

$$\boldsymbol{R} = \mathbf{1} + \sin\Theta\boldsymbol{\Gamma} + (1 - \cos\Theta)\boldsymbol{\Gamma}^2. \tag{4.5.5}$$

Now let $\boldsymbol{W}$ be the skew-symmetric matrix associated with $\boldsymbol{\Omega} = \Theta\boldsymbol{C}$, that is

$$\boldsymbol{W} = \Theta\boldsymbol{\Gamma} = \Theta(\boldsymbol{B}\otimes\boldsymbol{A} - \boldsymbol{A}\otimes\boldsymbol{B}). \tag{4.5.6}$$

By definition,

$$\exp\boldsymbol{W} = \mathbf{1} + \boldsymbol{W} + \frac{1}{2!}\boldsymbol{W}^2 + \frac{1}{3!}\boldsymbol{W}^3 + \frac{1}{4!}\boldsymbol{W}^4 + \cdots. \tag{4.5.7}$$

But, using the Cayley–Hamilton theorem, we note that

$$\boldsymbol{W}^3 + \Theta^2\boldsymbol{W} = \mathbf{0}, \tag{4.5.8}$$

and hence

$$\begin{gathered}\boldsymbol{W} = \Theta\boldsymbol{\Gamma}, \quad \boldsymbol{W}^2 = \Theta^2\boldsymbol{\Gamma}^2, \quad \boldsymbol{W}^3 = -\Theta^3\boldsymbol{\Gamma},\\ \boldsymbol{W}^4 = -\Theta^4\boldsymbol{\Gamma}^2, \quad \boldsymbol{W}^5 = \Theta^5\boldsymbol{\Gamma}, \ldots,\end{gathered} \tag{4.5.9}$$

so that

$$\begin{aligned}\exp\boldsymbol{W} = \mathbf{1} &+ \left(1 - \frac{\Theta^3}{3!} + \frac{\Theta^5}{5!} - \cdots\right)\boldsymbol{\Gamma}\\ &+ \left(\frac{\Theta^2}{2!} - \frac{\Theta^4}{4!} + \frac{\Theta^6}{6!} - \cdots\right)\boldsymbol{\Gamma}^2.\end{aligned} \tag{4.5.10}$$

Using the Taylor series of $\sin\Theta$ and $\cos\Theta$, and comparing with (4.5.5), we finally obtain

$$\boldsymbol{R} = \mathbf{1} + \sin\Theta\boldsymbol{\Gamma} + (1 - \cos\Theta)\boldsymbol{\Gamma}^2 = \exp\boldsymbol{W} = \exp(\Theta\boldsymbol{\Gamma}), \tag{4.5.11}$$

which shows that the proper orthogonal matrix $\boldsymbol{R}$ is the exponential of the skew-symmetric matrix $\boldsymbol{W}$.

(b) We now show that $\boldsymbol{R}$ given by (4.5.2) may be written as the exponential of a skew-symmetric matrix $\boldsymbol{W}$ (associated with an isotropic bivector).

Let $\boldsymbol{W}$ be the skew-symmetric matrix associated with the isotropic bivector $\boldsymbol{\Omega} = -\mathrm{i}\boldsymbol{D}'$. From 4.4 (case (b)), we know that

$$\boldsymbol{W} = \boldsymbol{D}'\otimes\boldsymbol{C} - \boldsymbol{C}\otimes\boldsymbol{D}'. \tag{4.5.12}$$

Also, because $\boldsymbol{D}'\cdot\boldsymbol{D}' = \boldsymbol{C}\cdot\boldsymbol{D}' = 0$, it is easily seen that

$$\boldsymbol{W}^2 = -\boldsymbol{D}'\otimes\boldsymbol{D}', \quad \boldsymbol{W}^3 = \boldsymbol{W}^4 = \cdots = \mathbf{0}. \tag{4.5.13}$$

Thus, using the definition (4.5.7) of the exponential of $\boldsymbol{W}$, we may write the spectral decomposition (4.5.2) of $\boldsymbol{R}$ as

$$\boldsymbol{R} = \mathbf{1} + \boldsymbol{W} + \tfrac{1}{2}\boldsymbol{W}^2 = \exp \boldsymbol{W}. \tag{4.5.14}$$

This shows that the proper orthogonal matrix $\boldsymbol{R}$ is the exponential of the skew-symmetric matrix $\boldsymbol{W}$.

Exercises 4.4

1. Check that $\tilde{\boldsymbol{R}} = \exp \tilde{\boldsymbol{W}}$, where $\tilde{\boldsymbol{R}}$ and $\tilde{\boldsymbol{W}}$ are given respectively by (4.3.7) with $\varepsilon = +1$, and (4.4.6).
2. Check that $\tilde{\boldsymbol{R}} = \exp \tilde{\boldsymbol{W}}$, where $\tilde{\boldsymbol{R}}$ and $\tilde{\boldsymbol{W}}$ are given respectively by (4.3.23) with $\varepsilon = +1$, and (4.4.9).
3. Check that $\tilde{\boldsymbol{R}}' = \exp \tilde{\boldsymbol{W}}'$, where $\tilde{\boldsymbol{R}}'$ and $\tilde{\boldsymbol{W}}'$ are given respectively by (4.3.29) with $\varepsilon = +1$, and (4.4.13).

4.6 Further links between skew-symmetric and orthogonal matrices

Here we exclude the case when the complex orthogonal 3×3 matrix $\boldsymbol{R}$ has both the eigenvalues $+1$ and -1. We then show that a proper orthogonal matrix $\boldsymbol{R}$ may be written as

$$\boldsymbol{R} = (\mathbf{1} - \tfrac{1}{2}\boldsymbol{W}')^{-1}(\mathbf{1} + \tfrac{1}{2}\boldsymbol{W}'), \tag{4.6.1}$$

where $\boldsymbol{W}'$ is an appropriate 3×3 skew-symmetric matrix. Hence an improper orthogonal matrix $\boldsymbol{R}$ may be written as

$$\boldsymbol{R} = (\mathbf{1} + \tfrac{1}{2}\boldsymbol{W}')^{-1}(\tfrac{1}{2}\boldsymbol{W}' - \mathbf{1}). \tag{4.6.2}$$

We already know that a complex proper orthogonal matrix is either of the form (4.5.1) or of the form (4.5.2).

(a) We first show that $\boldsymbol{R}$ given by (4.5.1) with $\Theta \neq \pi$ may be written in the form (4.6.1).

Let $\boldsymbol{X}$ be an arbitrary bivector, and $\boldsymbol{Y} = \boldsymbol{R}\boldsymbol{X}$ be its transform by the matrix $\boldsymbol{R}$ given by (4.5.1). Since (4.5.1) may also be written under the form (4.5.5), where $\boldsymbol{\Gamma}$ is the skew-symmetric matrix associated with the bivector $\boldsymbol{C}$, we have

$$\boldsymbol{Y} = \boldsymbol{R}\boldsymbol{X} = \cos\Theta\, \boldsymbol{X} + \sin\Theta\, \boldsymbol{C} \times \boldsymbol{X} + (1 - \cos\Theta)(\boldsymbol{C}\cdot\boldsymbol{X})\boldsymbol{C}. \tag{4.6.3}$$

In passing, we note that for $\boldsymbol{R}$ real, $\boldsymbol{C}$ is a real unit vector, Θ is real, and (4.6.3) is a usual formula for the transform of a vector $\boldsymbol{X}$ by a rotation $\boldsymbol{R}$ of axis along $\boldsymbol{C}$ and of angle Θ.

Using the usual formulae

$$\cos\Theta = \frac{1-\tan^2\frac{1}{2}\Theta}{1+\tan^2\frac{1}{2}\Theta}, \quad \sin\Theta = \frac{2\tan\frac{1}{2}\Theta}{1+\tan^2\frac{1}{2}\Theta}, \tag{4.6.4}$$

and introducing the bivector $\boldsymbol{\Omega}'$ defined by

$$\boldsymbol{\Omega}' = 2\left(\tan\frac{\Theta}{2}\right)\boldsymbol{C}, \tag{4.6.5}$$

we may write (4.6.3) as

$$(1+\tfrac{1}{4}\boldsymbol{\Omega}'\cdot\boldsymbol{\Omega}')\boldsymbol{Y} = (1-\tfrac{1}{4}\boldsymbol{\Omega}'\cdot\boldsymbol{\Omega}')\boldsymbol{X} + \boldsymbol{\Omega}'\times\boldsymbol{X} + \tfrac{1}{2}(\boldsymbol{\Omega}'\cdot\boldsymbol{X})\boldsymbol{\Omega}'. \tag{4.6.6}$$

Of course, we have to assume that $\Theta \neq \pi$ in order to avoid an infinite $\boldsymbol{\Omega}'$, which means that -1 is not an eigenvalue of the matrix $\boldsymbol{R}$ given by (4.5.1) or (4.5.5). In passing, we also note that, when $\boldsymbol{C}$ is a real unit vector and Θ is real, (4.6.6) is a usual formula expressing that $\boldsymbol{Y}$ is the transform of $\boldsymbol{X}$ by the rotation of axis along $\boldsymbol{C}$ and of angle Θ.

Taking the cross product of (4.6.6) with $\boldsymbol{\Omega}'$ yields

$$\begin{aligned}(1+\tfrac{1}{4}\boldsymbol{\Omega}'\cdot\boldsymbol{\Omega}')\boldsymbol{\Omega}'\times\boldsymbol{Y} &= -(\boldsymbol{\Omega}'\cdot\boldsymbol{\Omega}')\boldsymbol{X}\\ &\quad + (1-\tfrac{1}{4}\boldsymbol{\Omega}'\cdot\boldsymbol{\Omega}')\boldsymbol{\Omega}'\times\boldsymbol{X} + (\boldsymbol{\Omega}'\cdot\boldsymbol{X})\boldsymbol{\Omega}'.\end{aligned} \tag{4.6.7}$$

Hence, combining (4.6.6) and (4.6.7), we obtain

$$\boldsymbol{Y} - \tfrac{1}{2}\boldsymbol{\Omega}'\times\boldsymbol{Y} = \boldsymbol{X} + \tfrac{1}{2}\boldsymbol{\Omega}'\times\boldsymbol{X}. \tag{4.6.8}$$

Let $\boldsymbol{W}'$ be the skew-symmetric matrix associated with the bivector $\boldsymbol{\Omega}'$, that is

$$\boldsymbol{W}' = 2\left(\tan\frac{\Theta}{2}\right)\boldsymbol{\Gamma}, \tag{4.6.9}$$

where $\boldsymbol{\Gamma}$ is the skew-symmetric matrix associated with the bivector $\boldsymbol{C}$. Then, (4.6.8) also reads

$$(\mathbf{1}-\tfrac{1}{2}\boldsymbol{W}')\boldsymbol{Y} = (\mathbf{1}+\tfrac{1}{2}\boldsymbol{W}')\boldsymbol{X}, \tag{4.6.10}$$

and thus,

$$\boldsymbol{Y} = \boldsymbol{R}\boldsymbol{X} = (\mathbf{1}-\tfrac{1}{2}\boldsymbol{W}')^{-1}(\mathbf{1}+\tfrac{1}{2}\boldsymbol{W}')\boldsymbol{X}. \tag{4.6.11}$$

Because this is valid for any bivector $\boldsymbol{X}$, the proper orthogonal matrix $\boldsymbol{R}$ may be written in the form (4.6.1), with $\boldsymbol{W}'$ given by (4.6.9).

(b) We now show that $\boldsymbol{R}$ given by (4.5.2) may be written in the form (4.6.1).

In this case we know from section 4.5 that $\boldsymbol{R}$ may be written as

$$\boldsymbol{R} = \mathbf{1} + \boldsymbol{W} + \tfrac{1}{2}\boldsymbol{W}^2, \tag{4.6.12}$$

where $\boldsymbol{W}$ is the skew-symmetric matrix associated with the isotropic bivector $\boldsymbol{\Omega} = -\mathrm{i}\boldsymbol{D}'$. But, because $\boldsymbol{W}^3 = \mathbf{0}$, we have

$$(\mathbf{1} - \tfrac{1}{2}\boldsymbol{W})(\mathbf{1} + \tfrac{1}{2}\boldsymbol{W} + \tfrac{1}{4}\boldsymbol{W}^2) = \mathbf{1}, \tag{4.6.13}$$

and hence

$$\begin{aligned}(\mathbf{1} - \tfrac{1}{2}\boldsymbol{W})^{-1}(\mathbf{1} + \tfrac{1}{2}\boldsymbol{W}) &= (\mathbf{1} + \tfrac{1}{2}\boldsymbol{W} + \tfrac{1}{4}\boldsymbol{W}^2)(\mathbf{1} + \tfrac{1}{2}\boldsymbol{W}) \\ &= \mathbf{1} + \boldsymbol{W} + \tfrac{1}{2}\boldsymbol{W}^2. \end{aligned} \tag{4.6.14}$$

Thus, the proper orthogonal matrix $\boldsymbol{R}$ may be written in the form (4.6.1), where $\boldsymbol{W}' = \boldsymbol{W}$ is the skew-symmetric matrix associated with the isotropic bivector $\boldsymbol{\Omega}' = \boldsymbol{\Omega} = -\mathrm{i}\boldsymbol{D}'$.

Remark From (4.6.1), the skew-symmetric matrix $\boldsymbol{W}'$ may be easily obtained in terms of the proper orthogonal matrix $\boldsymbol{R}$ (it is assumed that -1 is not an eigenvalue of $\boldsymbol{R}$). We have

$$\tfrac{1}{2}\boldsymbol{W}' = (\boldsymbol{R} - \mathbf{1})(\boldsymbol{R} + \mathbf{1})^{-1}. \tag{4.6.15}$$

Exercises 4.5

1. Let

$$\boldsymbol{R} = \begin{bmatrix} 1 & \frac{1}{2}\mathrm{i} & \frac{1}{2} \\ -\mathrm{i} & \frac{3}{2} & \frac{1}{2}\mathrm{i} \\ -1 & -\mathrm{i} & 1 \end{bmatrix}.$$

Check that it is a proper orthogonal matrix.
Find $\boldsymbol{W}'$ and $\boldsymbol{\Omega}'$.

2. Let

$$\boldsymbol{R} = \begin{bmatrix} 2 & \sqrt{2}(-1+\mathrm{i}) & 1+2\mathrm{i} \\ \sqrt{2}(1+\mathrm{i}) & -1 & \sqrt{2}(-1+\mathrm{i}) \\ 1-2\mathrm{i} & \sqrt{2}(1+\mathrm{i}) & 2 \end{bmatrix}$$

(see exercises 4.1.2 and 4.2.2)
Find $\boldsymbol{W}'$ and $\boldsymbol{\Omega}'$.

5

Ellipsoids

Here we first present some useful bivector identities involving a 3×3 matrix g. This matrix may be real or complex. The case when the matrix g is assumed to be real, nonsingular and symmetric is important for applications in this chapter, because we consider the properties of ellipsoids and of pairs of concentric ellipsoids.

5.1 Further bivector identities

The adjugate matrix, g^*, of a 3×3 matrix g, is the matrix formed by the cofactors of the elements of the matrix g:

$$g_{ij}^* = \text{cofactor}(g_{ij}) = \tfrac{1}{2} e_{ikm} e_{jln} g_{kl} g_{mn}. \qquad (5.1.1)$$

In the special case when g is nonsingular, $\det(g) \neq 0$, then g^{-1} is the transpose of g^* divided by the determinant of g, so that

$$g^* = (\det g)(g^{-1})^{\mathrm{T}}. \qquad (5.1.2)$$

Let P and Q be any two bivectors or vectors. Then

$$gP \times gQ = g^*(P \times Q). \qquad (5.1.3)$$

This is because

$$\begin{aligned}[g^*(P \times Q)]_i &= \tfrac{1}{2} e_{ikm} e_{jln} g_{kl} g_{mn} e_{jrs} P_r Q_s \\ &= \tfrac{1}{2} e_{ikm} (\delta_{lr}\delta_{ns} - \delta_{ls}\delta_{nr}) g_{kl} g_{mn} P_r Q_s \\ &= e_{ikm} g_{kr} P_r g_{ms} Q_s = (gP \times gQ)_i.\end{aligned}$$

Because P and Q are arbitrary, so is $P \times Q$, and thus (5.1.3) may be used as a definition of the adjugate matrix g^* of the matrix g.

Also, if P, Q, R, T are any four bivectors, then

$$P \times g^*(Q \times R) = gQ(P^{\mathrm{T}} gR) - gR(P^{\mathrm{T}} gQ), \qquad (5.1.4)$$

$$(P \times Q)^{\mathrm{T}} g^*(R \times T) = (P^{\mathrm{T}} gR)(Q^{\mathrm{T}} gT) - (P^{\mathrm{T}} gT)(Q^{\mathrm{T}} gR). \qquad (5.1.5)$$

These follow from (5.1.3). Indeed, using (5.1.3),

$$\begin{aligned} \boldsymbol{P} \times \boldsymbol{g}^*(\boldsymbol{Q} \times \boldsymbol{R}) &= \boldsymbol{P} \times (\boldsymbol{gQ} \times \boldsymbol{gR}) \\ &= \boldsymbol{gQ}\{\boldsymbol{P}\cdot(\boldsymbol{gR})\} - \boldsymbol{gR}\{\boldsymbol{P}\cdot(\boldsymbol{gQ})\}, \end{aligned}$$

which gives (5.1.4). Also, using (5.1.3)

$$\begin{aligned} (\boldsymbol{P} \times \boldsymbol{Q})^{\mathrm{T}}\boldsymbol{g}^*(\boldsymbol{R} \times \boldsymbol{T}) &= (\boldsymbol{P} \times \boldsymbol{Q})\cdot(\boldsymbol{gR} \times \boldsymbol{gT}) \\ &= \{\boldsymbol{P}\cdot(\boldsymbol{gR})\}\{\boldsymbol{Q}\cdot(\boldsymbol{gT})\} - \{\boldsymbol{P}\cdot(\boldsymbol{gT})\}\{\boldsymbol{Q}\cdot(\boldsymbol{gR})\}, \end{aligned}$$

which gives (5.1.5).

Exercises 5.1

1. For any matrix $\boldsymbol{g}$, prove that $\boldsymbol{g}^*\boldsymbol{g}^{\mathrm{T}} = \boldsymbol{g}^{\mathrm{T}}\boldsymbol{g}^* = (\det \boldsymbol{g})\mathbf{1}$.
2. For any two matrices $\boldsymbol{g}_1, \boldsymbol{g}_2$, prove that $(\boldsymbol{g}_1\boldsymbol{g}_2)^* = \boldsymbol{g}_1^*\boldsymbol{g}_2^*$.
3. For any bivector $\boldsymbol{A}$, prove that $\{(\boldsymbol{A}\cdot\boldsymbol{A})\mathbf{1} - \boldsymbol{A} \otimes \boldsymbol{A}\}^* = (\boldsymbol{A}\cdot\boldsymbol{A})\boldsymbol{A} \otimes \boldsymbol{A}$.
4. For any skew-symmetric matrix $\boldsymbol{W}$, prove that $\boldsymbol{W}^* = \boldsymbol{\Omega} \otimes \boldsymbol{\Omega}$, where $\boldsymbol{\Omega}$ is the bivector associated with $\boldsymbol{W}$ (see section 4.4).
5. For any matrix $\boldsymbol{g}$, prove that $\operatorname{tr}(\boldsymbol{g}^*) = \frac{1}{2}\{(\operatorname{tr} \boldsymbol{g})^2 - \operatorname{tr}(\boldsymbol{g}^2)\}$, $\det(\boldsymbol{g}^*) = (\det \boldsymbol{g})^2$.

5.2 Diametral planes and conjugate diameters

Let $\boldsymbol{\alpha}$ be a 3×3 real symmetric positive definite matrix. Then

$$\boldsymbol{x}^{\mathrm{T}}\boldsymbol{\alpha}\boldsymbol{x} = 1, \tag{5.2.1}$$

is the equation of an ellipsoid centred at the origin of coordinates.

Consider the points of intersection with the ellipsoid $\boldsymbol{x}^{\mathrm{T}}\boldsymbol{\alpha}\boldsymbol{x} = 1$ of the straight line $\boldsymbol{x} = \boldsymbol{a} + t\boldsymbol{b}$, $(-\infty \leqslant t \leqslant \infty)$, through the point $\boldsymbol{a}$ parallel to $\boldsymbol{b}$. The corresponding values of t are the roots of the quadratic

$$(\boldsymbol{b}^{\mathrm{T}}\boldsymbol{\alpha}\boldsymbol{b})t^2 + 2(\boldsymbol{b}^{\mathrm{T}}\boldsymbol{\alpha}\boldsymbol{a})t + \boldsymbol{a}^{\mathrm{T}}\boldsymbol{\alpha}\boldsymbol{a} = 0. \tag{5.2.2}$$

These roots will be equal and opposite if and only if

$$\boldsymbol{b}^{\mathrm{T}}\boldsymbol{\alpha}\boldsymbol{a} = 0, \tag{5.2.3}$$

which means that $\boldsymbol{a}$ will be the mid point of the chord of intersection. All points such as $\boldsymbol{a}$, bisecting chords parallel to $\boldsymbol{b}$, lie on the plane

$$\boldsymbol{b}^{\mathrm{T}}\boldsymbol{\alpha}\boldsymbol{x} = 0, \quad b_i\alpha_{ij}x_j = 0. \tag{5.2.4}$$

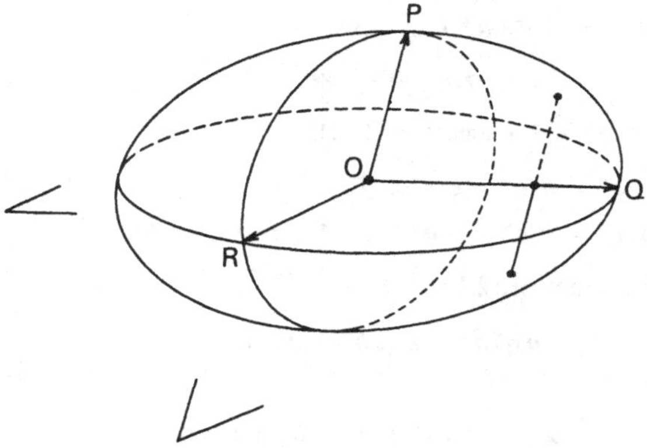

Figure 5.1 *Conjugate diameters with respect to an ellipsoid.*

This central plane which bisects all chords of the ellipsoid parallel to $\boldsymbol{b}$, is parallel to the tangent planes at the extremities of the diameter which is parallel to $\boldsymbol{b}$. It is called the **diametral plane** of $\boldsymbol{b}$ with respect to the ellipsoid. Also, the direction of $\boldsymbol{b}$ and the planes parallel to the diametral plane of $\boldsymbol{b}$, are said to be **conjugate** with respect to the ellipsoid.

The symmetry of (5.2.3) shows that if the point Q (say) is on the diametral plane of OP, then P will be on the diametral plane of OQ (Figure 5.1). Also, suppose OR is the line of intersection of the diametral planes of OP and OQ. Then, since the diametral planes of OP and OQ pass through OR, the diametral plane of OR will pass through P and Q and is therefore the plane POQ. Thus the plane through any two of the three lines OP, OQ, OR is diametral to the third. Three such planes are said to be **conjugate** and their lines of intersection are **conjugate diameters** with respect to the ellipsoid.

Let P, Q, R be three points on the ellipsoid $\boldsymbol{x}^{\mathrm{T}}\boldsymbol{\alpha}\boldsymbol{x} = 1$ at the extremities of three conjugate diameters. Let $\boldsymbol{OP} = \boldsymbol{a}, \boldsymbol{OQ} = \boldsymbol{b}, \boldsymbol{OR} = \boldsymbol{c}$. Then as each one lies on the diametral planes of the other two, we have

$$\boldsymbol{a}^{\mathrm{T}}\boldsymbol{\alpha}\boldsymbol{b} = \boldsymbol{b}^{\mathrm{T}}\boldsymbol{\alpha}\boldsymbol{c} = \boldsymbol{c}^{\mathrm{T}}\boldsymbol{\alpha}\boldsymbol{a} = 0, \tag{5.2.5}$$

and because each point is on the ellipsoid

$$\boldsymbol{a}^{\mathrm{T}}\boldsymbol{\alpha}\boldsymbol{a} = \boldsymbol{b}^{\mathrm{T}}\boldsymbol{\alpha}\boldsymbol{b} = \boldsymbol{c}^{\mathrm{T}}\boldsymbol{\alpha}\boldsymbol{c} = 1. \tag{5.2.6}$$

Thus the three vectors $\boldsymbol{a}^*, \boldsymbol{b}^*, \boldsymbol{c}^*$, defined by

$$\boldsymbol{a}^* = \boldsymbol{\alpha a}, \quad \boldsymbol{b}^* = \boldsymbol{\alpha b}, \quad \boldsymbol{c}^* = \boldsymbol{\alpha c}, \tag{5.2.7}$$

are reciprocal to $\boldsymbol{a}, \boldsymbol{b}, \boldsymbol{c}$ (section 2.11):

$$\boldsymbol{a}\cdot\boldsymbol{a}^* = \boldsymbol{b}\cdot\boldsymbol{b}^* = \boldsymbol{c}\cdot\boldsymbol{c}^* = 1,$$
$$\boldsymbol{a}\cdot\boldsymbol{b}^* = \boldsymbol{b}\cdot\boldsymbol{c}^* = \boldsymbol{c}\cdot\boldsymbol{a}^* = \boldsymbol{a}\cdot\boldsymbol{c}^* = \boldsymbol{b}\cdot\boldsymbol{a}^* = \boldsymbol{c}\cdot\boldsymbol{b}^* = 0.$$

Recalling that (section 2.13)

$$\boldsymbol{a}\otimes\boldsymbol{a}^* + \boldsymbol{b}\otimes\boldsymbol{b}^* + \boldsymbol{c}\otimes\boldsymbol{c}^* = \mathbf{1},$$

we note that

$$\boldsymbol{\alpha} = \boldsymbol{a}^*\otimes\boldsymbol{a}^* + \boldsymbol{b}^*\otimes\boldsymbol{b}^* + \boldsymbol{c}^*\otimes\boldsymbol{c}^*, \tag{5.2.8}$$

because this satisfies (5.2.5) and (5.2.6) identically. Further, the inverse of $\boldsymbol{\alpha}$ is given by

$$\boldsymbol{\alpha}^{-1} = \boldsymbol{a}\otimes\boldsymbol{a} + \boldsymbol{b}\otimes\boldsymbol{b} + \boldsymbol{c}\otimes\boldsymbol{c}. \tag{5.2.9}$$

The equation of the ellipsoid may be written

$$1 = \boldsymbol{x}^{\mathrm{T}}\boldsymbol{\alpha x} = (\boldsymbol{x}\cdot\boldsymbol{a}^*)^2 + (\boldsymbol{x}\cdot\boldsymbol{b}^*)^2 + (\boldsymbol{x}\cdot\boldsymbol{c}^*)^2. \tag{5.2.10}$$

If we write

$$\boldsymbol{x} = x'_1\hat{\mathbf{a}} + x'_2\hat{\mathbf{b}} + x'_3\hat{\mathbf{c}}, \tag{5.2.11}$$

where $\hat{\mathbf{a}}, \hat{\mathbf{b}}, \hat{\mathbf{c}}$ are unit vectors along the three conjugate diameters, so that x'_1, x'_2, x'_3 are the (oblique) coordinates measured along $\hat{\mathbf{a}}, \hat{\mathbf{b}}, \hat{\mathbf{c}}$, then (5.2.10) becomes

$$\frac{x_1'^2}{a^2} + \frac{x_2'^2}{b^2} + \frac{x_3'^2}{c^2} = 1. \tag{5.2.12}$$

If n_i are the components of a unit vector, then

$$\boldsymbol{x} = n_1\boldsymbol{a} + n_2\boldsymbol{b} + n_3\boldsymbol{c}, \tag{5.2.13}$$

is a point on the ellipsoid because

$$\boldsymbol{x}\cdot\boldsymbol{a}^* = n_1, \quad \boldsymbol{x}\cdot\boldsymbol{b}^* = n_2, \quad \boldsymbol{x}\cdot\boldsymbol{c}^* = n_3, \tag{5.2.14}$$

and thus, because $n_i n_i = 1$, it follows that (5.2.10) is satisfied.

Without loss we may write the position vector of a point on the ellipsoid as

$$\boldsymbol{x} = \cos\theta\cos\phi\boldsymbol{a} + \cos\theta\sin\phi\boldsymbol{b} + \sin\theta\boldsymbol{c}, \quad 0 \leqslant \theta \leqslant \pi, \quad 0 \leqslant \phi \leqslant 2\pi. \tag{5.2.15}$$

This is the parametric form of the equation of an ellipsoid of which $\boldsymbol{a}, \boldsymbol{b}, \boldsymbol{c}$ are three conjugate radii.

5.3 Polar reciprocal of an ellipsoid with respect to a sphere

The polar reciprocal of an ellipsoid with respect to the unit sphere is the envelope of the polar planes of the points on the ellipsoid with respect to the sphere.

A typical point $\hat{\boldsymbol{x}}$ (say) on the ellipsoid (5.2.15) may be described in terms of the parameters (θ, ϕ). Now the polar plane of the point $\hat{\boldsymbol{x}}(\theta, \phi)$ on the ellipsoid with respect to the unit sphere $\boldsymbol{x}^{\mathrm{T}}\boldsymbol{x} = 1$, is $\hat{\boldsymbol{x}}^{\mathrm{T}}\boldsymbol{x} = 1$, or

$$(\cos\theta\cos\phi\boldsymbol{a}^{\mathrm{T}} + \cos\theta\sin\phi\boldsymbol{b}^{\mathrm{T}} + \sin\theta\boldsymbol{c}^{\mathrm{T}})\boldsymbol{x} = 1. \tag{5.3.1}$$

The envelope of this two parameter family of planes is the solution of (5.3.1) and

$$(-\sin\theta\cos\phi\boldsymbol{a}^{\mathrm{T}} - \sin\theta\sin\phi\boldsymbol{b}^{\mathrm{T}} + \cos\theta\boldsymbol{c}^{\mathrm{T}})\boldsymbol{x} = 0, \tag{5.3.2}$$

$$(-\cos\theta\sin\phi\boldsymbol{a}^{\mathrm{T}} + \cos\theta\cos\phi\boldsymbol{b}^{\mathrm{T}})\boldsymbol{x} = 0. \tag{5.3.3}$$

The solution of these three equations is

$$\boldsymbol{x} = \cos\theta\cos\phi\boldsymbol{a}^{*} + \cos\theta\sin\phi\boldsymbol{b}^{*} + \sin\theta\boldsymbol{c}^{*}. \tag{5.3.4}$$

It is an ellipsoid with conjugate radii $\boldsymbol{a}^{*}, \boldsymbol{b}^{*}, \boldsymbol{c}^{*}$.

Indeed, the polar reciprocal of the ellipsoid $\boldsymbol{x}^{\mathrm{T}}\boldsymbol{\alpha}\boldsymbol{x} = 1$ with respect to the unit sphere is the ellipsoid $\boldsymbol{x}^{\mathrm{T}}\boldsymbol{\alpha}^{-1}\boldsymbol{x} = 1$.

5.4 Invariants of the ellipsoid

Because $\boldsymbol{\alpha}$ is symmetric positive definite, it possesses a unique positive definite square root $\boldsymbol{\alpha}^{1/2}$, so that $(\boldsymbol{\alpha}^{1/2})^2 = \boldsymbol{\alpha}$. Then, from (5.2.5) and (5.2.6), the vectors $\boldsymbol{a}', \boldsymbol{b}', \boldsymbol{c}'$, defined by

$$\boldsymbol{a}' = \boldsymbol{\alpha}^{1/2}\boldsymbol{a}, \quad \boldsymbol{b}' = \boldsymbol{\alpha}^{1/2}\boldsymbol{b}, \quad \boldsymbol{c}' = \boldsymbol{\alpha}^{1/2}\boldsymbol{c}, \tag{5.4.1}$$

are mutually orthogonal unit vectors.

The equation of the ellipsoid may be written

$$1 = \boldsymbol{x}^{\mathrm{T}}\boldsymbol{\alpha}\boldsymbol{x} = (\boldsymbol{\alpha}^{1/2}\boldsymbol{x})\cdot(\boldsymbol{\alpha}^{1/2}\boldsymbol{x}). \tag{5.4.2}$$

If $\boldsymbol{x}$ is a point on the ellipsoid, then $\boldsymbol{x}' = \boldsymbol{\alpha}^{1/2}\boldsymbol{x}$, is of unit length, so that $\boldsymbol{x}'$ lies on a unit sphere. Thus, $\boldsymbol{\alpha}^{1/2}$ transforms the ellipsoid into a sphere, three conjugate radii of the ellipsoid transforming into three mutually orthogonal radii of the sphere.

We note that the sum of the squares of the lengths of the three conjugate radii $\boldsymbol{a}, \boldsymbol{b}$ and $\boldsymbol{c}$ is

$$\begin{aligned}\boldsymbol{a}\cdot\boldsymbol{a}+\boldsymbol{b}\cdot\boldsymbol{b}+\boldsymbol{c}\cdot\boldsymbol{c} &= \boldsymbol{a}'^{\mathrm{T}}\boldsymbol{\alpha}^{-1}\boldsymbol{a}'+\boldsymbol{b}'^{\mathrm{T}}\boldsymbol{\alpha}^{-1}\boldsymbol{b}'+\boldsymbol{c}'^{\mathrm{T}}\boldsymbol{\alpha}^{-1}\boldsymbol{c}'\\ &= \alpha_{ij}^{-1}(a_i'a_j'+b_i'b_j'+c_i'c_j')\\ &= \alpha_{ij}^{-1}(\delta_{ij}) = \mathrm{tr}(\boldsymbol{\alpha}^{-1}),\end{aligned} \tag{5.4.3}$$

which is invariant.

Thus the sum of the squares of the lengths of any three conjugate radii is invariant.

The vector area of the parallelogram formed by the pair of conjugate radii $\boldsymbol{a}$ and $\boldsymbol{b}$ is

$$\begin{aligned}\boldsymbol{a}\times\boldsymbol{b} &= (\boldsymbol{\alpha}^{-1/2}\boldsymbol{a}')\times(\boldsymbol{\alpha}^{-1/2}\boldsymbol{b}')\\ &= \det(\boldsymbol{\alpha}^{-1/2})\boldsymbol{\alpha}^{1/2}(\boldsymbol{a}'\times\boldsymbol{b}')\\ &= \pm\det(\boldsymbol{\alpha}^{-1/2})\boldsymbol{\alpha}^{1/2}\boldsymbol{c}',\end{aligned} \tag{5.4.4}$$

on using (5.1.2) and (5.1.3). Hence

$$\begin{aligned}&(\boldsymbol{a}\times\boldsymbol{b})\cdot(\boldsymbol{a}\times\boldsymbol{b})+(\boldsymbol{b}\times\boldsymbol{c})\cdot(\boldsymbol{b}\times\boldsymbol{c})+(\boldsymbol{c}\times\boldsymbol{a})\cdot(\boldsymbol{c}\times\boldsymbol{a})\\ &\quad= \det(\boldsymbol{\alpha}^{-1})(\boldsymbol{a}'^{\mathrm{T}}\boldsymbol{\alpha}\boldsymbol{a}'+\boldsymbol{b}'^{\mathrm{T}}\boldsymbol{\alpha}\boldsymbol{b}'+\boldsymbol{c}'^{\mathrm{T}}\boldsymbol{\alpha}\boldsymbol{c}')\\ &\quad= (\det\boldsymbol{\alpha})^{-1}\,\mathrm{tr}\,\boldsymbol{\alpha} = \tfrac{1}{2}\{(\mathrm{tr}\,\boldsymbol{\alpha}^{-1})^2-\mathrm{tr}(\boldsymbol{\alpha}^{-2})\},\end{aligned} \tag{5.4.5}$$

which is invariant.

Thus the sum of the squares of the areas of the parallelograms formed by any three pairs of conjugate radii in the ellipsoid is invariant.

Finally the squared volume of the parallelepiped formed by the three conjugate radii $\boldsymbol{a}, \boldsymbol{b}$ and $\boldsymbol{c}$ is

$$\begin{aligned}\{(\boldsymbol{a}\times\boldsymbol{b})\cdot\boldsymbol{c}\}^2 &= \det(\boldsymbol{\alpha}^{-1})\{(\boldsymbol{\alpha}^{1/2}\boldsymbol{c}')\cdot(\boldsymbol{\alpha}^{-1/2}\boldsymbol{c}')\}^2\\ &= \det(\boldsymbol{\alpha}^{-1}),\end{aligned} \tag{5.4.6}$$

and is also invariant.

Thus the volume of the parallelepiped formed by any three conjugate radii of an ellipsoid is invariant.

5.5 Isotropy with respect to a metric

Let $\boldsymbol{g}$ be a real symmetric positive definite 3×3 matrix. With this matrix is associated the ellipsoid $\boldsymbol{x}^{\mathrm{T}}\boldsymbol{g}\boldsymbol{x}=1$ which we call the **$\boldsymbol{g}$-metric ellipsoid**, or, for brevity, the $\boldsymbol{g}$-ellipsoid.

A bivector $\boldsymbol{P}$ is said to be isotropic with respect to the metric $\boldsymbol{g}$ when

$$\boldsymbol{P}^{\mathrm{T}}\boldsymbol{g}\boldsymbol{P}=0. \tag{5.5.1}$$

Here we present a geometrical interpretation of this condition.

If we write $\boldsymbol{P}=\boldsymbol{P}^{+}+\mathrm{i}\boldsymbol{P}^{-}$, then from (5.5.1)

$$\boldsymbol{P}^{+\mathrm{T}}\boldsymbol{g}\boldsymbol{P}^{-}=0, \tag{5.5.2a}$$

$$\boldsymbol{P}^{+\mathrm{T}}\boldsymbol{g}\boldsymbol{P}^{+}=\boldsymbol{P}^{-\mathrm{T}}\boldsymbol{g}\boldsymbol{P}^{-}. \tag{5.5.2b}$$

By (5.5.2a), $\boldsymbol{P}^{+}$ and $\boldsymbol{P}^{-}$ are along conjugate directions of the $\boldsymbol{g}$-ellipsoid. By (5.5.2b), the ratio of the lengths of the radii of the $\boldsymbol{g}$-ellipsoid along $\boldsymbol{P}^{+}$ and $\boldsymbol{P}^{-}$, respectively, is equal to the ratio of the lengths of the radii of the ellipse of $\boldsymbol{P}$ along $\boldsymbol{P}^{+}$ and $\boldsymbol{P}^{-}$, respectively. Thus it follows (section 1.7) that if (5.5.1) is satisfied then the ellipse of $\boldsymbol{P}$ is similar and similarly situated to the ellipse in which the plane of $\boldsymbol{P}$ cuts the $\boldsymbol{g}$-ellipsoid.

If, in addition to the condition of isotropy (5.5.1), we require that

$$\boldsymbol{P}^{\mathrm{T}}\boldsymbol{g}\bar{\boldsymbol{P}}=2, \tag{5.5.3}$$

then

$$\boldsymbol{P}^{+\mathrm{T}}\boldsymbol{g}\boldsymbol{P}^{+}=\boldsymbol{P}^{-\mathrm{T}}\boldsymbol{g}\boldsymbol{P}^{-}=1. \tag{5.5.4}$$

Then, $\boldsymbol{P}^{+}$ and $\boldsymbol{P}^{-}$ are radii of the $\boldsymbol{g}$-ellipsoid. Hence, if both (5.5.1) and (5.5.3) hold, then the ellipse of $\boldsymbol{P}$ (centred at the origin of the coordinates) is a central section of the $\boldsymbol{g}$-ellipsoid.

5.6 Central circular sections of an ellipsoid

A section of an ellipsoid by a plane through its centre is generally an ellipse. Exceptionally the central section may be a circle. It is shown that if all three semi-axes of the ellipsoid have different lengths, then there are two central circular sections, whilst for a spheroid (ellipsoid of revolution) there is only one central circular section. Of course, for a sphere every central section is a circle.

Consider the ellipsoid whose equation referred to cartesian coordinates along its principal axes is

$$\frac{x^2}{a^2}+\frac{y^2}{b^2}+\frac{z^2}{c^2}=1, \quad 0<a\leqslant b\leqslant c. \tag{5.6.1}$$

The points of its intersection with the concentric sphere of radius R, $x^2+y^2+z^2=R^2$, lie on the cone

$$\frac{x^2}{a^2}+\frac{y^2}{b^2}+\frac{z^2}{c^2}=\frac{x^2+y^2+z^2}{R^2}. \tag{5.6.2}$$

This cone degenerates into a pair of real planes provided the matrix $\boldsymbol{\beta}$ given by

$$\boldsymbol{\beta} = \operatorname{diag}\left(\frac{1}{a^2} - \frac{1}{R^2}, \frac{1}{b^2} - \frac{1}{R^2}, \frac{1}{c^2} - \frac{1}{R^2}\right)$$

is such that $\det \boldsymbol{\beta} = 0$ and further that the non-zero eigenvalues of $\boldsymbol{\beta}$ have opposite signs. Thus we must have $R^2 = b^2$, and the planes are

$$\sqrt{b^2 - a^2}\,cx = \pm\sqrt{c^2 - b^2}\,az, \tag{5.6.3}$$

the circular sections each having radius b.

Alternatively, the equation of the ellipsoid may be written $\boldsymbol{x}^{\mathrm{T}}\boldsymbol{\alpha}\boldsymbol{x} = 1$ where $\boldsymbol{\alpha}$ is positive definite, with the spectral decomposition and Hamilton cyclic form (sections 2.3 and 2.4) given by

$$\boldsymbol{\alpha} = \lambda \mathbf{s} \otimes \mathbf{s} + \mu \mathbf{t} \otimes \mathbf{t} + \nu \mathbf{u} \otimes \mathbf{u}, \tag{5.6.4}$$

$$\boldsymbol{\alpha} = \mu \mathbf{1} + \frac{\lambda - \nu}{2}(\mathbf{h} \otimes \mathbf{k} + \mathbf{k} \otimes \mathbf{h}), \tag{5.6.5}$$

with $\lambda > \mu > \nu > 0$. Here $\mathbf{s}, \mathbf{t}, \mathbf{u}$ are unit vectors along the principal axes of the ellipsoid and $\mathbf{h}$ and $\mathbf{k}$ are unit vectors given by

$$\begin{aligned} \sqrt{\lambda - \nu}\,\mathbf{h} &= \sqrt{\lambda - \mu}\,\mathbf{s} + \sqrt{\mu - \nu}\,\mathbf{u}, \\ \sqrt{\lambda - \nu}\,\mathbf{k} &= \sqrt{\lambda - \mu}\,\mathbf{s} - \sqrt{\mu - \nu}\,\mathbf{u}. \end{aligned} \tag{5.6.6}$$

Then, from (5.6.5), the equation $\boldsymbol{x}^{\mathrm{T}}\boldsymbol{\alpha}\boldsymbol{x} = 1$ may be written

$$1 = \mu \boldsymbol{x} \cdot \boldsymbol{x} + (\lambda - \nu)(\mathbf{h} \cdot \boldsymbol{x})(\mathbf{k} \cdot \boldsymbol{x}), \tag{5.6.7}$$

so that the points of intersection of the ellipsoid with the plane $\mathbf{h} \cdot \boldsymbol{x} = 0$ satisfy

$$1 = \mu \boldsymbol{x} \cdot \boldsymbol{x}, \quad 0 = \mathbf{h} \cdot \boldsymbol{x}. \tag{5.6.8}$$

This intersection is thus a circle of radius $\mu^{-1/2} = b$. Similarly the plane $\mathbf{k} \cdot \boldsymbol{x} = 0$ is a plane of a central circular section.

If the ellipsoid is a spheroid, either

$$\frac{x^2}{a^2} + \frac{y^2 + z^2}{b^2} = 1, \quad a < b, \tag{5.6.9}$$

and the plane of the central circular section is $x = 0$, or

$$\frac{x^2 + y^2}{a^2} + \frac{z^2}{c^2} = 1, \quad a < c, \tag{5.6.10}$$

and the plane of the central circular section is $z = 0$.

5.6.1 *Bivector approach*

Alternatively, using bivectors, suppose that the bivector $\boldsymbol{A}$ is such that

$$\boldsymbol{A}\cdot\boldsymbol{A}=0, \quad \boldsymbol{A}^{\mathrm{T}}\boldsymbol{\alpha}\boldsymbol{A}=0. \tag{5.6.11}$$

The first condition means that the ellipse of $\boldsymbol{A}$ is a circle, and the second that it is similar and similarly situated to a central section of the ellipsoid $\boldsymbol{x}^{\mathrm{T}}\boldsymbol{\alpha}\boldsymbol{x}=1$. Thus (5.6.11) means that the ellipse of $\boldsymbol{A}$ is similar and similarly situated to a central circular section of this ellipsoid. Using (5.6.5) we note that (5.6.11) may also be written

$$\boldsymbol{A}\cdot\boldsymbol{A}=0, \quad (\mathbf{h}\cdot\boldsymbol{A})(\mathbf{k}\cdot\boldsymbol{A})=0. \tag{5.6.12}$$

Hence either $\mathbf{h}\cdot\boldsymbol{A}=0$ or $\mathbf{k}\cdot\boldsymbol{A}=0$, and thus $\mathbf{h}$ and $\mathbf{k}$ are along the normals to the central circular sections.

If, in addition to (5.6.11), we require that

$$\boldsymbol{A}^{\mathrm{T}}\boldsymbol{\alpha}\bar{\boldsymbol{A}}=2, \tag{5.6.13}$$

then the ellipse of $\boldsymbol{A}$ is actually a central circular section of the ellipsoid. Using (5.6.5) and (5.6.12), the additional condition (5.6.13) may also be written

$$\mu(\boldsymbol{A}\cdot\bar{\boldsymbol{A}})=2. \tag{5.6.14}$$

The bivectors $\boldsymbol{A}$ satisfying (5.6.11) and (5.6.13), or equivalently (5.6.12) and (5.6.14), are $\boldsymbol{A}_h$, $\boldsymbol{A}_k$ (say) given by

$$\boldsymbol{A}_h=\mu^{-1/2}\left\{\left(\frac{\mu-\nu}{\lambda-\nu}\right)^{1/2}\mathbf{s}+\mathrm{i}\mathbf{t}-\left(\frac{\lambda-\mu}{\lambda-\nu}\right)^{1/2}\mathbf{u}\right\},$$

$$\boldsymbol{A}_k=\mu^{-1/2}\left\{\left(\frac{\mu-\nu}{\lambda-\nu}\right)^{1/2}\mathbf{s}+\mathrm{i}\mathbf{t}+\left(\frac{\lambda-\mu}{\lambda-\nu}\right)^{1/2}\mathbf{u}\right\}. \tag{5.6.15}$$

The case when the ellipsoid is a spheroid, $\lambda>\mu=\nu$ or $\lambda=\mu>\nu$, is dealt with similarly. For the spheroids (5.6.9) or (5.6.10), the bivector $\boldsymbol{A}$ whose ellipse is the central circular section of the spheroid, is respectively $\boldsymbol{A}_x$ or $\boldsymbol{A}_z$ given by

$$\boldsymbol{A}_x=b(\mathrm{i}\mathbf{t}+\mathbf{u}), \tag{5.6.16a}$$

$$\boldsymbol{A}_z=a(\mathbf{s}+\mathrm{i}\mathbf{t}), \tag{5.6.16b}$$

with $b=\nu^{-1/2}$ and $a=\lambda^{-1/2}$.

Exercises 5.2

1. Let $\boldsymbol{x}^{\mathrm{T}}\boldsymbol{\alpha}_1\boldsymbol{x}=1$ and $\boldsymbol{x}^{\mathrm{T}}\boldsymbol{\alpha}_2\boldsymbol{x}=1$ be two ellipsoids. Prove that a necessary and sufficient condition that the two ellipsoids have the same planes of central circular sections is that there exists numbers $c_1 \neq 0$, $c_2 \neq 0$, c_0, such that $c_1\boldsymbol{\alpha}_1 - c_2\boldsymbol{\alpha}_2 = c_0\mathbf{1}$.
2. Suppose $\boldsymbol{\alpha}$ is given by (5.6.4) and $\lambda > \mu > \nu > 0$, and $\boldsymbol{A}_h$, $\boldsymbol{A}_k$ are given by (5.6.15). Show that

$$\begin{aligned}\boldsymbol{\alpha} &= (\lambda+\nu-\mu)\mathbf{1}\\ &\quad+\frac{\mu}{4}(\lambda-\mu)\{\boldsymbol{A}_h\otimes\boldsymbol{A}_k+\boldsymbol{A}_k\otimes\boldsymbol{A}_h+\bar{\boldsymbol{A}}_h\otimes\bar{\boldsymbol{A}}_k+\bar{\boldsymbol{A}}_k\otimes\bar{\boldsymbol{A}}_h\}\\ &\quad+\frac{\mu}{4}(\mu-\nu)\{\bar{\boldsymbol{A}}_h\otimes\boldsymbol{A}_k+\boldsymbol{A}_k\otimes\bar{\boldsymbol{A}}_h+\boldsymbol{A}_h\otimes\bar{\boldsymbol{A}}_k+\bar{\boldsymbol{A}}_k\otimes\boldsymbol{A}_h\}.\end{aligned}$$

   Find the corresponding expression for a spheroid.
3. Let $\boldsymbol{\alpha}$ be given by (5.6.5) with $\lambda > \mu > \nu$. Show that

$$2\boldsymbol{\alpha}\mathbf{h} = (\lambda+\nu)\mathbf{h} + (\lambda-\nu)\mathbf{k}, \quad 2\boldsymbol{\alpha}\mathbf{k} = (\lambda-\nu)\mathbf{h} + (\lambda+\nu)\mathbf{k}.$$

4. Let $\boldsymbol{\alpha}$ be given by (5.6.5) with $\lambda > \mu > \nu$. Show that

$$\begin{aligned}\mu\boldsymbol{\alpha}^{-1} &= \mathbf{1} - \frac{(\lambda-\nu)^2}{4}(\boldsymbol{\alpha}^{-1}\mathbf{h}\otimes\boldsymbol{\alpha}^{-1}\mathbf{h}+\boldsymbol{\alpha}^{-1}\mathbf{k}\otimes\boldsymbol{\alpha}^{-1}\mathbf{k})\\ &\quad-\frac{\lambda^2-\nu^2}{4}(\boldsymbol{\alpha}^{-1}\mathbf{h}\otimes\boldsymbol{\alpha}^{-1}\mathbf{k}+\boldsymbol{\alpha}^{-1}\mathbf{k}\otimes\boldsymbol{\alpha}^{-1}\mathbf{h}).\end{aligned}$$

5. Let $\boldsymbol{\alpha}$ be given by (5.6.5) with $\lambda > \mu > \nu$. Derive the identity

$$\begin{aligned}&(\lambda+\nu)\mathbf{1} - \tfrac{1}{2}(\lambda-\nu)(\mathbf{h}\otimes\mathbf{k}+\mathbf{k}\otimes\mathbf{h})\\ &= (\lambda+\nu)\mu\boldsymbol{\alpha}^{-1} + \tfrac{1}{2}\lambda\nu(\lambda-\nu)(\boldsymbol{\alpha}^{-1}\mathbf{h}\otimes\boldsymbol{\alpha}^{-1}\mathbf{k}+\boldsymbol{\alpha}^{-1}\mathbf{k}\otimes\boldsymbol{\alpha}^{-1}\mathbf{h}).\end{aligned}$$

5.7 Principal axes of a central section of an ellipsoid

The determination of the principal axes, both in magnitude and direction, of a central section of an ellipsoid is considered here. We note two areas in which this problem arises.

In classical optics (Chapter 9), the properties of a magnetically isotropic but electrically anisotropic crystal are described in terms of a positive definite symmetric tensor $\boldsymbol{\kappa}$ – the electric permittivity. Associated with this is the Fresnel ellipsoid $\kappa_{ij}x_ix_j = 1$. In general, for any direction of propagation, $\boldsymbol{n}$, two waves may propagate. Their electric displacement amplitudes are along the principal axes of the central section of the Fresnel ellipsoid by the plane $\boldsymbol{n}\cdot\boldsymbol{x}=0$, and their speeds of propagation are proportional to the lengths of the semi-axes of this section.

The propagation of homogeneous waves in a direction $\boldsymbol{n}$ in a homogeneous anisotropic incompressible linearly elastic material (Chapter 10) is described in terms of an acoustical tensor $\boldsymbol{Q}(\boldsymbol{n})$ which is positive definite and symmetric. Since the material is incompressible the waves are transverse. The displacement amplitudes of the two waves propagating along $\boldsymbol{n}$ lie along the principal axes of the central section of the ellipsoid $Q_{ij}(\boldsymbol{n})x_i x_j=1$ by the plane $\boldsymbol{n}\cdot\boldsymbol{x}=0$ and the corresponding squared speeds are proportional to the lengths of the principal semi-axes of the section.

The problem is to determine the magnitude and direction of the principal semi-axes of the central section of the ellipsoid $\alpha_{ij}x_i x_j=1$ by the plane $n_i x_i=0$. It is assumed that the eigenvalues and the eigenvectors of the positive definite symmetric matrix $\boldsymbol{\alpha}$ are known.

5.7.1 *Ellipsoid of three axes*

Suppose that $\boldsymbol{\alpha}$ has the spectral representation (5.6.4) with $\lambda>\mu>\nu>0$. From (5.6.6), we note that the eigenvectors $\boldsymbol{s}$ and $\boldsymbol{u}$ are along $\mathbf{h}\pm\mathbf{k}$, and also note, on evaluating $\mathbf{h}\cdot\mathbf{k}$, that

$$2\mu=\lambda+\nu-(\lambda-\nu)\mathbf{h}\cdot\mathbf{k}. \tag{5.7.1}$$

Also

$$\alpha_{ij}x_i x_j=\mu\boldsymbol{x}\cdot\boldsymbol{x}+(\lambda-\nu)(\mathbf{h}\cdot\boldsymbol{x})(\mathbf{k}\cdot\boldsymbol{x}). \tag{5.7.2}$$

We refer to the directions of the unit vectors $\mathbf{h}$ and $\mathbf{k}$ as the 'optic axes' of the ellipsoid.

Now we examine the central section of the ellipsoid by the plane $\boldsymbol{n}\cdot\boldsymbol{x}=0$.

Together with $\boldsymbol{n}\cdot\boldsymbol{x}=0$, the equation of this central section may be written $\boldsymbol{x}^{\mathrm{T}}\boldsymbol{\alpha}^*\boldsymbol{x}=1$, where

$$\alpha^*_{ij}=(\delta_{ip}-n_i n_p)\alpha_{pq}(\delta_{qj}-n_q n_j). \tag{5.7.3}$$

In fact $\boldsymbol{x}^{\mathrm{T}}\boldsymbol{\alpha}^*\boldsymbol{x}=1$ is the equation of an elliptical cylinder whose base is this section and whose generators are straight lines parallel to $\boldsymbol{n}$. From (5.6.5),

$$\begin{aligned}\alpha^*_{ij}&=\mu(\delta_{ij}-n_i n_j)+\tfrac{1}{2}(\lambda-\nu)(h^*_i k^*_j+h^*_j k^*_i)\\&=\mu(\delta_{ij}-n_i n_j)+\tfrac{1}{2}(\lambda-\nu)\sin\phi_1\sin\phi_2(\hat{h}^*_i\hat{k}^*_j+\hat{h}^*_j\hat{k}^*_i),\end{aligned} \tag{5.7.4}$$

where $\mathbf{h}^*, \mathbf{k}^*$ are the orthogonal projections of the vectors $\mathbf{h}, \mathbf{k}$, on the plane $\boldsymbol{n}\cdot\boldsymbol{x} = 0$, given by

$$\mathbf{h}^* = \mathbf{h} - (\mathbf{h}\cdot\boldsymbol{n})\boldsymbol{n}, \quad \mathbf{k}^* = \mathbf{k} - (\mathbf{k}\cdot\boldsymbol{n})\boldsymbol{n}. \tag{5.7.5}$$

Also, $\hat{\mathbf{h}}^*, \hat{\mathbf{k}}^*$ are unit vectors along $\mathbf{h}^*, \mathbf{k}^*$, given by

$$\mathbf{h}^* = \sin\phi_1 \hat{\mathbf{h}}^*, \quad \mathbf{k}^* = \sin\phi_2 \hat{\mathbf{k}}^*. \tag{5.7.6}$$

Here $\boldsymbol{n}$ makes the angle ϕ_1 with $\mathbf{h}$ and the angle ϕ_2 with $\mathbf{k}$.

We have to distinguish between two cases.

*(a) General case: **n** not in plane containing both optic axes* For the moment we assume that $\boldsymbol{n}$ is not coplanar with $\mathbf{h}$ and $\mathbf{k}$, so that $\mathbf{h}^*$ and $\mathbf{k}^*$ are not in the same direction.

From the form of the expression (5.7.4) it is clear that $\boldsymbol{\alpha}^*$ has the eigenvector $\boldsymbol{n}$, with zero eigenvalue, and the eigenvectors $\hat{\mathbf{h}}^* \pm \hat{\mathbf{k}}^*$ with eigenvalues $\gamma^\pm$ (say). Indeed, from (5.7.4)

$$\alpha^*_{ij}(\hat{h}^*_j \pm \hat{k}^*_j) = \gamma^\pm(\hat{h}^*_i \pm \hat{k}^*_i), \tag{5.7.7}$$

where, using (5.7.1),

$$\begin{aligned}
\gamma^\pm &= \mu + \tfrac{1}{2}(\lambda - \nu)\sin\phi_1 \sin\phi_2(\hat{\mathbf{h}}^*\cdot\hat{\mathbf{k}}^* \pm 1) \\
&= \mu + \tfrac{1}{2}(\lambda - \nu)\{\mathbf{h}\cdot\mathbf{k} - \cos(\phi_1 \pm \phi_2)\} \\
&= \tfrac{1}{2}(\lambda + \nu) - \tfrac{1}{2}(\lambda - \nu)\cos(\phi_1 \pm \phi_2), \\
\hat{\mathbf{h}}^* \pm \hat{\mathbf{k}}^* &= (\sin\phi_1)^{-1}\{\mathbf{h} - (\mathbf{h}\cdot\hat{\mathbf{n}})\hat{\mathbf{n}}\} \pm (\sin\phi_2)^{-1}\{\boldsymbol{k} - (\boldsymbol{k}\cdot\hat{\mathbf{n}})\hat{\mathbf{n}}\}.
\end{aligned} \tag{5.7.8}$$

Thus the principal axes of the section are along $\hat{\mathbf{h}}^* \pm \hat{\mathbf{k}}^*$ and the corresponding lengths of the principal semi-axes of the section are $(\gamma^\pm)^{-1/2}$.

Geometrically, the principal axes of the section by a given central plane are along the internal and external bisectors of the angle between the projections of the optic axes onto this plane.

*(b) Special case: **n** lying in plane containing both optic axes* If $\boldsymbol{n}$ lies in the plane of $\mathbf{h}$ and $\mathbf{k}$, the plane containing the optic axes, then either $\hat{\mathbf{h}}^* = -\hat{\mathbf{k}}^*$, or $\hat{\mathbf{h}}^* = \hat{\mathbf{k}}^*$. If $\hat{\mathbf{h}}^* = -\hat{\mathbf{k}}^*$, then equation (5.7.4) reads

$$\alpha^*_{ij} = \mu(\delta_{ij} - n_i n_j) - (\lambda - \nu)\sin\phi_1 \sin\phi_2 \hat{h}^*_i \hat{h}^*_j, \tag{5.7.9}$$

and we note that $\hat{\mathbf{h}}^*$ and $\hat{\mathbf{h}}^* \times \boldsymbol{n}$ are eigenvectors of $\boldsymbol{\alpha}^*$:

$$\alpha^*_{ij}\hat{h}^*_j = \{\mu - (\lambda - \nu)\sin\phi_1 \sin\phi_2\}\hat{h}^*_i, \tag{5.7.10}$$

$$\alpha^*_{ij}(\hat{\mathbf{h}}^* \times \boldsymbol{n})_j = \mu(\hat{\mathbf{h}}^* \times \boldsymbol{n})_i. \tag{5.7.11}$$

We now show that these values for the eigenvalues are consistent with (5.7.8). Since $\mathbf{h}, \mathbf{k}$ and $\boldsymbol{n}$ are coplanar, with $\hat{\mathbf{h}}^* = -\hat{\mathbf{k}}^*$,

$$\mathbf{h}\cdot\mathbf{k} = \cos(\phi_1 + \phi_2), \tag{5.7.12}$$

and from (5.7.1),

$$\mu = \tfrac{1}{2}(\lambda + \nu) - \tfrac{1}{2}(\lambda - \nu)\cos(\phi_1 + \phi_2) = \gamma^+, \tag{5.7.13}$$

$$\mu - (\lambda - \nu)\sin\phi_1 \sin\phi_2 = \tfrac{1}{2}(\lambda + \nu) - \tfrac{1}{2}(\lambda - \nu)\cos(\phi_1 - \phi_2) = \gamma^-,$$

which are the expressions (5.7.8) for the eigenvalues.

If $\hat{\mathbf{h}}^* = \hat{\mathbf{k}}^*$, then $\mathbf{h}\cdot\mathbf{k} = \cos(\phi_1 - \phi_2)$, and the eigenvectors of $\boldsymbol{\alpha}^*$ are again $\hat{\mathbf{h}}^* \times \boldsymbol{n}$, with eigenvalue $\mu = \gamma^-$, and $\hat{\mathbf{h}}^*$, with eigenvalue $\mu + (\lambda - \nu)\sin\phi_1 \sin\phi_2 = \gamma^+$.

Thus $\gamma^\pm$ are the eigenvalues in every case. If $\mathbf{h}, \mathbf{k}$ and $\boldsymbol{n}$ are not coplanar, then the principal axes of the central section are along $\hat{\mathbf{h}}^* \pm \hat{\mathbf{k}}^*$. However, if $\mathbf{h}, \mathbf{k}$ and $\boldsymbol{n}$ are coplanar, the principal axes of the section are along $\hat{\mathbf{h}}^*$ and $\hat{\mathbf{h}}^* \times \boldsymbol{n}$, or equivalently, along $\hat{\mathbf{k}}^*$ and $\hat{\mathbf{k}}^* \times \boldsymbol{n}$.

Suppose $\boldsymbol{n}$ is along an optic axis, say along $\mathbf{k}$, so that $\phi_2 = 0$. Then $\gamma^+ = \gamma^- = \tfrac{1}{2}(\lambda + \nu) - \tfrac{1}{2}(\lambda - \nu)\cos\phi_1 = \mu$ by (5.7.1), because now $\cos\phi_1 = \boldsymbol{n}\cdot\mathbf{h} = \mathbf{k}\cdot\mathbf{h}$, and the central section is a circle as it ought to be, because $\boldsymbol{n}$ is along an optic axis.

5.7.2 *Spheroid*

Suppose the ellipsoid is a spheroid. Let $\lambda = \mu$. Then from (5.6.4),

$$\alpha_{ij} = \lambda\delta_{ij} + (\nu - \lambda)u_i u_j, \tag{5.7.14}$$

$$\alpha^*_{ij} = \lambda(\delta_{ij} - n_i n_j) + (\nu - \lambda)u^*_i u^*_j, \tag{5.7.15}$$

where $\boldsymbol{u}^* = \boldsymbol{u} - (\boldsymbol{u}\cdot\boldsymbol{n})\boldsymbol{n}$. In this case the optic axis is along $\boldsymbol{u}$, and $\phi_1 = \phi_2 = \phi$ (say). The eigenvectors of $\boldsymbol{\alpha}^*$ are $\boldsymbol{n}$, with eigenvalue zero, $\boldsymbol{u}^*$ with eigenvalue γ given by

$$\begin{aligned} \gamma &= \lambda + (\nu - \lambda)\boldsymbol{u}^*\cdot\boldsymbol{u}^* \\ &= \lambda + (\nu - \lambda)\sin^2\phi \\ &= \tfrac{1}{2}(\lambda + \nu) - \tfrac{1}{2}(\lambda - \nu)\cos 2\phi, \end{aligned} \tag{5.7.16}$$

and $\boldsymbol{u}^* \times \boldsymbol{n} (= \boldsymbol{u} \times \boldsymbol{n})$ with eigenvalue λ.

Note that the expression for γ may be obtained from γ^+ (5.7.8) by letting $\phi_1, \phi_2 \to \phi$. Thus the expressions (5.7.8) for the eigenvalues are also valid in this case. The directions of the principal axes of the section, however, are along $\boldsymbol{u}^*$, the projection of the optic axis onto the plane normal to $\boldsymbol{n}$, and $\boldsymbol{u}^* \times \boldsymbol{n}$.

Remark *The use of Lagrange multipliers* The problem of determining the lengths of the principal axes of a central section of the ellipsoid $\boldsymbol{x}^{\mathrm{T}}\boldsymbol{\alpha}\boldsymbol{x} = 1$ may also be tackled using Lagrange multipliers. Thus the problem is to determine the extrema of $L^2 = x_i x_i$ subject to the constraints

$$x_i \alpha_{ij} x_j = 1, \quad n_i x_i = 0, \tag{5.7.17}$$

where $\mathbf{n}$ is a unit vector. Then, using Lagrange multipliers p and q, we write

$$L^2 = x_i x_i + p(x_i \alpha_{ij} x_j - 1) + 2q n_i x_i. \tag{5.7.18}$$

Then, $\partial L^2/\partial \boldsymbol{x} = 0$ gives

$$x_i + p\alpha_{ij} x_j + q n_i = 0. \tag{5.7.19}$$

On multiplying (5.7.19) by x_i and by n_i, and using (5.7.17), we obtain $p = -x_i x_i = -L^2$ and $q = L^2 n_i \alpha_{ij} x_j$. Hence (5.7.19) yields the eigenvalue problem

$$\{L^{-2}\delta_{ij} + (n_i n_p - \delta_{ip})\alpha_{pj}\} x_j = 0, \tag{5.7.20}$$

or equivalently, using (5.7.3), because $n_i x_i = 0$,

$$\{L^{-2}\delta_{ij} - \alpha^*_{ij}\} x_j = 0. \tag{5.7.21}$$

Thus the values of L^{-2} corresponding to the extrema are the roots of

$$\det(L^{-2}\delta_{ij} - \alpha^*_{ij}) = 0, \tag{5.7.22}$$

and because $\det \boldsymbol{\alpha}^* = 0$, one root is zero. Also because $\boldsymbol{\alpha}^*$ is real and symmetric, the two other roots are the real solutions of a quadratic in L^{-2}, and the corresponding eigenvectors are orthogonal.

The speeds and amplitude vectors of the two homogeneous electromagnetic waves propagating in the direction $\boldsymbol{n}$ in an electrically anisotropic but magnetically isotropic medium are determined through an eigenvalue problem of the form of (5.7.21) (Chapter 9).

5.8 Orthogonality of bivectors with respect to a metric

In section 2.4, we considered the geometrical implications of the orthogonality of two bivectors. Here we give a geometrical interpretation of the orthogonality of a pair of bivectors $\boldsymbol{P}, \boldsymbol{Q}$ with respect to a real symmetric positive definite 3×3 matrix $\boldsymbol{g}$: $\boldsymbol{P}^{\mathrm{T}}\boldsymbol{g}\boldsymbol{Q} = 0$.

Associated with $\boldsymbol{g}$ is the $\boldsymbol{g}$-metric ellipsoid and associated with each bivector is an ellipse. We consider separately the cases when the ellipses of $\boldsymbol{P}$ and $\boldsymbol{Q}$ are coplanar (section 5.8.1) and not coplanar (section 5.8.2).

Because

$$\boldsymbol{P}^{\mathrm{T}}\boldsymbol{g}\boldsymbol{Q} = 0, \tag{5.8.1}$$

it follows that

$$\begin{aligned} \boldsymbol{P}^{+\mathrm{T}}\boldsymbol{g}\boldsymbol{Q}^{+} - \boldsymbol{P}^{-\mathrm{T}}\boldsymbol{g}\boldsymbol{Q}^{-} &= 0, \\ \boldsymbol{P}^{+\mathrm{T}}\boldsymbol{g}\boldsymbol{Q}^{-} + \boldsymbol{P}^{-\mathrm{T}}\boldsymbol{g}\boldsymbol{Q}^{+} &= 0. \end{aligned} \tag{5.8.2}$$

The directional ellipses associated with $\boldsymbol{P}$ and $\boldsymbol{Q}$ are defined by the parametric equations

$$\boldsymbol{x} = \boldsymbol{P}^{+} \cos\theta + \boldsymbol{P}^{-} \sin\theta, \quad \boldsymbol{x} = \boldsymbol{Q}^{+} \cos\theta + \boldsymbol{Q}^{-} \sin\theta. \tag{5.8.3}$$

5.8.1 P and Q coplanar

If $\boldsymbol{y}$ denotes the position vector of a generic point of the common plane of $\boldsymbol{P}$ and $\boldsymbol{Q}$, then

$$\boldsymbol{y}^{\mathrm{T}}\boldsymbol{g}\boldsymbol{y} = 1, \tag{5.8.4}$$

is the equation of an ellipse – the '$\boldsymbol{g}$-metric ellipse' – which is the section of the $\boldsymbol{g}$-metric ellipsoid by the plane of $\boldsymbol{P}$ and $\boldsymbol{Q}$. We note the following properties.

Property 5.1

The ellipse of $\boldsymbol{Q}$ is similar and similarly situated to the polar reciprocal of the ellipse of $\boldsymbol{P}$ with respect to the metric ellipse. The ellipses of $\boldsymbol{P}$ and $\boldsymbol{Q}$ are described in the same sense.

Proof The polar reciprocal of the ellipse of $\boldsymbol{P}$ with respect to the metric ellipse is by definition the envelope of the polars of all the points of the ellipse of $\boldsymbol{P}$ with respect to the metric ellipse. Its parametric equation, with θ as parameter, may be obtained by solving for $\boldsymbol{y}$ the system

$$(\boldsymbol{P}^+\cos\theta + \boldsymbol{P}^-\sin\theta)^{\mathrm{T}}\boldsymbol{g}\boldsymbol{y} = 1, \qquad (5.8.5a)$$

$$(-\boldsymbol{P}^+\sin\theta + \boldsymbol{P}^-\cos\theta)^{\mathrm{T}}\boldsymbol{g}\boldsymbol{y} = 0. \qquad (5.8.5b)$$

Indeed (5.8.5a) is the equation of the polar of the point $(\boldsymbol{P}^+\cos\theta + \boldsymbol{P}^-\sin\theta)$ with respect to the metric ellipse, and (5.8.5b) is its derivative with respect to θ.

We introduce in the plane of $\boldsymbol{P}$ the pair of vectors $(\boldsymbol{P}_*^+, \boldsymbol{P}_*^-)$ reciprocal to the pair $(\boldsymbol{P}^+, \boldsymbol{P}^-)$ with respect to the metric $\boldsymbol{g}$. Thus

$$\begin{aligned} \boldsymbol{P}^{-\mathrm{T}}\boldsymbol{g}\boldsymbol{P}_*^+ &= 0, \quad \boldsymbol{P}^{+\mathrm{T}}\boldsymbol{g}\boldsymbol{P}_*^+ = 1, \\ \boldsymbol{P}^{+\mathrm{T}}\boldsymbol{g}\boldsymbol{P}_*^- &= 0, \quad \boldsymbol{P}^{-\mathrm{T}}\boldsymbol{g}\boldsymbol{P}_*^- = 1. \end{aligned} \qquad (5.8.6)$$

The solution of (5.8.5) for $\boldsymbol{y}$ is easily checked to be

$$\boldsymbol{y} = \boldsymbol{P}_*^+\cos\theta + \boldsymbol{P}_*^-\sin\theta, \qquad (5.8.7)$$

so that the polar reciprocal of the ellipse of $\boldsymbol{P}$ with respect to the metric ellipse is the ellipse associated with $\boldsymbol{P}_* = \boldsymbol{P}_*^+ + \mathrm{i}\boldsymbol{P}_*^-$ (Figure 5.2).

From (5.8.6) we have that

$$\boldsymbol{P}^{\mathrm{T}}\boldsymbol{g}\boldsymbol{P}_* = 0. \qquad (5.8.8)$$

Then, since $\boldsymbol{P}_*$ and $\boldsymbol{Q}$ are both in the plane of $\boldsymbol{P}$, equation (5.8.8) together with (5.8.1) implies that $\boldsymbol{Q}$ and $\boldsymbol{P}_*$ are two parallel bivectors. This means that the ellipse of $\boldsymbol{Q}$ is similar and similarly situated to the ellipse of $\boldsymbol{P}_*$.

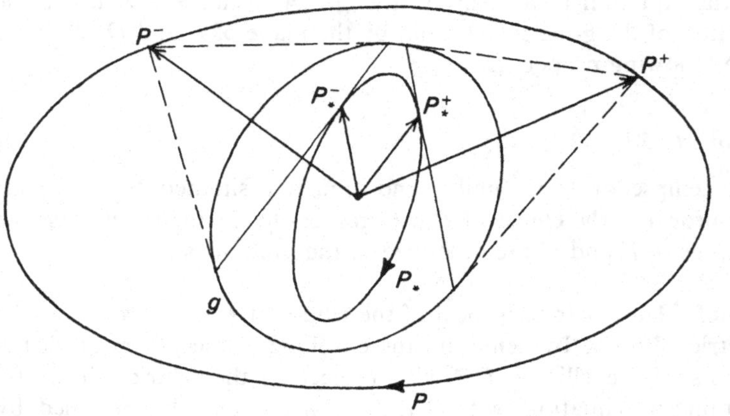

Figure 5.2 *Orthogonality with respect to a metric: coplanar bivectors.*

Also, the pairs of coplanar vectors $(\boldsymbol{P}^+, \boldsymbol{P}^-)$ and $(\boldsymbol{Q}^+, \boldsymbol{Q}^-)$ have the same orientation, because, from (5.1.5)

$$\begin{aligned}&(\det \boldsymbol{g})(\boldsymbol{P}^+ \times \boldsymbol{P}^-)^{\mathrm{T}}\boldsymbol{g}^{-1}(\boldsymbol{Q}^+ \times \boldsymbol{Q}^-)\\ &\quad = (\boldsymbol{P}^{+\mathrm{T}}\boldsymbol{g}\boldsymbol{Q}^+)(\boldsymbol{P}^{-\mathrm{T}}\boldsymbol{g}\boldsymbol{Q}^-) - (\boldsymbol{P}^{+\mathrm{T}}\boldsymbol{g}\boldsymbol{Q}^-)(\boldsymbol{P}^{-\mathrm{T}}\boldsymbol{g}\boldsymbol{Q}^+)\\ &\quad = (\boldsymbol{P}^{+\mathrm{T}}\boldsymbol{g}\boldsymbol{Q}^+)^2 + (\boldsymbol{P}^{+\mathrm{T}}\boldsymbol{g}\boldsymbol{Q}^-)^2 > 0, \end{aligned} \qquad (5.8.9)$$

on using (5.8.2). Thus the two ellipses are described in the same sense.

Thus we have established Property 5.1.

Property 5.2

The ellipses of $\boldsymbol{P}$ and $\boldsymbol{Q}$ and the $\boldsymbol{g}$-metric ellipse have a pair of common conjugate directions.

Proof We know (section 1.7) that the ellipses of $\boldsymbol{P}$ and $\boldsymbol{Q}$ have a pair of common conjugate directions. Let $\boldsymbol{u}$ and $\boldsymbol{v}$ be along these directions with their extremities on the $\boldsymbol{g}$-metric ellipse:

$$\boldsymbol{u}^{\mathrm{T}}\boldsymbol{g}\boldsymbol{u} = \boldsymbol{v}^{\mathrm{T}}\boldsymbol{g}\boldsymbol{v} = 1. \qquad (5.8.10)$$

Then, $\boldsymbol{P}$ and $\boldsymbol{Q}$ may be written

$$\boldsymbol{P} = \mathrm{e}^{\mathrm{i}\phi}(a\boldsymbol{u} + \mathrm{i}b\boldsymbol{v}), \quad \boldsymbol{Q} = \mathrm{e}^{\mathrm{i}\psi}(c\boldsymbol{u} + \mathrm{i}d\boldsymbol{v}), \qquad (5.8.11)$$

where a, b, c and d are some real scalars. Introducing this into (5.8.1) yields

$$ac - bd = 0, \qquad (5.8.12a)$$

$$(ad + bc)\boldsymbol{u}^{\mathrm{T}}\boldsymbol{g}\boldsymbol{v} = 0. \qquad (5.8.12b)$$

But, using (5.8.12a), we note that

$$ad + bc = \frac{a}{d}(c^2 + d^2) = \frac{d}{a}(a^2 + b^2) \neq 0. \qquad (5.8.13)$$

Hence, from (5.8.12b)

$$\boldsymbol{u}^{\mathrm{T}}\boldsymbol{g}\boldsymbol{v} = 0, \qquad (5.8.14)$$

showing that $\boldsymbol{u}$ and $\boldsymbol{v}$ are also conjugate directions of the $\boldsymbol{g}$-metric ellipse.

Equations of the ellipses using their common conjugate directions Let Ox and Oy be oblique axes along the common conjugate directions

$\boldsymbol{u}, \boldsymbol{v}$ of the three ellipses, with the lengths of $\boldsymbol{u}$ and $\boldsymbol{v}$ being taken as units of length on the Ox and Oy axes respectively. The equation of the $\boldsymbol{g}$-metric ellipse is then

$$x^2 + y^2 = 1, \tag{5.8.15}$$

whilst the equations of the ellipses of $\boldsymbol{P}$ and $\boldsymbol{Q}$ are

$$\frac{x^2}{a^2} + \frac{y^2}{b^2} = 1, \quad \frac{x^2}{c^2} + \frac{y^2}{d^2} = 1, \tag{5.8.16}$$

where a, b, c, d satisfy (5.8.12a). Hence, with an arbitrary $k > 0$,

$$\frac{x^2}{a^2} + \frac{y^2}{b^2} = k, \quad \frac{x^2}{b^2} + \frac{y^2}{a^2} = k, \tag{5.8.17}$$

are the equations of ellipses similar and similarly situated to the ellipses of $\boldsymbol{P}$ and $\boldsymbol{Q}$ respectively. Figure 5.3(a), (b) represents these

(a)

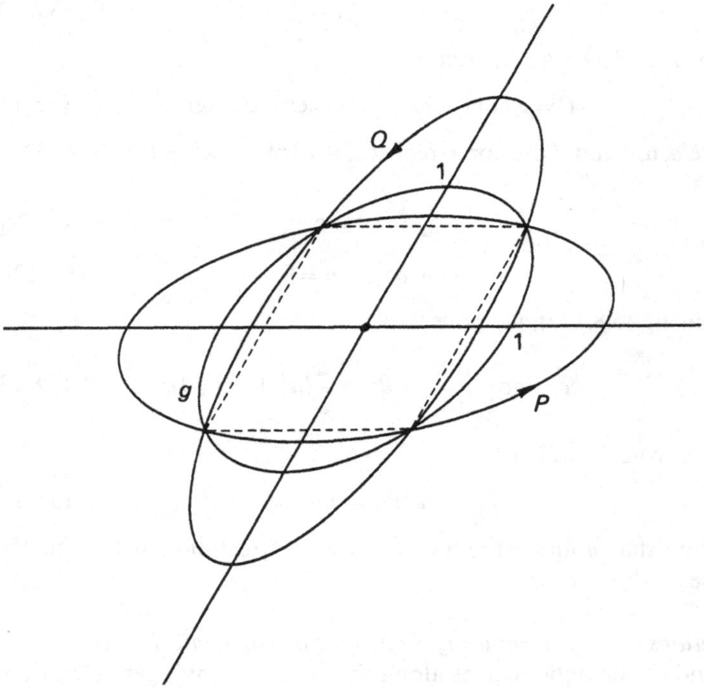

(b)

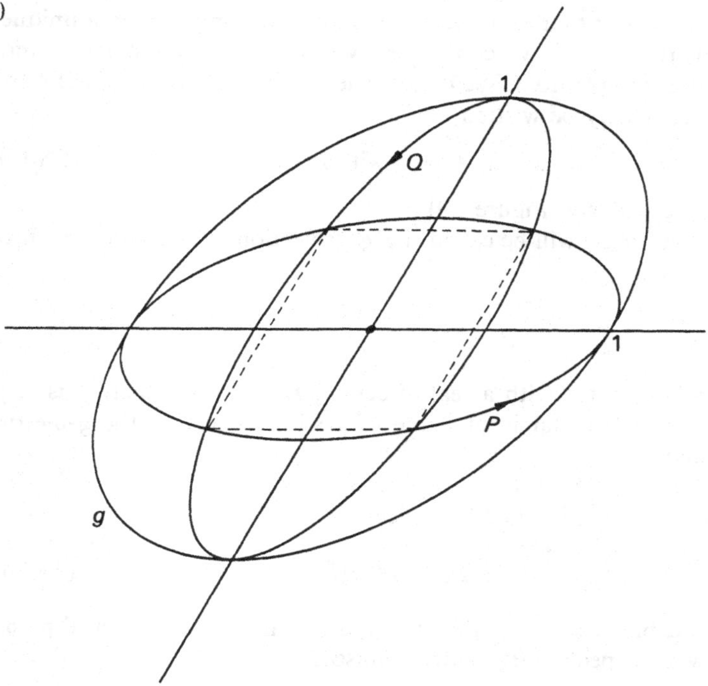

Figure 5.3 *Orthogonality with respect to a metric: ellipses referred to their common conjugate directions. (a)* $k=(a^2+b^2)(2a^2b^2)^{-1}$. *(b)* $k=a^{-2}$.

ellipses and the $\boldsymbol{g}$-metric ellipse when $k=(a^2+b^2)(2a^2b^2)^{-1}$ and $k=a^{-2}$.

5.8.2 *P and Q not coplanar*

Let us now assume that the ellipses associated with the bivectors $\boldsymbol{P}$ and $\boldsymbol{Q}$ are not in the same plane.

It is convenient to introduce here the concept of orthogonal projection with respect to the metric $\boldsymbol{g}$ or '$\boldsymbol{g}$-projection'. Let a be a given plane through the origin and let $\boldsymbol{n}$ be a vector along the direction conjugate to a with respect to the metric ellipsoid: a is the diametral plane of $\boldsymbol{n}$, and its equation is

$$\boldsymbol{n}^{\mathrm{T}}\boldsymbol{g}\boldsymbol{x}=0. \qquad (5.8.18)$$

It is easily seen that every vector $\boldsymbol{x}$ may be decomposed in a unique way as the sum of two component vectors, one in the plane a, and the other parallel to $\boldsymbol{n}$. We denote the component in the plane a by $\boldsymbol{x}_a$. Then $\boldsymbol{x}$ may be written

$$\boldsymbol{x} = \boldsymbol{x}_a + \lambda \boldsymbol{n}, \quad \text{with } \boldsymbol{n}^{\mathrm{T}} \boldsymbol{g} \boldsymbol{x}_a = 0, \tag{5.8.19}$$

for some scalar λ (Figure 5.4).

The vector $\boldsymbol{x}_a$ will be called the '$\boldsymbol{g}$-projection' of the vector $\boldsymbol{x}$ onto the plane a.

Property 5.3

If $\boldsymbol{P}$ is a bivector with a real direction, $\boldsymbol{P} = \mu \boldsymbol{s}$ (say), then $\boldsymbol{Q}$ is any bivector in the diametral plane of $\boldsymbol{s}$ with respect to the $\boldsymbol{g}$-metric ellipsoid.

Proof In this case, (5.8.1) reduces to

$$\boldsymbol{s}^{\mathrm{T}} \boldsymbol{g} \boldsymbol{Q}^+ = \boldsymbol{s}^{\mathrm{T}} \boldsymbol{g} \boldsymbol{Q}^- = 0, \tag{5.8.20}$$

which expresses the fact that $\boldsymbol{Q}^+$ and $\boldsymbol{Q}^-$ are in the diametral plane of $\boldsymbol{s}$ with respect to the metric ellipsoid.

Property 5.4

If neither $\boldsymbol{P}$ nor $\boldsymbol{Q}$ is a bivector with a real direction, the plane of the ellipse of $\boldsymbol{Q}$ may not contain the conjugate direction to the plane of the ellipse of $\boldsymbol{P}$ with respect to the metric ellipsoid.

Proof Let $\boldsymbol{n} \neq \boldsymbol{0}$ be a vector in the conjugate direction to the plane a of $\boldsymbol{P}^+$ and $\boldsymbol{P}^-$ with respect to the metric ellipsoid:

$$\boldsymbol{P}^{+\mathrm{T}} \boldsymbol{g} \boldsymbol{n} = \boldsymbol{P}^{-\mathrm{T}} \boldsymbol{g} \boldsymbol{n} = 0. \tag{5.8.21}$$

We now show that $\boldsymbol{n}$ may not lie in the plane of $\boldsymbol{Q}^+$ and $\boldsymbol{Q}^-$. For, suppose that $\boldsymbol{n}$ lies in this plane. Then, for some real scalars ν and γ $(\nu^2 + \gamma^2 \neq 0)$,

$$\boldsymbol{n} = \nu \boldsymbol{Q}^+ + \gamma \boldsymbol{Q}^-, \tag{5.8.22}$$

and (5.8.21) then reads

$$\nu \boldsymbol{P}^{+\mathrm{T}} \boldsymbol{g} \boldsymbol{Q}^+ + \gamma \boldsymbol{P}^{+\mathrm{T}} \boldsymbol{g} \boldsymbol{Q}^- = \nu \boldsymbol{P}^{-\mathrm{T}} \boldsymbol{g} \boldsymbol{Q}^+ + \gamma \boldsymbol{P}^{-\mathrm{T}} \boldsymbol{g} \boldsymbol{Q}^- = 0. \tag{5.8.23}$$

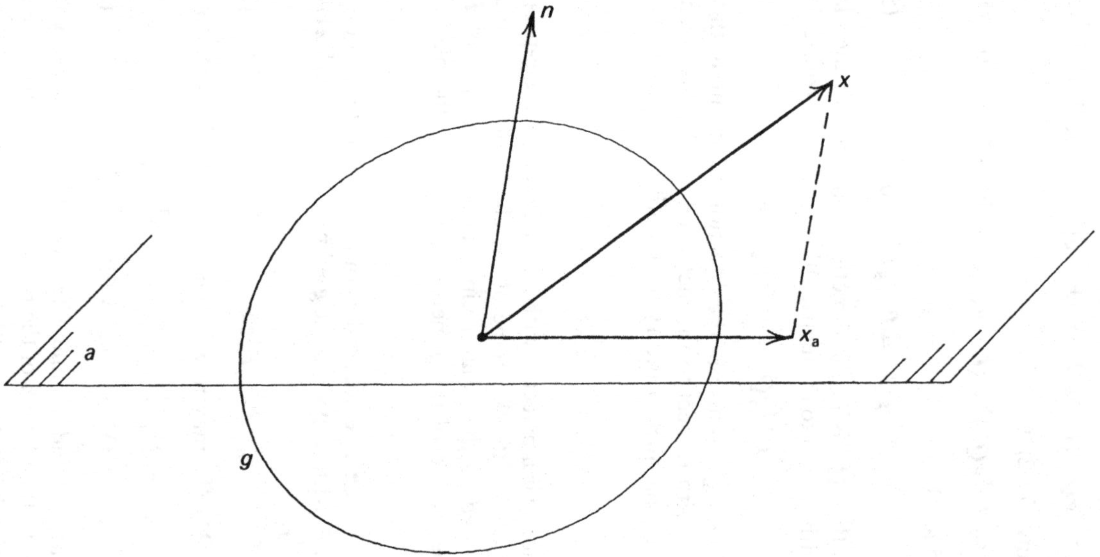

Figure 5.4 *The 'g–projection'.*

Eliminating v and γ from these equations gives

$$(\boldsymbol{P}^{+\mathrm{T}}\boldsymbol{g}\boldsymbol{Q}^{+})(\boldsymbol{P}^{-\mathrm{T}}\boldsymbol{g}\boldsymbol{Q}^{-})=(\boldsymbol{P}^{+\mathrm{T}}\boldsymbol{g}\boldsymbol{Q}^{-})(\boldsymbol{P}^{-\mathrm{T}}\boldsymbol{g}\boldsymbol{Q}^{+}), \tag{5.8.24}$$

and then, using (5.8.2), we obtain

$$(\boldsymbol{P}^{+\mathrm{T}}\boldsymbol{g}\boldsymbol{Q}^{+})^2+(\boldsymbol{P}^{+\mathrm{T}}\boldsymbol{g}\boldsymbol{Q}^{-})^2=(\boldsymbol{P}^{-\mathrm{T}}\boldsymbol{g}\boldsymbol{Q}^{-})^2+(\boldsymbol{P}^{-\mathrm{T}}\boldsymbol{g}\boldsymbol{Q}^{+})^2=0. \tag{5.8.25}$$

Now, from (5.8.25)

$$\boldsymbol{P}^{+\mathrm{T}}\boldsymbol{g}\boldsymbol{Q}^{+}=\boldsymbol{P}^{-\mathrm{T}}\boldsymbol{g}\boldsymbol{Q}^{+}=0, \tag{5.8.26}$$

which means that $\boldsymbol{Q}^{+}$ is in the direction conjugate to the plane of $\boldsymbol{P}^{+}$ and $\boldsymbol{P}^{-}$ with respect to the metric ellipsoid. Again, from (5.8.25),

$$\boldsymbol{P}^{+\mathrm{T}}\boldsymbol{g}\boldsymbol{Q}^{-}=\boldsymbol{P}^{-\mathrm{T}}\boldsymbol{g}\boldsymbol{Q}^{-}=0, \tag{5.8.27}$$

which means that $\boldsymbol{Q}^{-}$ is also in the direction conjugate to the plane of $\boldsymbol{P}^{+}$ and $\boldsymbol{P}^{-}$ with respect to the metric ellipsoid. This means that $\boldsymbol{Q}^{+}$ and $\boldsymbol{Q}^{-}$ are parallel, contrary to hypothesis.

Property 5.5

If neither $\boldsymbol{P}$ nor $\boldsymbol{Q}$ is a bivector with a real direction, the $\boldsymbol{g}$-projection of the ellipse of $\boldsymbol{Q}$ onto the plane of the ellipse of $\boldsymbol{P}$ is similar and similarly situated to the polar reciprocal of the ellipse of $\boldsymbol{P}$ with respect to the section of the $\boldsymbol{g}$-metric ellipsoid by the plane of the bivector $\boldsymbol{P}$.

Proof Let $\boldsymbol{n}\neq\boldsymbol{0}$ be a vector in the conjugate direction to the plane a of $\boldsymbol{P}^{+}$ and $\boldsymbol{P}^{-}$ with respect to the $\boldsymbol{g}$-ellipsoid, so that (5.8.21) holds, or equivalently,

$$\boldsymbol{P}^{\mathrm{T}}\boldsymbol{g}\boldsymbol{n}=0. \tag{5.8.28}$$

The vectors $\boldsymbol{Q}^{+}$, $\boldsymbol{Q}^{-}$ may now be decomposed as in (5.8.19):

$$\boldsymbol{Q}^{+}=\boldsymbol{Q}_a^{+}+\lambda^{+}\boldsymbol{n},\quad\text{with}\quad \boldsymbol{n}^{\mathrm{T}}\boldsymbol{g}\boldsymbol{Q}_a^{+}=0, \tag{5.8.29}$$

$$\boldsymbol{Q}^{-}=\boldsymbol{Q}_a^{-}+\lambda^{-}\boldsymbol{n},\quad\text{with}\quad \boldsymbol{n}^{\mathrm{T}}\boldsymbol{g}\boldsymbol{Q}_a^{-}=0, \tag{5.8.30}$$

for some scalars λ^{+} and λ^{-}, and where $\boldsymbol{Q}_a^{+}, \boldsymbol{Q}_a^{-}$ are the $\boldsymbol{g}$-projections of $\boldsymbol{Q}^{+}, \boldsymbol{Q}^{-}$ onto the plane a of the bivector $\boldsymbol{P}$. In terms of bivectors, (5.8.29) and (5.8.30) may be written

$$\boldsymbol{Q}=\boldsymbol{Q}_a+\lambda\boldsymbol{n},\quad\text{with}\quad \boldsymbol{n}^{\mathrm{T}}\boldsymbol{g}\boldsymbol{Q}_a=0, \tag{5.8.31}$$

where $\boldsymbol{Q}_a=\boldsymbol{Q}_a^{+}+\mathrm{i}\boldsymbol{Q}_a^{-}$ and $\lambda=\lambda^{+}+\mathrm{i}\lambda^{-}$. Introducing (5.8.31) into

(5.8.1) and using (5.8.28), we note that (5.8.1) is equivalent to

$$P^{\mathrm{T}} g Q_a = 0. \tag{5.8.32}$$

The bivectors $\boldsymbol{P}$ and $\boldsymbol{Q}_a$ entering (5.8.32) have their ellipses in the same plane a, so that they obey Property 5.1 where the metric ellipse is the elliptical section of the $\boldsymbol{g}$-metric ellipsoid by the common plane a. Thus the ellipse of $\boldsymbol{Q}_a$ is similar and similarly situated to the polar reciprocal of the ellipse of $\boldsymbol{P}$ with respect to the elliptical section of the metric ellipsoid by the plane a. This completes the proof since the ellipse of $\boldsymbol{Q}_a$ is the $\boldsymbol{g}$-projection of the ellipse of $\boldsymbol{Q}$ onto the plane a.

5.9 A matrix identity

Let $\boldsymbol{\alpha}$ and $\boldsymbol{\beta}$ be any two real, positive definite, symmetric 3×3 matrices. Here we derive an identity relating $\boldsymbol{\alpha}$ and $\boldsymbol{\beta}$ and the eigenvectors of $\boldsymbol{\beta}^{-1}\boldsymbol{\alpha}$.

5.9.1 General case. Three different eigenvalues

Let

$$\boldsymbol{\alpha r} = \lambda \boldsymbol{\beta r}, \quad \boldsymbol{\alpha s} = \mu \boldsymbol{\beta s}, \quad \boldsymbol{\alpha t} = \nu \boldsymbol{\beta t}, \quad \lambda > \mu > \nu > 0, \tag{5.9.1}$$

where λ, μ, ν are the roots of the cubic equation in γ:

$$\det(\boldsymbol{\alpha} - \gamma \boldsymbol{\beta}) = 0. \tag{5.9.2}$$

Thus, $\boldsymbol{r}, \boldsymbol{s}, \boldsymbol{t}$ are the eigenvectors of the matrix $\boldsymbol{\alpha}$ with respect to the matrix $\boldsymbol{\beta}$, and λ, μ, γ are the corresponding eigenvalues. Because we have assumed that the three roots are different, it is easily seen that

$$\boldsymbol{r}^{\mathrm{T}}\boldsymbol{\alpha s} = \boldsymbol{r}^{\mathrm{T}}\boldsymbol{\beta s} = \boldsymbol{r}^{\mathrm{T}}\boldsymbol{\alpha t} = \boldsymbol{r}^{\mathrm{T}}\boldsymbol{\beta t} = \boldsymbol{s}^{\mathrm{T}}\boldsymbol{\alpha t} = \boldsymbol{s}^{\mathrm{T}}\boldsymbol{\beta t} = 0, \tag{5.9.3}$$

$$\lambda = \frac{\boldsymbol{r}^{\mathrm{T}}\boldsymbol{\alpha r}}{\boldsymbol{r}^{\mathrm{T}}\boldsymbol{\beta r}}, \quad \mu = \frac{\boldsymbol{s}^{\mathrm{T}}\boldsymbol{\alpha s}}{\boldsymbol{s}^{\mathrm{T}}\boldsymbol{\beta s}}, \quad \nu = \frac{\boldsymbol{t}^{\mathrm{T}}\boldsymbol{\alpha t}}{\boldsymbol{t}^{\mathrm{T}}\boldsymbol{\beta t}}, \tag{5.9.4}$$

and

$$\boldsymbol{\beta}^{-1}\boldsymbol{\alpha r} = \lambda \boldsymbol{r}, \quad \boldsymbol{\beta}^{-1}\boldsymbol{\alpha s} = \mu \boldsymbol{s}, \quad \boldsymbol{\beta}^{-1}\boldsymbol{\alpha t} = \nu \boldsymbol{t}. \tag{5.9.5}$$

Note that (5.9.3) means that $\boldsymbol{r}, \boldsymbol{s}, \boldsymbol{t}$ are along three conjugate diameters with respect to both the $\boldsymbol{\alpha}$ and $\boldsymbol{\beta}$-ellipsoids. We define $\boldsymbol{r}_*, \boldsymbol{s}_*, \boldsymbol{t}_*$ through

$$\boldsymbol{r}_* = \frac{\boldsymbol{\alpha r}}{\boldsymbol{r}^{\mathrm{T}}\boldsymbol{\alpha r}} = \frac{\boldsymbol{\beta r}}{\boldsymbol{r}^{\mathrm{T}}\boldsymbol{\beta r}}, \quad \boldsymbol{s}_* = \frac{\boldsymbol{\alpha s}}{\boldsymbol{s}^{\mathrm{T}}\boldsymbol{\alpha s}} = \frac{\boldsymbol{\beta s}}{\boldsymbol{s}^{\mathrm{T}}\boldsymbol{\beta s}}, \quad \boldsymbol{t}_* = \frac{\boldsymbol{\alpha t}}{\boldsymbol{t}^{\mathrm{T}}\boldsymbol{\alpha t}} = \frac{\boldsymbol{\beta t}}{\boldsymbol{t}^{\mathrm{T}}\boldsymbol{\beta t}}. \tag{5.9.6}$$

Then

$$\boldsymbol{r}\cdot\boldsymbol{r}_* = \boldsymbol{s}\cdot\boldsymbol{s}_* = \boldsymbol{t}\cdot\boldsymbol{t}_* = 1,$$
$$\boldsymbol{r}\cdot\boldsymbol{s}_* = \boldsymbol{r}\cdot\boldsymbol{t}_* = \boldsymbol{s}\cdot\boldsymbol{r}_* = \boldsymbol{s}\cdot\boldsymbol{t}_* = \boldsymbol{t}\cdot\boldsymbol{r}_* = \boldsymbol{t}\cdot\boldsymbol{s}_* = 0. \tag{5.9.7}$$

Thus $\boldsymbol{r}, \boldsymbol{s}, \boldsymbol{t}$ and $\boldsymbol{r}_*, \boldsymbol{s}_*, \boldsymbol{t}_*$ are reciprocal. We note from (5.9.6) that

$$\boldsymbol{r}_*^{\mathrm{T}}\boldsymbol{\alpha}^{-1}\boldsymbol{s}_* = \boldsymbol{r}_*^{\mathrm{T}}\boldsymbol{\beta}^{-1}\boldsymbol{s}_* = \boldsymbol{r}_*^{\mathrm{T}}\boldsymbol{\alpha}^{-1}\boldsymbol{t}_* = \boldsymbol{r}_*^{\mathrm{T}}\boldsymbol{\beta}^{-1}\boldsymbol{t}_* = \boldsymbol{s}_*^{\mathrm{T}}\boldsymbol{\alpha}^{-1}\boldsymbol{t}_* = \boldsymbol{s}_*^{\mathrm{T}}\boldsymbol{\beta}^{-1}\boldsymbol{t}_* = 0, \tag{5.9.8}$$

and

$$(\boldsymbol{r}^{\mathrm{T}}\boldsymbol{\beta}\boldsymbol{r})(\boldsymbol{r}_*^{\mathrm{T}}\boldsymbol{\beta}^{-1}\boldsymbol{r}_*) = 1, \quad (\boldsymbol{r}^{\mathrm{T}}\boldsymbol{\alpha}\boldsymbol{r})(\boldsymbol{r}_*^{\mathrm{T}}\boldsymbol{\alpha}^{-1}\boldsymbol{r}_*) = 1, \text{ etc.} \tag{5.9.9}$$

Also,

$$\boldsymbol{r}_*^{\mathrm{T}}\boldsymbol{\beta}^{-1}\boldsymbol{\alpha} = \lambda\boldsymbol{r}_*^{\mathrm{T}}, \quad \boldsymbol{s}_*^{\mathrm{T}}\boldsymbol{\beta}^{-1}\boldsymbol{\alpha} = \mu\boldsymbol{s}_*^{\mathrm{T}}, \quad \boldsymbol{t}_*^{\mathrm{T}}\boldsymbol{\beta}^{-1}\boldsymbol{\alpha} = \nu\boldsymbol{t}_*^{\mathrm{T}}. \tag{5.9.10}$$

Note that (5.9.8) means that $\boldsymbol{r}_*, \boldsymbol{s}_*, \boldsymbol{t}_*$ are along three conjugate diameters with respect to both the $\boldsymbol{\alpha}^{-1}$ and $\boldsymbol{\beta}^{-1}$-ellipsoids. From (5.9.5) and (5.9.10) it is clear that $(\boldsymbol{r}, \boldsymbol{s}, \boldsymbol{t})$ are right eigenvectors of $\boldsymbol{\beta}^{-1}\boldsymbol{\alpha}$ with eigenvalues (λ, μ, ν) respectively, and $(\boldsymbol{r}_*, \boldsymbol{s}_*, \boldsymbol{t}_*)$ are left eigenvectors of $\boldsymbol{\beta}^{-1}\boldsymbol{\alpha}$ with eigenvalues (λ, μ, ν) respectively.

The spectral decomposition of $\boldsymbol{\beta}^{-1}\boldsymbol{\alpha}$ is therefore

$$\begin{aligned}\boldsymbol{\beta}^{-1}\boldsymbol{\alpha} &= \lambda\boldsymbol{r}\otimes\boldsymbol{r}_* + \mu\boldsymbol{s}\otimes\boldsymbol{s}_* + \nu\boldsymbol{t}\otimes\boldsymbol{t}_* \\ &= \lambda\boldsymbol{r}\otimes\boldsymbol{r}_* + \mu(\mathbf{1} - \boldsymbol{r}\otimes\boldsymbol{r}_* - \boldsymbol{t}\otimes\boldsymbol{t}_*) + \nu\boldsymbol{t}\otimes\boldsymbol{t}_* \\ &= \mu\mathbf{1} + (\lambda-\mu)\boldsymbol{r}\otimes\boldsymbol{r}_* - (\mu-\nu)\boldsymbol{t}\otimes\boldsymbol{t}_*.\end{aligned} \tag{5.9.11}$$

Thus, using (5.9.6) and (5.9.9), we have

$$\boldsymbol{\alpha} = \mu\boldsymbol{\beta} + (\lambda-\mu)(\boldsymbol{r}^{\mathrm{T}}\boldsymbol{\beta}\boldsymbol{r})\boldsymbol{r}_*\otimes\boldsymbol{r}_* - (\mu-\nu)(\boldsymbol{t}^{\mathrm{T}}\boldsymbol{\beta}\boldsymbol{t})\boldsymbol{t}_*\otimes\boldsymbol{t}_*, \tag{5.9.12}$$

$$\boldsymbol{\beta}^{-1} = \mu\boldsymbol{\alpha}^{-1} + (\lambda-\mu)(\boldsymbol{r}_*^{\mathrm{T}}\boldsymbol{\alpha}^{-1}\boldsymbol{r}_*)\boldsymbol{r}\otimes\boldsymbol{r} - (\mu-\nu)(\boldsymbol{t}_*^{\mathrm{T}}\boldsymbol{\alpha}^{-1}\boldsymbol{t}_*)\boldsymbol{t}\otimes\boldsymbol{t}. \tag{5.9.13}$$

We may write the identities (5.9.12) and (5.9.13) in Hamilton's form:

$$\boldsymbol{\alpha} = \mu\boldsymbol{\beta} + \tfrac{1}{2}(\boldsymbol{n}_+^*\otimes\boldsymbol{n}_-^* + \boldsymbol{n}_-^*\otimes\boldsymbol{n}_+^*), \tag{5.9.14}$$

where

$$\begin{aligned}\boldsymbol{n}_\pm^* &= \sqrt{(\lambda-\mu)(\boldsymbol{r}^{\mathrm{T}}\boldsymbol{\beta}\boldsymbol{r})}\,\boldsymbol{r}_* \pm \sqrt{(\mu-\nu)(\boldsymbol{t}^{\mathrm{T}}\boldsymbol{\beta}\boldsymbol{t})}\,\boldsymbol{t}_* \\ &= \left(\frac{\lambda-\mu}{\boldsymbol{r}_*^{\mathrm{T}}\boldsymbol{\beta}^{-1}\boldsymbol{r}_*}\right)^{1/2}\boldsymbol{r}_* \pm \left(\frac{\mu-\nu}{\boldsymbol{t}_*^{\mathrm{T}}\boldsymbol{\beta}^{-1}\boldsymbol{t}_*}\right)^{1/2}\boldsymbol{t}_*;\end{aligned} \tag{5.9.15}$$

and

$$\boldsymbol{\beta}^{-1} = \mu\boldsymbol{\alpha}^{-1} + \tfrac{1}{2}(\boldsymbol{n}_+\otimes\boldsymbol{n}_- + \boldsymbol{n}_-\otimes\boldsymbol{n}_+), \tag{5.9.16}$$

where

$$\begin{aligned} \boldsymbol{n}_{\pm} &= \sqrt{(\lambda-\mu)(\boldsymbol{r}_*^{\mathrm{T}}\boldsymbol{\alpha}^{-1}\boldsymbol{r}_*)}\,\boldsymbol{r} \pm \sqrt{(\mu-\nu)(\boldsymbol{t}_*^{\mathrm{T}}\boldsymbol{\alpha}^{-1}\boldsymbol{t}_*)}\,\boldsymbol{t} \\ &= \left(\frac{\lambda-\mu}{\boldsymbol{r}^{\mathrm{T}}\boldsymbol{\alpha}\boldsymbol{r}}\right)^{1/2}\boldsymbol{r} \pm \left(\frac{\mu-\nu}{\boldsymbol{t}^{\mathrm{T}}\boldsymbol{\alpha}\boldsymbol{t}}\right)^{1/2}\boldsymbol{t}. \end{aligned} \tag{5.9.17}$$

Note that

$$\boldsymbol{n}_+^{\mathrm{T}}\boldsymbol{\alpha}\boldsymbol{n}_+ = \boldsymbol{n}_-^{\mathrm{T}}\boldsymbol{\alpha}\boldsymbol{n}_- = \lambda - \nu = \boldsymbol{n}_+^{*\mathrm{T}}\boldsymbol{\beta}^{-1}\boldsymbol{n}_+^* = \boldsymbol{n}_-^{*\mathrm{T}}\boldsymbol{\beta}^{-1}\boldsymbol{n}_-^*. \tag{5.9.18}$$

5.9.2 *Special case. Two eigenvalues equal*

Suppose now that two eigenvalues are equal, say $\lambda = \mu \neq \nu > 0$. Then, having found $\boldsymbol{r}$ and $\boldsymbol{t}$ satisfying $\boldsymbol{\alpha r} = \lambda\boldsymbol{\beta r}$, $\boldsymbol{\alpha t} = \nu\boldsymbol{\beta t}$, the vector $\boldsymbol{s}$ is chosen so that $\boldsymbol{s}^{\mathrm{T}}\boldsymbol{\alpha r} = 0$ (and therefore also $\boldsymbol{s}^{\mathrm{T}}\boldsymbol{\beta r} = 0$), and $\boldsymbol{s}^{\mathrm{T}}\boldsymbol{\alpha t} = 0$ (and therefore also $\boldsymbol{s}^{\mathrm{T}}\boldsymbol{\beta t} = 0$). Thus, for some scalar ε,

$$\begin{aligned} \boldsymbol{s} &= \varepsilon(\boldsymbol{\beta r} \times \boldsymbol{\beta t}) = \varepsilon\det(\boldsymbol{\beta})\boldsymbol{\beta}^{-1}(\boldsymbol{r} \times \boldsymbol{t}) \\ &= \varepsilon(\boldsymbol{\alpha r} \times \boldsymbol{\alpha t})(\lambda\nu)^{-1} = \varepsilon\det(\boldsymbol{\alpha})\boldsymbol{\alpha}^{-1}(\boldsymbol{r} \times \boldsymbol{t})(\lambda\nu)^{-1}, \end{aligned} \tag{5.9.19}$$

on using (5.1.3), and so

$$(\lambda\nu)\boldsymbol{\alpha s} = \frac{\det(\boldsymbol{\alpha})}{\det(\boldsymbol{\beta})}\boldsymbol{\beta s} = \lambda^2\nu\boldsymbol{\beta s}, \tag{5.9.20}$$

or

$$\boldsymbol{\alpha s} = \lambda\boldsymbol{\beta s}. \tag{5.9.21}$$

Then, if $\boldsymbol{r}_*, \boldsymbol{s}_*, \boldsymbol{t}_*$ are as given in (5.9.6), they form a set reciprocal to $\boldsymbol{r}, \boldsymbol{s}, \boldsymbol{t}$, and

$$\begin{aligned} \boldsymbol{\beta}^{-1}\boldsymbol{\alpha} &= \lambda(\boldsymbol{r} \otimes \boldsymbol{r}_* + \boldsymbol{s} \otimes \boldsymbol{s}_*) + \nu\boldsymbol{t} \otimes \boldsymbol{t}_* \\ &= \lambda\mathbf{1} - (\lambda - \nu)\boldsymbol{t} \otimes \boldsymbol{t}_*. \end{aligned} \tag{5.9.22}$$

Thus we have the identities,

$$\begin{aligned} \boldsymbol{\alpha} &= \lambda\boldsymbol{\beta} - (\lambda - \nu)(\boldsymbol{t}^{\mathrm{T}}\boldsymbol{\beta t})\boldsymbol{t}_* \otimes \boldsymbol{t}_*, \\ \boldsymbol{\beta}^{-1} &= \lambda\boldsymbol{\alpha}^{-1} - (\lambda - \nu)(\boldsymbol{t}_*^{\mathrm{T}}\boldsymbol{\alpha}^{-1}\boldsymbol{t}_*)\boldsymbol{t} \otimes \boldsymbol{t}. \end{aligned} \tag{5.9.23}$$

Similarly, if $\lambda \neq \mu = \nu > 0$, then

$$\begin{aligned} \boldsymbol{\alpha} &= \nu\boldsymbol{\beta} + (\lambda - \nu)(\boldsymbol{r}^{\mathrm{T}}\boldsymbol{\beta r})\boldsymbol{r}_* \otimes \boldsymbol{r}_*, \\ \boldsymbol{\beta}^{-1} &= \nu\boldsymbol{\alpha}^{-1} + (\lambda - \nu)(\boldsymbol{r}_*^{\mathrm{T}}\boldsymbol{\alpha}^{-1}\boldsymbol{r}_*)\boldsymbol{r} \otimes \boldsymbol{r}. \end{aligned} \tag{5.9.24}$$

5.10 Two concentric ellipsoids. Similar and similarly situated central sections

Here we consider the pair of concentric ellipsoids

$$\boldsymbol{x}^{\mathrm{T}}\boldsymbol{\alpha}\boldsymbol{x} = 1, \tag{5.10.1a}$$

$$\boldsymbol{x}^{\mathrm{T}}\boldsymbol{\beta}\boldsymbol{x} = 1, \tag{5.10.1b}$$

where $\boldsymbol{\alpha}$ and $\boldsymbol{\beta}$ are real, positive definite, symmetric matrices.

First we assume that the eigenvalues of $\boldsymbol{\alpha}$ with respect to $\boldsymbol{\beta}$, the roots for γ of $\det(\boldsymbol{\alpha}-\gamma\boldsymbol{\beta})=0$, are all different: $\lambda>\mu>\nu>0$. Then, using (5.9.14)

$$\boldsymbol{x}^{\mathrm{T}}\boldsymbol{\alpha}\boldsymbol{x} = \mu\boldsymbol{x}^{\mathrm{T}}\boldsymbol{\beta}\boldsymbol{x} + (\boldsymbol{n}_+^*\cdot\boldsymbol{x})(\boldsymbol{n}_-^*\cdot\boldsymbol{x}). \tag{5.10.2}$$

Thus, on the plane $\boldsymbol{n}_+^*\cdot\boldsymbol{x}=0$, we have

$$\boldsymbol{x}^{\mathrm{T}}\boldsymbol{\alpha}\boldsymbol{x} = \mu\boldsymbol{x}^{\mathrm{T}}\boldsymbol{\beta}\boldsymbol{x}. \tag{5.10.3}$$

Since this is valid for all $\boldsymbol{x}$ satisfying $\boldsymbol{n}_+^*\cdot\boldsymbol{x}=0$ it follows, on comparing coefficients, that the central section of the ellipsoid (5.10.1a) by the plane $\boldsymbol{n}_+^*\cdot\boldsymbol{x}=0$ is an ellipse, similar, and similarly situated to the central section of the ellipsoid (5.10.1b) by the plane $\boldsymbol{n}_+^*\cdot\boldsymbol{x}=0$, the similarity ratio being μ. In the same way the central sections of the two ellipsoids by the plane $\boldsymbol{n}_-^*\cdot\boldsymbol{x}=0$ are also two similar and similarly situated ellipses.

There are only two such pairs of similar and similarly situated ellipses, because we have already proved in the case when one of the ellipsoids is a sphere that there are only two central circular sections of the ellipsoid.

Next suppose $\lambda=\mu\neq\nu>0$. Then, using (5.9.23)

$$\boldsymbol{x}^{\mathrm{T}}\boldsymbol{\alpha}\boldsymbol{x} = \lambda\boldsymbol{x}^{\mathrm{T}}\boldsymbol{\beta}\boldsymbol{x} - (\lambda-\nu)(\boldsymbol{t}^{\mathrm{T}}\boldsymbol{\beta}\boldsymbol{t})(\boldsymbol{t}_*\cdot\boldsymbol{x})^2 \tag{5.10.4}$$

and hence on the central plane $\boldsymbol{t}_*\cdot\boldsymbol{x}=0$, the central sections of the two ellipsoids are similar and similarly situated. There is only one such pair because in the case when one ellipsoid is a spheroid and the other a sphere, the ellipsoid has only one central circular section.

If $\lambda\neq\mu=\nu>0$, then, also, there is only one pair of similar and similarly situated central elliptical sections. They lie on the plane $\boldsymbol{r}_*\cdot\boldsymbol{x}=0$.

5.10.1 Bivector approach

There is a more direct approach using bivectors. Suppose the plane of the ellipse of the bivector $\boldsymbol{A}$ intersects the ellipsoids (5.10.1) in a pair of similar and similarly situated ellipses, which are also similar and similarly situated to the ellipse of $\boldsymbol{A}$. Then

$$\boldsymbol{A}^{\mathrm{T}}\boldsymbol{\alpha}\boldsymbol{A} = 0, \quad \text{and} \quad \boldsymbol{A}^{\mathrm{T}}\boldsymbol{\beta}\boldsymbol{A} = 0. \tag{5.10.5}$$

In the case of three different eigenvalues, using the identity (5.9.14), these may be written

$$(\boldsymbol{n}_+^* \cdot \boldsymbol{A})(\boldsymbol{n}_-^* \cdot \boldsymbol{A}) = 0, \quad \boldsymbol{A}^{\mathrm{T}}\boldsymbol{\beta}\boldsymbol{A} = 0, \tag{5.10.6}$$

or equivalently,

$$(\boldsymbol{n}_+^* \cdot \boldsymbol{A})(\boldsymbol{n}_-^* \cdot \boldsymbol{A}) = 0, \quad \boldsymbol{A}^{+}\boldsymbol{\alpha}\boldsymbol{A} = 0. \tag{5.10.7}$$

Thus $\boldsymbol{n}_+^*$ and $\boldsymbol{n}_-^*$ are the two possible normals to the plane of $\boldsymbol{A}$.

The case of two equal eigenvalues may be dealt with similarly.

6

Homogeneous and inhomogeneous plane waves

In this chapter, we introduce homogeneous and inhomogeneous plane waves. We show how such wave solutions may be obtained for some simple model linear partial differential equations with constant coefficients.

The propagation of homogeneous waves is described in terms of a single direction: the direction of propagation. The field is homogeneous in any plane orthogonal to this direction. However, for inhomogeneous waves, two different directions are of importance: the propagation direction which is normal to the planes of constant phase, and the attenuation direction which is normal to the planes of constant amplitude.

For inhomogeneous waves, we show how the results of Chapter 2 may be used to introduce a systematic method (Hayes, 1984) for deriving all possible inhomogeneous plane wave solutions. We call it the 'directional ellipse method', or 'DE method'.

6.1 Time harmonic homogeneous plane waves. No attenuation

6.1.1 Scalar equation

We consider the scalar wave equation

$$\nabla^2 u = \frac{1}{c^2}\frac{\partial^2 u}{\partial t^2}, \tag{6.1.1}$$

for the real field $u = u(\boldsymbol{x}, t)$ where c is a constant. Unattenuated time harmonic homogeneous plane waves are solutions of the form

$$u = \{A \exp \mathrm{i}(\boldsymbol{k}\cdot\boldsymbol{x} - \omega t)\}^{+}. \tag{6.1.2}$$

Here A is called the 'complex amplitude' and is a constant. It may be written $A = ae^{i\varphi}$, where a is called the 'real amplitude' and φ the 'phase shift'. The vector $\boldsymbol{k}$ is called the 'wave vector' and is a constant real vector, and ω is called the 'angular frequency' and is also assumed to be real. Then (6.1.2) describes an infinite train of waves propagating without damping in the direction of $\boldsymbol{k}$, with phase speed $v = \omega|\boldsymbol{k}|^{-1}$, period $2\pi\omega^{-1}$ and wave length $2\pi|\boldsymbol{k}|^{-1}$.

Because (6.1.1) is linear with real coefficients, the real solutions of the form (6.1.2) are obtained by seeking for u complex solutions of the form

$$U = A \exp i(\boldsymbol{k}\cdot\boldsymbol{x} - \omega t), \tag{6.1.3}$$

and taking their real parts. Inserting (6.1.3) into (6.1.1) leads to the propagation condition

$$\left(\boldsymbol{k}\cdot\boldsymbol{k} - \frac{\omega^2}{c^2}\right)A = 0. \tag{6.1.4}$$

Hence, in order to obtain nontrivial solutions ($A \neq 0$), the wave vector $\boldsymbol{k}$ and the angular frequency ω must satisfy

$$\boldsymbol{k}\cdot\boldsymbol{k} = \frac{\omega^2}{c^2}, \tag{6.1.5}$$

which is called the 'dispersion relation'. Thus

$$v = \pm c. \tag{6.1.6}$$

If instead of (6.1.1) we consider the 'anisotropic' wave equation

$$\alpha_{ij}\frac{\partial^2 u}{\partial x_i \partial x_j} = \frac{1}{c^2}\frac{\partial^2 u}{\partial t^2}, \tag{6.1.7}$$

where $\boldsymbol{\alpha} = (\alpha_{ij})$ is a real positive definite symmetric constant matrix, then insertion of (6.1.3) into (6.1.7) now gives the dispersion relation

$$\boldsymbol{k}^{\mathrm{T}}\boldsymbol{\alpha}\boldsymbol{k} = \frac{\omega^2}{c^2}. \tag{6.1.8}$$

Thus, writing

$$\boldsymbol{k} = k\mathbf{n}, \quad k^2 = \boldsymbol{k}\cdot\boldsymbol{k}, \quad \mathbf{n}\cdot\mathbf{n} = 1, \tag{6.1.9}$$

we obtain

$$v^2 = c^2(\mathbf{n}^{\mathrm{T}}\boldsymbol{\alpha}\mathbf{n}). \tag{6.1.10}$$

Because $\boldsymbol{\alpha}$ is positive definite, it follows that (6.1.10) yields two real and opposite values of v for any given unit vector $\mathbf{n}$. Obviously, the unit vector $\mathbf{n}$ in the propagation direction may be arbitrarily prescribed and the corresponding phase velocities are then determined through (6.1.10).

Of course if $\boldsymbol{\alpha}$ is not positive definite, then (6.1.7) is no longer hyperbolic and for some choices of $\mathbf{n}$ there may be zero or purely imaginary solutions of (6.1.8) for v. Suppose for example that $\boldsymbol{\alpha} = \mathrm{diag}(1, 4, -9)$. Then, for $\mathbf{n} = (n_1, n_2, 0)$, v^2 is real and positive and the corresponding solutions describe propagating waves, whereas if $\mathbf{n} = (0, 0, 1)$, then v^2 is negative (v purely imaginary) and the corresponding solutions $u = \exp(\pm 3kct)\{A \exp ikx_3\}^+$ do not describe propagating waves.

Exercises 6.1

1. Derive the dispersion relations for the equations
   (a) $\partial_t^2 u - \alpha^2 \nabla^2 u + \beta^2 u = 0$;
   (b) $\partial_t^2 u - \alpha^2 \nabla^2 u = \beta^2 \nabla^2 \partial_t^2 u$;
   (c) $\nabla^2 U = -\mathrm{i}\alpha \partial_t U$, where U is complex.
2. For the same equations, obtain the squared phase speed v^2 as a function of the angular frequency ω.

6.1.2 *Vector equation*

We now consider the vector wave equation

$$\nabla^2 \boldsymbol{u} = \frac{1}{c^2} \frac{\partial^2 \boldsymbol{u}}{\partial t^2}, \tag{6.1.11}$$

for the real vector field $\boldsymbol{u} = \boldsymbol{u}(\boldsymbol{x}, t)$, and we seek solutions of the form

$$\boldsymbol{u} = \{\boldsymbol{A} \exp \mathrm{i}(\boldsymbol{k} \cdot \boldsymbol{x} - \omega t)\}^+, \tag{6.1.12}$$

where $\boldsymbol{A} = \boldsymbol{A}^+ + \mathrm{i}\boldsymbol{A}^-$ is now a constant bivector, called the 'amplitude bivector'. The procedure of section 6.1 now leads to the propagation condition

$$\left(\boldsymbol{k} \cdot \boldsymbol{k} - \frac{\omega^2}{c^2}\right) \boldsymbol{A} = 0, \tag{6.1.13}$$

and hence to the dispersion relation (6.1.5) as in the case of the scalar wave equation. When $\boldsymbol{k}$ and ω satisfy this relation, the amplitude bivector $\boldsymbol{A}$ may be arbitrary.

Recalling the results of section 2.2, we note from (6.1.12) that when t is varied at fixed $\boldsymbol{x}$, the extremity of the vector $\boldsymbol{u}(\boldsymbol{x}, t)$ describes the ellipse associated with the bivector $\boldsymbol{A}$ (centred at $\boldsymbol{x}$). This ellipse is called the 'polarization ellipse' of the plane wave (6.1.12).

When $\boldsymbol{A} \times \bar{\boldsymbol{A}} = 0$, the bivector $\boldsymbol{A}$ has a real direction and the ellipse degenerates into a segment of a straight line (section 2.2, remark (c)). Then the wave is said to be 'linearly polarized', and the direction of polarization is the real direction of the bivector $\boldsymbol{A}$. When $\boldsymbol{A} \cdot \boldsymbol{A} = 0$, the bivector $\boldsymbol{A}$ is isotropic and the ellipse is a circle (section 2.2, remark (b)). Then the wave is said to be 'circularly polarized'. In all other cases, the wave is said to be 'elliptically polarized'.

Let us now consider the general linear anisotropic vector wave equation

$$\alpha_{ijpq} \frac{\partial^2 u_j}{\partial x_p \partial x_q} = \frac{1}{c^2} \frac{\partial^2 u_i}{\partial t^2}, \tag{6.1.14}$$

where α_{ijpq} is a fourth order constant tensor symmetric in (pq). Seeking solutions of the form (6.1.12), we now find the propagation condition

$$\left(\alpha_{ijpq} k_p k_q - \frac{\omega^2}{c^2} \delta_{ij} \right) A_j = 0, \tag{6.1.15}$$

or, equivalently on using (6.1.9)

$$\{\boldsymbol{Q}(\mathbf{n}) - v^2 \mathbf{1}\} \boldsymbol{A} = \mathbf{0}, \tag{6.1.16}$$

where $\boldsymbol{Q}(\mathbf{n})$ is the real matrix defined by

$$Q_{ij}(\mathbf{n}) = c^2 \alpha_{ijpq} n_p n_q. \tag{6.1.17}$$

Obviously, the unit vector $\mathbf{n}$ may be prescribed arbitrarily and the corresponding values of v^2 and of $\boldsymbol{A}$ are the eigenvalues and the eigenbivectors of the real matrix $\boldsymbol{Q}(\mathbf{n})$. Of course, unattenuated time harmonic plane waves are only obtained for real positive eigenvalues v^2.

If α_{ijpq} is symmetric in (ij), the matrix $\boldsymbol{Q}(\mathbf{n})$ is then real and symmetric for every $\mathbf{n}$, and the eigenvalues are real. If, in addition, $\boldsymbol{Q}(\mathbf{n})$ is positive definite for every $\mathbf{n}$, the three eigenvalues v^2 yield propagating waves. However, if $\boldsymbol{Q}(\mathbf{n})$ has negative eigenvalues, the corresponding solutions do not describe propagating waves.

Exercise 6.2

1. Find the propagation condition for the anisotropic vector wave equation

$$\nabla^2 u_i - \partial_i(\nabla \cdot \boldsymbol{u}) - \beta_{ij}\partial_t^2 u_j = 0,$$

where β_{ij} is a real symmetric positive definite matrix. Show that for any prescribed $\mathbf{n}$, one value of v^2 is zero and that the two others are positive.

6.2 Attenuated and damped homogeneous plane waves

Consider now the scalar equation

$$\nabla^2 u = \frac{1}{c^2}\left(\frac{\partial^2 u}{\partial t^2} + \alpha \frac{\partial u}{\partial t}\right), \tag{6.2.1}$$

where α is a positive constant.

'Attenuated' time harmonic homogeneous plane waves are solutions of the form (6.1.2), where $\boldsymbol{k} = k\mathbf{n} = (k^+ + ik^-)\mathbf{n}$ is a bivector with real direction $\mathbf{n}(\mathbf{n}\cdot\mathbf{n} = 1)$, and ω is a real number. For these, (6.1.2) becomes

$$u = \exp(-k^-\mathbf{n}\cdot\boldsymbol{x})\{A \exp i(k^+\mathbf{n}\cdot\boldsymbol{x} - \omega t)\}^+. \tag{6.2.2}$$

This describes an infinite train of waves propagating in the direction of $\mathbf{n}$ with phase speed $\omega(k^+)^{-1}$ and period $2\pi\omega^{-1}$. The field $u(\boldsymbol{x}, t)$ is no longer periodic in space. The 'attenuation coefficient' is k^-, so that the amplitude at position $\boldsymbol{x} + (2\pi(k)^{-1})\mathbf{n}$ is $e^{-2\pi}$ times its value at position $\boldsymbol{x}$.

'Damped' homogeneous plane waves are solutions of the form (6.1.2), where $\boldsymbol{k} = k\mathbf{n}$, $(\mathbf{n}\cdot\mathbf{n} = 1)$ is a real vector, and $\omega = \omega^+ + i\omega^-$ is a complex number. For these, (6.1.2) becomes

$$u = \exp(\omega^- t)\{A \exp i(k\mathbf{n}\cdot\boldsymbol{x} - \omega^+ t)\}^+. \tag{6.2.3}$$

This describes an infinite train of waves propagating in the direction of $\mathbf{n}$ with phase speed $\omega^+ k^{-1}$ and wavelength $2\pi k^{-1}$. The field $u(\boldsymbol{x}, t)$ is no longer periodic in time. The 'damping factor' is $(-\omega^-)$, so that the amplitude at time $t + 2\pi(-\omega^-)^{-1}$ is $e^{-2\pi}$ times its value at time t.

Inserting (6.1.3) into (6.2.1) yields the dispersion relation

$$c^2 \boldsymbol{k}\cdot\boldsymbol{k} = \omega^2 + i\omega\alpha. \tag{6.2.4}$$

If $\alpha \neq 0$, it is clear that there is no possibility of solutions to (6.2.4) unless $\boldsymbol{k}$ or ω is complex.

Assuming ω real and $\boldsymbol{k} = k\mathbf{n}$, with k complex (attenuated waves), we have

$$k^{+2} - k^{-2} = \frac{\omega^2}{c^2}, \quad k^+k^- = \tfrac{1}{2}\omega\alpha, \tag{6.2.5}$$

and k is given (up to a sign) by the square root of the complex number $(\omega^2 + \mathrm{i}\omega\alpha)c^{-2}$. The complex wave speeds $v = \omega k^{-1}$ are given by

$$v^2 = c^2\left(1 + \mathrm{i}\frac{\alpha}{\omega}\right)^{-1}. \tag{6.2.6}$$

Assuming now $\boldsymbol{k} = k\mathbf{n}$ real, and $\omega = \omega^+ + \mathrm{i}\omega^-$ complex, with $\omega^+ \neq 0$ (damped waves), we have

$$\omega^- = -\tfrac{1}{2}\alpha, \quad \omega^{+2} = c^2k^2 - \tfrac{1}{4}\alpha^2. \tag{6.2.7}$$

We note that in order to obtain solutions with $\omega^+ \neq 0$, we have to require that $k^2 > \frac{1}{4}\alpha^2c^{-2}$. Indeed when $k^2 \leqslant \frac{1}{4}\alpha^2c^{-2}$ the solutions ω of (6.2.4) are purely imaginary ($\boldsymbol{k}$ being real). The complex wave speeds $v = \omega k^{-1}$ are given by

$$v = \pm(c^2 - \alpha^2/4k^2)^{1/2} - \mathrm{i}\alpha/2k. \tag{6.2.8}$$

Remarks

(a) Complex eigenvalues $v^2 = \omega^2k^{-2}$ of the eigenvalue problem (6.1.16) may now be interpreted. Corresponding to these we have attenuated plane wave solutions (ω real, k complex) or damped plane wave solutions (ω complex, k real) of the anisotropic vector wave equation (6.1.14).

(b) Homogeneous plane waves with both attenuation and damping (homogeneous 'complex exponential solutions') may also be considered. For these, $\boldsymbol{k} = (k^+ + \mathrm{i}k^-)\mathbf{n}$ and $\omega = \omega^+ + \mathrm{i}\omega^-$, so that the field $u(\boldsymbol{x}, t)$ is given by

$$u = \exp -(k^-\mathbf{n}\cdot\boldsymbol{x} - \omega^-t)\{A\exp\mathrm{i}(k^+\mathbf{n}\cdot\boldsymbol{x} - \omega^+t)\}^+. \tag{6.2.9}$$

Here the field $u(\boldsymbol{x}, t)$ is neither periodic in time nor in space. By taking $\omega^+ = \pm ck^+$, $\omega^- = \pm ck^-$ it is seen that such solutions exist even for the classical wave equation (6.1.1).

In order to express that the real exponential represents a

decay (and not an amplification) of the amplitude, we require that $\exp -(k^- \mathbf{n}\cdot\boldsymbol{x} - \omega^- t)$ decreases with time at fixed phase (i.e. when $k^+\mathbf{n}\cdot\boldsymbol{x} - \omega^+ t = \text{constant}$). This will be the case if

$$\frac{k^-\omega^+ - \omega^- k^+}{k^+} \geqslant 0. \tag{6.2.10}$$

In the special case when $\omega^- = 0$ (attenuated waves) this reduces to $\omega k^-(k^+)^{-1} \geqslant 0$, and in the special case when $k^- = 0$ (damped waves) to $\omega^- \leqslant 0$. For (6.2.1) it follows from (6.2.5) and (6.2.7) that these conditions are satisfied when $\alpha \geqslant 0$.

Exercises 6.3

1. Derive attenuated time harmonic homogeneous plane wave solutions for the equation ('heat equation')

$$\nabla^2 u = \alpha \partial_t u, \quad (\alpha > 0).$$

Show that damped homogeneous plane waves (with nonzero phase speed) are not possible.

2. Find the dispersion relation for the equation ('telegraph equation')

$$\nabla^2 u = \frac{1}{c^2}(\partial_t^2 u + 2\alpha\partial_t u + \alpha^2 u), \quad (\alpha > 0).$$

Derive attenuated time harmonic homogeneous plane waves and damped homogeneous plane waves for this equation. Check the decay condition.

6.3 Time harmonic inhomogeneous plane waves

6.3.1 Scalar equation

Here we consider solutions of (6.1.1) or (6.1.7) of the form

$$u = \{A \exp \mathrm{i}\omega(\boldsymbol{S}\cdot\boldsymbol{x} - t)\}^+, \tag{6.3.1}$$

where $\boldsymbol{S} = \boldsymbol{S}^+ + \mathrm{i}\boldsymbol{S}^-$ is a bivector, called the 'slowness bivector'. As in (6.1.2), A is a complex constant and ω a real constant. The bivector $\boldsymbol{K} = \omega\boldsymbol{S}$ is called the 'wave bivector'. Writing $A = a\mathrm{e}^{\mathrm{i}\varphi}$, with a and φ real, we note that (6.3.1) may be written in the form

$$u = a\exp(-\omega\boldsymbol{S}^-\cdot\boldsymbol{x})\cos\{\omega(\boldsymbol{S}^+\cdot\boldsymbol{x} - t) + \varphi\}. \tag{6.3.2}$$

Of course, φ may be taken to be zero by changing the time origin. The planes

$$\boldsymbol{S}^+ \cdot \boldsymbol{x} = \text{constant} \tag{6.3.3}$$

are called the 'planes of constant phase', and the planes

$$\boldsymbol{S}^- \cdot \boldsymbol{x} = \text{constant} \tag{6.3.4}$$

are called the 'planes of constant amplitude'.

A solution of the form (6.3.1) is called an 'inhomogeneous plane wave' or an 'evanescent wave' when $\boldsymbol{S}^-$ is not parallel to $\boldsymbol{S}^+$, that is when the planes of constant amplitude are different from the planes of equal phase (when $\boldsymbol{S}^-$ is parallel to $\boldsymbol{S}^+$, we have $\boldsymbol{K} = \omega \boldsymbol{S} = (k^+ + ik^-)\mathbf{n}$, and we retrieve attenuated homogeneous plane waves). It describes an infinite train of waves propagating along $\boldsymbol{S}^+$ with phase speed $|\boldsymbol{S}^+|^{-1}$, and attenuated in the direction of $\boldsymbol{S}^-$. We call $|\boldsymbol{S}^-|$ the 'attenuation' factor. It is because the field $u(\boldsymbol{x}, t)$ is not homogeneous in a plane of equal phase (decay or amplification with distance along the projection of $\boldsymbol{S}^-$) that the wave is called 'inhomogeneous'.

Inserting $A \exp i\omega(\boldsymbol{S}\cdot\boldsymbol{x} - t)$ into the isotropic wave equation (6.1.1) leads to

$$\boldsymbol{S}\cdot\boldsymbol{S} = \frac{1}{c^2}. \tag{6.3.5}$$

Hence, taking the real and imaginary parts of this, we obtain

$$\boldsymbol{S}^+\cdot\boldsymbol{S}^+ - \boldsymbol{S}^-\cdot\boldsymbol{S}^- = \frac{1}{c^2}, \quad \boldsymbol{S}^+\cdot\boldsymbol{S}^- = 0. \tag{6.3.6}$$

Thus, in this case, the planes of constant phase have to be orthogonal to the planes of constant amplitude.

Inserting now $A \exp i\omega(\boldsymbol{S}\cdot\boldsymbol{x} - t)$ into the anisotropic wave equation (6.1.7) leads to

$$\boldsymbol{S}^{\mathrm{T}}\boldsymbol{\alpha}\boldsymbol{S} = \frac{1}{c^2}, \tag{6.3.7}$$

and thus

$$\boldsymbol{S}^{+\mathrm{T}}\boldsymbol{\alpha}\boldsymbol{S}^+ - \boldsymbol{S}^{-\mathrm{T}}\boldsymbol{\alpha}\boldsymbol{S}^- = \frac{1}{c^2}, \tag{6.3.8a}$$

$$\boldsymbol{S}^{+\mathrm{T}}\boldsymbol{\alpha}\boldsymbol{S}^- = 0. \tag{6.3.8b}$$

From (6.3.8b), it follows that in this case the normals to the planes of constant phase and constant amplitude have to be along conjugate directions with respect to the ellipsoid $\alpha_{ij}x_ix_j = 1$.

For homogeneous waves the direction $\mathbf{n}$ of the wave vector may be prescribed arbitrarily, and, corresponding to each prescribed $\mathbf{n}$, the wave speed v is determined. For time-harmonic inhomogeneous waves, it is now clear that the directions of $\boldsymbol{K}^+ = \omega \boldsymbol{S}^+$ and $\boldsymbol{K}^+ = \omega \boldsymbol{S}^-$ cannot in general be both arbitrarily prescribed. There is, however, a systematic procedure due to Hayes (1984) for analysing conditions of the type (6.3.5) or (6.3.7), and which generalizes the procedure used for homogeneous waves. We call it the 'directional ellipse method', or 'DE method', and describe it here.

Let us consider the ellipse associated with the bivector $\boldsymbol{S}$. Let us now introduce the ellipse that is similar and similarly situated to the ellipse of $\boldsymbol{S}$ and whose minor semi-axis is of unit length (Figure 6.1). Let $\hat{\mathbf{m}}$, $\hat{\mathbf{n}}$ be real unit vectors along the major and minor axes of this ellipse such that the orientation from $\hat{\mathbf{m}}$ to $\hat{\mathbf{n}}$ is the same as the orientation from $\boldsymbol{S}^+$ to $\boldsymbol{S}^-$, and let $m \geqslant 1$ be the major semi-axis of this ellipse:

$$\hat{\mathbf{m}} \cdot \hat{\mathbf{m}} = \hat{\mathbf{n}} \cdot \hat{\mathbf{n}} = 1, \quad \hat{\mathbf{m}} \cdot \hat{\mathbf{n}} = 0, \quad m \geqslant 1. \tag{6.3.9}$$

This ellipse is associated with the bivector $\boldsymbol{C}$ defined by

$$\boldsymbol{C} = m\hat{\mathbf{m}} + \mathrm{i}\hat{\mathbf{n}}. \tag{6.3.10}$$

Hence the slowness bivector $\boldsymbol{S}$ may be written

$$\boldsymbol{S} = N\boldsymbol{C}, \quad N = T e^{\mathrm{i}\phi}, \tag{6.3.11}$$

where N is a complex number (of modulus T and argument ϕ).

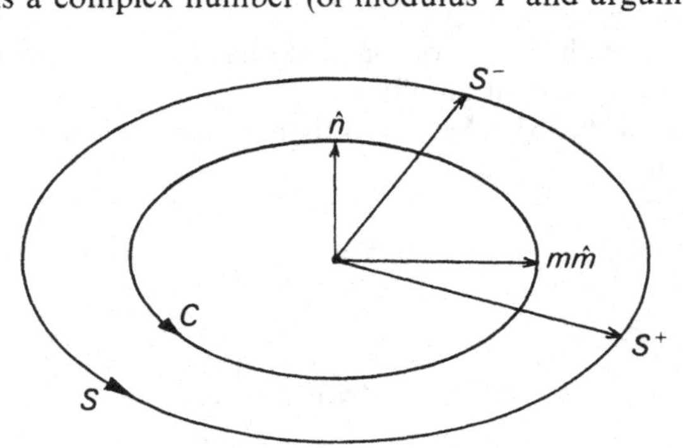

Figure 6.1 *The bivector* $\boldsymbol{C}$ *and the DE method.*

The DE method is as follows: prescribe $\boldsymbol{C}=m\hat{\mathbf{m}}+i\hat{\mathbf{n}}$ arbitrarily (thus prescribe arbitrarily $\hat{\mathbf{m}}$, $\hat{\mathbf{n}}$ and m satisfying (6.3.9)), and corresponding to each prescribed $\boldsymbol{C}$, determine the complex number N. Thus, for inhomogeneous waves, $\boldsymbol{C}$ (ellipse with minor semi-axis of unit length) plays the role of the propagation direction $\mathbf{n}$, and the complex number N^{-1} the role of the wave speed v. When N is determined, $\boldsymbol{S}^+$ and $\boldsymbol{S}^-$ are known as a pair of conjugate radii of an ellipse similar and similarly situated to the ellipse of $\boldsymbol{C}$ (Theorem 2.1). Thus, knowledge of N enables us to determine the planes of constant phase and the planes of constant amplitude, the phase speed and the attenuation factor.

From (6.3.10) and (6.3.11), we have

$$\begin{aligned}\boldsymbol{S}^+ &= T(m\hat{\mathbf{m}}\cos\phi-\hat{\mathbf{n}}\sin\phi), & |\boldsymbol{S}^+| &= T(m^2\cos^2\phi+\sin^2\phi)^{1/2},\\ \boldsymbol{S}^- &= T(m\hat{\mathbf{m}}\sin\phi+\hat{\mathbf{n}}\cos\phi), & |\boldsymbol{S}^-| &= T(m^2\sin^2\phi+\cos^2\phi)^{1/2}.\end{aligned} \tag{6.3.12}$$

Thus the angle θ (say), between the planes of constant phase and the planes of constant amplitude, is given in terms of m and ϕ by

$$\tan\theta=\frac{m}{(m^2-1)\cos\phi\sin\phi}=\frac{\tan 2\beta}{\sin 2\phi}, \tag{6.3.13}$$

where

$$\tan\beta=\frac{1}{m},\quad 0\leqslant\beta\leqslant\frac{\pi}{4}. \tag{6.3.14}$$

It is clear that these planes are orthogonal if N is real or purely imaginary, so that either $\sin\phi=0$ or $\cos\phi=0$, and also when $m=1$ because in this case the ellipse of $\boldsymbol{C}$ is a circle and then all pairs of conjugate radii are orthogonal. We also note from (6.3.13) that 2ϕ and 2β may be interpreted as the sides of a right-angled spherical triangle, with angle θ opposite to the side 2β (Appendix, formula (A.15) and Figure A.2, with $b=2\phi$, $c=2\beta$, $\hat{C}=\theta$).

Let us now apply the DE method to the relations (6.3.5) and (6.3.7), obtained respectively from the isotropic and anisotropic scalar wave equation.

(a) Isotropic scalar wave equation Introducing (6.3.11) into (6.3.5) yields

$$N^{-2}=c^2\boldsymbol{C}\cdot\boldsymbol{C}=c^2(m^2-1), \tag{6.3.15}$$

and hence the modulus T, and the argument ϕ, of N are given by

$$T = c^{-1}(m^2 - 1)^{-1/2}, \quad \phi = 0, \pi. \tag{6.3.16}$$

Of course here we have to assume $m \neq 1$ in order to obtain a finite N. The slowness bivector is then given by

$$\boldsymbol{S} = \pm \frac{m\hat{\mathbf{m}} + \mathrm{i}\hat{\mathbf{n}}}{c(m^2 - 1)^{1/2}}. \tag{6.3.17}$$

The orthogonal unit vectors $\hat{\mathbf{m}}, \hat{\mathbf{n}}$ and the real number $m > 1$ being arbitrarily prescribed, (6.3.17) gives all the bivectors $\boldsymbol{S}$ satisfying (6.3.5), and hence all the vectors $\boldsymbol{S}^+$ and $\boldsymbol{S}^-$ satisfing (6.3.6). Without loss of generality, the '+' sign may be chosen in (6.3.17). Clearly here the planes of constant phase are orthogonal to the planes of constant amplitude. The general time harmonic inhomogeneous plane wave solution of the wave equation (6.1.1) may thus be written

$$u = a \exp\left\{ -\omega \frac{\hat{\mathbf{n}} \cdot \boldsymbol{x}}{c(m^2 - 1)^{1/2}} \right\} \cos\left\{ \omega \left(\frac{m\hat{\mathbf{m}} \cdot \boldsymbol{x}}{c(m^2 - 1)^{1/2}} - t \right) \right\}. \tag{6.3.18}$$

This describes a train of waves propagating along $\hat{\mathbf{m}}$ with phase speed $c(m^2 - 1)^{1/2} m^{-1}$, whose amplitude decays along $\hat{\mathbf{n}}$, the attenuation factor being $c^{-1}(m^2 - 1)^{-1/2}$. The orthogonal unit vectors $\hat{\mathbf{m}}, \hat{\mathbf{n}}$ and the real numbers a and $m > 1$ are arbitrary.

Also, we note that if $m = 1$, the bivector $\boldsymbol{C}$ defined by (6.3.10) is isotropic, and thus, for any finite N, $\boldsymbol{S} = N\boldsymbol{C}$ is also isotropic: $\boldsymbol{S} \cdot \boldsymbol{S} = 0$. Then, clearly the relation (6.3.5) is not satisfied and there is no propagating solution in this case. What we do have, however, are static exponential solutions

$$u = \{A \exp \mathrm{i}\omega N \boldsymbol{C} \cdot \boldsymbol{x}\}^+, \tag{6.3.19}$$

where ωN is now an arbitrary complex number. These are solutions of the Laplace equation $\nabla^2 u = 0$, and, because they are time-independent, they are also solutions of the isotropic wave equation (6.1.1).

(b) Anisotropic scalar wave equation Introducing (6.3.11) into (6.3.7) yields

$$N^{-2} = c^2 \boldsymbol{C}^{\mathrm{T}} \boldsymbol{\alpha} \boldsymbol{C}. \tag{6.3.20}$$

Of course here we have to assume $\boldsymbol{C}^{\mathrm{T}} \boldsymbol{\alpha} \boldsymbol{C} \neq 0$ in order to obtain a finite N. Thus

$$T^{-2}\cos 2\phi = c^2(m^2\hat{\mathbf{m}}^{\mathrm{T}}\boldsymbol{\alpha}\hat{\mathbf{m}} - \hat{\mathbf{n}}^{\mathrm{T}}\boldsymbol{\alpha}\hat{\mathbf{n}}),$$
$$T^{-2}\sin 2\phi = -2c^2 m\hat{\mathbf{m}}^{\mathrm{T}}\boldsymbol{\alpha}\hat{\mathbf{n}}. \tag{6.3.21}$$

Hence, for arbitrarily prescribed $\boldsymbol{C} = m\hat{\mathbf{m}} + \mathrm{i}\hat{\mathbf{n}}$, T and ϕ are determined. Indeed

$$T^{-2} = c^2\{(m^2\hat{\mathbf{m}}^{\mathrm{T}}\boldsymbol{\alpha}\hat{\mathbf{m}} - \hat{\mathbf{n}}^{\mathrm{T}}\boldsymbol{\alpha}\hat{\mathbf{n}})^2 + 4m^2(\hat{\mathbf{m}}^{\mathrm{T}}\boldsymbol{\alpha}\hat{\mathbf{n}})^2\}^{1/2},$$

$$\tan 2\phi = -\frac{2m\hat{\mathbf{m}}^{\mathrm{T}}\boldsymbol{\alpha}\hat{\mathbf{n}}}{(m^2\hat{\mathbf{m}}^{\mathrm{T}}\boldsymbol{\alpha}\hat{\mathbf{m}} - \hat{\mathbf{n}}^{\mathrm{T}}\boldsymbol{\alpha}\hat{\mathbf{n}})}. \tag{6.3.22}$$

We note that if $\boldsymbol{C}^{\mathrm{T}}\boldsymbol{\alpha}\boldsymbol{C} = 0$, then, for any finite N, $\boldsymbol{S} = N\boldsymbol{C}$ satisfies $\boldsymbol{S}^{\mathrm{T}}\boldsymbol{\alpha}\boldsymbol{S} = 0$. Then clearly the relation (6.3.7) is not satisfied and there is no propagating solution in this case. Rather, (6.3.19) is now a solution of $\alpha_{ij}\partial^2 u/(\partial x_i \partial x_j) = 0$. We recall (see section 5.5) that $\boldsymbol{C}^{\mathrm{T}}\boldsymbol{\alpha}\boldsymbol{C} = 0$ means that the ellipse of the bivector $\boldsymbol{C}$ has been chosen to be similar and similarly situated to an elliptical section of the ellipsoid $\boldsymbol{x}^{\mathrm{T}}\boldsymbol{\alpha}\boldsymbol{x} = 1$. There is an infinity of such choices. For all other choices of $\boldsymbol{C}$, one can always determine propagating solutions.

Exercises 6.4

1. Consider the inhomogeneous plane wave solutions of the anisotropic wave equation (6.1.7). How must the ellipse of $\boldsymbol{C}$ be chosen in order to obtain inhomogeneous waves with the planes of constant phase orthogonal to the planes of constant amplitude?
2. Apply the DE method to equation $\partial_t^2 u - \alpha^2\nabla^2 u + \beta^2 u = 0$. Find the general form of the slowness bivector $\boldsymbol{S}$.
3. Apply the DE method to equation $\nabla^2 u = \alpha\partial_t u$ $(\alpha > 0)$. Find the general form of the slowness bivector $\boldsymbol{S}$.

6.3.2 *Vector equation*

For the vector wave equation (6.1.11) or (6.1.14), time harmonic inhomogeneous plane waves are solutions of the form

$$\boldsymbol{u} = \{\boldsymbol{A}\exp \mathrm{i}\omega(\boldsymbol{S}\cdot\boldsymbol{x} - t)\}^+ = \exp(-\omega\boldsymbol{S}^-\cdot\boldsymbol{x})\{\boldsymbol{A}\exp \mathrm{i}\omega(\boldsymbol{S}^+\cdot\boldsymbol{x} - t)\}^+. \tag{6.3.23}$$

Here two bivectors enter the expression for $\boldsymbol{u}$: the slowness bivector $\boldsymbol{S} = \boldsymbol{S}^+ + \mathrm{i}\boldsymbol{S}^-$ and the amplitude bivector $\boldsymbol{A} = \boldsymbol{A}^+ + \mathrm{i}\boldsymbol{A}^-$.

Inserting $\boldsymbol{A} \exp i\omega(\boldsymbol{S}\cdot\boldsymbol{x} - t)$ into the isotropic vector wave equation (6.1.11) leads to the propagation condition

$$\left(\boldsymbol{S}\cdot\boldsymbol{S} - \frac{1}{c^2}\right)\boldsymbol{A} = 0, \tag{6.3.24}$$

and hence to the relation (6.3.5) as in the case of the scalar equation. Thus applying the DE method, the general form of the slowness bivector is given by (6.3.17). The corresponding amplitude bivector $\boldsymbol{A}$ may be arbitrary.

We note from (6.3.1) that when t is varied at fixed $\boldsymbol{x}$, the extremity of the vector $\boldsymbol{u}(\boldsymbol{x}, t)$ describes an ellipse which is similar and similarly situated to the ellipse associated with the bivector $\boldsymbol{A}$ (centred at $\boldsymbol{x}$). The magnitude of this ellipse is the same for all positions $\boldsymbol{x}$ in a plane of constant amplitude, but varies for one plane of constant amplitude to another. For simplicity, the ellipse of $\boldsymbol{A}$ will still be called the 'polarization ellipse' of the inhomogeneous wave (6.3.23).

For all positions $\boldsymbol{x}$ in a plane of constant phase, the field $\boldsymbol{u}(\boldsymbol{x}, t)$ consists, at any time t, of parallel vectors of different magnitudes. With time, these vectors describe similar and similarly situated ellipses (centred at the different points $\boldsymbol{x}$ of the plane).

For all positions $\boldsymbol{x}$ in a plane of constant amplitude, the field $\boldsymbol{u}(\boldsymbol{x}, t)$ consists, at any time t, of vectors in different directions. With time, these vectors all describe the same ellipse up to a translation (centred at the different points $\boldsymbol{x}$ of the plane).

When $\boldsymbol{A}$ is chosen to be parallel to $\boldsymbol{S}$, so that $\boldsymbol{S} \times \boldsymbol{A} = 0$, the wave is said to be 'longitudinal'. This means (section 2.3) that the polarization ellipse is similar and similarly situated to the ellipse of $\boldsymbol{S}$ (or, equivalently, of $\boldsymbol{C}$). When $\boldsymbol{A}$ is chosen to be orthogonal to $\boldsymbol{S}$, so that $\boldsymbol{S}\cdot\boldsymbol{A} = 0$, the wave is said to be 'transverse'. This means that the projection of the polarization ellipse onto the plane of $\boldsymbol{S}$ is similar and similarly situated to the ellipse of $\boldsymbol{S}$ (or, equivalently, of $\boldsymbol{C}$) when rotated through a quadrant.

Let us now insert $\boldsymbol{A} \exp i\omega(\boldsymbol{S}\cdot\boldsymbol{x} - t)$ into the anisotropic vector wave equation (6.1.14). We obtain the propagation condition

$$\{\boldsymbol{Q}(\boldsymbol{S}) - \boldsymbol{1}\}\boldsymbol{A} = \boldsymbol{0}, \tag{6.3.25}$$

where $\boldsymbol{Q}(\boldsymbol{S})$ is the complex matrix defined by

$$Q_{ij}(\boldsymbol{S}) = c^2 \alpha_{ijpq} S_p S_q. \tag{6.3.26}$$

Applying the DE method, we write $\boldsymbol{S} = N\boldsymbol{C}, \boldsymbol{C}$ being of the form

(6.3.10), and (6.3.25) becomes

$$\{\boldsymbol{Q}(\boldsymbol{C}) - N^{-2}\mathbf{1}\}\boldsymbol{A} = \mathbf{0}. \tag{6.3.27}$$

Hence, for each prescribed bivector $\boldsymbol{C} = m\hat{\mathbf{m}} + i\hat{\mathbf{n}}$, the corresponding values of N^{-2} and of $\boldsymbol{A}$ are the eigenvalues and the eigenbivectors of the complex matrix $\boldsymbol{Q}(\boldsymbol{C})$. To obtain the special case of homogeneous waves, we take $\boldsymbol{C} = \mathbf{n}$ (real unit vector) and we retrieve the eigenvalue problem (6.1.16).

Assuming α_{ijpq} symmetric in (ij), the matrix $\boldsymbol{Q}(\boldsymbol{C})$ is complex symmetric. The properties of the eigenvalues and eigenbivectors of such matrices have been studied in Chapter 3. In particular, we note that a necessary and sufficient condition for obtaining a circularly polarized wave ($\boldsymbol{A}\cdot\boldsymbol{A} = 0$) is that the equation

$$\det\{\boldsymbol{Q}(\boldsymbol{C}) - N^{-2}\mathbf{1}\} = 0 \tag{6.3.28}$$

have a double or triple root for N^{-2}.

We also note that if, for a certain prescribed $\boldsymbol{C}$, $\boldsymbol{Q}(\boldsymbol{C})$ has a zero eigenvalue, then, for any finite N, we have $\{\boldsymbol{Q}(\boldsymbol{S})\}\boldsymbol{A} = \mathbf{0}$, where $\boldsymbol{A}$ is an eigenbivector corresponding to the zero eigenvalue. Then, clearly, the propagation condition (6.3.25) is not satisfied, and there is no propagating solution in this case. What we do have, however, are static exponential solutions

$$\boldsymbol{u} = \{\boldsymbol{A} \exp i\omega N\boldsymbol{C}\cdot\boldsymbol{x}\}^{+}, \tag{6.3.29}$$

where ωN is now an arbitrary complex number. These are solutions of

$$\alpha_{ijpq}\frac{\partial^2 u_j}{\partial x_p \partial x_q} = 0.$$

Exercises 6.5

1. Find the propagation condition for the equation $\partial_t\boldsymbol{u} + \nabla\times\boldsymbol{u} = \mathbf{0}$. Determine all the possible time-harmonic inhomogeneous plane wave solutions.
2. Find the propagation condition for the anisotropic vector wave equation

$$\nabla^2 u_i - \partial_i(\nabla\cdot\boldsymbol{u}) - \beta_{ij}\partial_t^2 u_j = 0,$$

where β_{ij} is a real symmetric positive definite constant matrix.

Find the time-harmonic inhomogeneous waves corresponding to any prescribed isotropic $\boldsymbol{C}(\boldsymbol{C}\cdot\boldsymbol{C}=0)$.

6.4 Damped inhomogeneous plane waves (complex exponential solutions)

Now we allow ω to be complex and consider waves of the form (6.3.1) with

$$\omega=\omega^{+}+\mathrm{i}\omega^{-}=\Omega \mathrm{e}^{\mathrm{i}\delta}, \tag{6.4.1}$$

where Ω is the modulus and δ is the argument of the complex number ω. Introducing the wave bivector $\boldsymbol{K}$ defined by

$$\boldsymbol{K}=\omega\boldsymbol{S}=\boldsymbol{K}^{+}+\mathrm{i}\boldsymbol{K}^{-}, \tag{6.4.2}$$

we note that the field u may now be written in the form

$$u=\exp-(\boldsymbol{K}^{-}\cdot\boldsymbol{x}-\omega^{-}t)\{A\exp \mathrm{i}(\boldsymbol{K}^{+}\cdot\boldsymbol{x}-\omega^{+}t)\}. \tag{6.4.3}$$

Using (6.3.11), we have

$$\boldsymbol{K}=\Omega T\mathrm{e}^{\mathrm{i}(\delta+\phi)}\boldsymbol{C}=\Omega T(\boldsymbol{h}^{+}+\mathrm{i}\boldsymbol{h}^{-}), \tag{6.4.4}$$

where

$$\boldsymbol{h}^{+}+\mathrm{i}\boldsymbol{h}^{-}=\mathrm{e}^{\mathrm{i}(\delta+\phi)}\boldsymbol{C}. \tag{6.4.5}$$

Thus, using (6.3.10),

$$\begin{aligned}\boldsymbol{h}^{+}&=\cos(\delta+\phi)m\hat{\mathbf{m}}-\sin(\delta+\phi)\hat{\mathbf{n}}, \quad \boldsymbol{K}^{+}=\Omega T\boldsymbol{h}^{+},\\ \boldsymbol{h}^{-}&=\cos(\delta+\phi)\hat{\mathbf{n}}+\sin(\delta+\phi)m\hat{\mathbf{m}}, \quad \boldsymbol{K}^{-}=\Omega T\boldsymbol{h}^{-}.\end{aligned} \tag{6.4.6}$$

The vectors $\boldsymbol{h}^{+}$ and $\boldsymbol{h}^{-}$ form a pair of conjugate radii of the ellipse of the bivector $\boldsymbol{C}$, and $\boldsymbol{K}^{+}$ and $\boldsymbol{K}^{-}$ which are along $\boldsymbol{h}^{+}$ and $\boldsymbol{h}^{-}$ form a pair of conjugate radii of an ellipse similar and similarly situated to the ellipse of $\boldsymbol{C}$ which is also the ellipse of the bivector $\boldsymbol{K}$.

Now, let

$$\begin{aligned}w^{+}(\boldsymbol{x},t)&=\boldsymbol{K}^{+}\cdot\boldsymbol{x}-\omega^{+}t=\Omega\{T\boldsymbol{h}^{+}\cdot\boldsymbol{x}-(\cos\delta)t\},\\ w^{-}(\boldsymbol{x},t)&=\boldsymbol{K}^{-}\cdot\boldsymbol{x}-\omega^{-}t=\Omega\{T\boldsymbol{h}^{-}\cdot\boldsymbol{x}-(\sin\delta)t\},\end{aligned} \tag{6.4.7}$$

so that (6.4.3) reads

$$u=\mathrm{e}^{-w^{-}}\{\mathrm{e}^{\mathrm{i}w^{+}}\}^{+}=\{\mathrm{e}^{\mathrm{i}(w^{+}+\mathrm{i}w^{-})}\}^{+}. \tag{6.4.8}$$

On the planes

$$w^{-}(\boldsymbol{x},t)=\boldsymbol{K}^{-}\cdot\boldsymbol{x}-\omega^{-}t=\text{constant}, \tag{6.4.9}$$

the amplitude is constant, the amplitude of the field u being propagated unchanged in the direction of $\boldsymbol{K}^- = \Omega T\boldsymbol{h}^-$, with velocity $\boldsymbol{v}_a$ given by

$$\boldsymbol{v}_a = \frac{\omega^-}{\boldsymbol{K}^-\cdot\boldsymbol{K}^-}\boldsymbol{K}^- = \frac{\sin\delta}{T(\boldsymbol{h}^-\cdot\boldsymbol{h}^-)}\boldsymbol{h}^-. \qquad (6.4.10)$$

The amplitude waves have neither wavelength nor period.

On the planes

$$w^+(\boldsymbol{x},t) = \boldsymbol{K}^+\cdot\boldsymbol{x} - \omega^+ t = \text{constant}, \qquad (6.4.11)$$

the phase is constant, the phase being propagated unchanged in the direction of $\boldsymbol{K}^+ = \Omega T\boldsymbol{h}^+$, with velocity $\boldsymbol{v}_p$ given by

$$\boldsymbol{v}_p = \frac{\omega^+}{\boldsymbol{K}^+\cdot\boldsymbol{K}^+}\boldsymbol{K}^+ = \frac{\cos\delta}{T(\boldsymbol{h}^+\cdot\boldsymbol{h}^+)}\boldsymbol{h}^+, \qquad (6.4.12)$$

the wavelength, λ, and the period, σ, being given by

$$\lambda = \frac{2\pi}{|\boldsymbol{K}^+|}, \quad \sigma = \frac{2\pi}{\omega^+}. \qquad (6.4.13)$$

The normal to the planes of constant phase and the normal to the planes of constant amplitude are along $\boldsymbol{h}^+$ and $\boldsymbol{h}^-$, respectively, conjugate radii of the ellipse of $\boldsymbol{C}$. The angle θ between these planes is given by

$$\tan\theta = \frac{m}{(m^2-1)\cos(\delta+\phi)\sin(\delta+\phi)} = \frac{\tan 2\beta}{\sin 2(\delta+\phi)}, \qquad (6.4.14)$$

with β defined by (6.3.14). For $m \neq 1$, these planes are orthogonal when either $\sin(\delta+\phi)=0$ or $\cos(\delta+\phi)=0$, so that ωN is either real or purely imaginary. As was done for (6.3.13), relation (6.4.14) may be interpreted as a formula of spherical trigonometry (Appendix, (A.15) and Figure A.2 with $b = 2(\delta+\phi)$, $c = 2\beta$, $\hat{C} = \theta$).

Damped inhomogeneous waves (complex exponential solutions) of given complex angular frequency ω may be systematically obtained by the DE method: first prescribe $\boldsymbol{C} = m\hat{\boldsymbol{m}} + \mathrm{i}\hat{\boldsymbol{n}}$, and then find the complex number $N = T\mathrm{e}^{\mathrm{i}\phi}$. Knowledge of N enables us to determine $\boldsymbol{K} = \omega N\boldsymbol{C}$ and thus the amplitude and phase velocities $\boldsymbol{v}_a$, $\boldsymbol{v}_p$.

As an example, consider the heat equation

$$\nabla^2 u = \alpha\partial_t u, \quad (\alpha > 0). \qquad (6.4.15)$$

Inserting $A \exp i\omega(\boldsymbol{S}\cdot\boldsymbol{x} - t)$ into (6.4.15) leads to the complex dispersion relation

$$\boldsymbol{K}\cdot\boldsymbol{K} = \omega^2 \boldsymbol{S}\cdot\boldsymbol{S} = i\alpha\omega. \tag{6.4.16}$$

Then, writing $\omega = \Omega e^{i\delta}$ and $\boldsymbol{S} = N\boldsymbol{C}$, yields

$$N^{-2} = \frac{\Omega(m^2 - 1)}{\alpha} \exp i\left(\delta - \frac{\pi}{2}\right), \tag{6.4.17}$$

and hence, the modulus T and the argument ϕ of N are given by

$$T^2 = \frac{\alpha}{\Omega(m^2 - 1)}, \quad \phi = \frac{\pi}{4} - \frac{\delta}{2}, \quad \text{or} \quad \frac{5\pi}{4} - \frac{\delta}{2}. \tag{6.4.18}$$

Thus, from (6.4.4), we obtain

$$\boldsymbol{K} = \pm\left(\frac{\alpha\Omega}{m^2 - 1}\right)^{1/2} \exp i\left(\frac{\delta}{2} + \frac{\pi}{4}\right)\boldsymbol{C}, \tag{6.4.19}$$

so that

$$\boldsymbol{K}^+ = \pm\left(\frac{\alpha\Omega}{m^2 - 1}\right)^{1/2} \left\{\cos\left(\frac{\delta}{2} + \frac{\pi}{4}\right)m\hat{\mathbf{m}} - \sin\left(\frac{\delta}{2} + \frac{\pi}{4}\right)\hat{\mathbf{n}}\right\}, \tag{6.4.20}$$

and

$$\boldsymbol{K}^- = \pm\left(\frac{\alpha\Omega}{m^2 - 1}\right)^{1/2} \left\{\cos\left(\frac{\delta}{2} + \frac{\pi}{4}\right)\hat{\mathbf{n}} + \sin\left(\frac{\delta}{2} + \frac{\pi}{4}\right)m\hat{\mathbf{m}}\right\}. \tag{6.4.21}$$

From (6.4.14), we note that the angle, θ, between the planes of equal phase and the planes of equal amplitude is given by

$$\tan\theta = \frac{m}{(m^2 - 1)\cos\delta} = \frac{\tan 2\beta}{\cos\delta}. \tag{6.4.22}$$

Remarks

(a) The angular frequency $\omega = \Omega e^{i\delta}$ may be prescribed. Thus, we note from (6.4.14), that by choosing δ appropriately there is some scope for changing the angle, θ, between the planes of constant phase and constant amplitude. The qualification 'some' is used because it depends on the prescribed value of m. If $m = 1$ ($\boldsymbol{C}$ isotropic), then $\theta = \frac{1}{2}\pi$ irrespective of the value of δ. However, for given $m \neq 1$, the acute angle between the two families of planes may be changed between the smallest value 2β corresponding to $\delta + \phi = \frac{1}{4}\pi$ or $\frac{5}{4}\pi$, and the largest value $\frac{1}{2}\pi$ corresponding to $\delta + \phi = 0$ or $\frac{1}{2}\pi$. We note that for $\delta + \phi = \frac{1}{4}\pi$ or $\frac{5}{4}\pi$, we have $\boldsymbol{h}^+ + \mathrm{i}\boldsymbol{h}^- = \pm\exp(\mathrm{i}\frac{1}{4}\pi)\boldsymbol{C}$, so that the normals $\boldsymbol{h}^+$ and $\boldsymbol{h}^-$ to the planes of constant phase and constant amplitude are along the equi-conjugate radii of the ellipse of the bivector $\boldsymbol{C}$. The most acute angle θ is the angle 2β between the equi-conjugate radii of the ellipse of the bivector $\boldsymbol{C}$.

(b) For the example of the heat equation (6.4.15), we note that if the complex angular frequency ω is not prescribed, then the wave bivector $\boldsymbol{K}$ may be arbitrary. Indeed, the complex dispersion relation (6.4.16) yields a value of ω for every given bivector $\boldsymbol{K}$.

This situation is typical for damped inhomogeneous plane waves. As another example, consider the anisotropic vector wave equation (6.1.14). The propagation condition (6.3.25) obtained in the context of time-harmonic inhomogeneous plane waves remains valid for damped waves because it does not involve ω. For prescribed complex ω and $\boldsymbol{C}$, the values of N are obtained from the eigenvalue problem (6.3.27). We now note that introducing $\boldsymbol{K} = \omega\boldsymbol{S}$, the propagation condition (6.3.25) reads

$$\{\boldsymbol{Q}(\boldsymbol{K}) - \omega^2\mathbf{1}\}\boldsymbol{A} = \mathbf{0}. \tag{6.4.23}$$

Hence, if ω is not prescribed, $\boldsymbol{K}$ may be any bivector and the eigenvalue problem (6.4.23) yields values of ω for every given bivector $\boldsymbol{K}$. If, for a given $\boldsymbol{K}$, $\boldsymbol{Q}(\boldsymbol{K})$ has a zero eigenvalue, we have a corresponding static exponential solution

$$\boldsymbol{u} = \{\boldsymbol{A}\exp\mathrm{i}\boldsymbol{K}\cdot\boldsymbol{x}\}^+. \tag{6.4.24}$$

Exercises 6.6

1. For the equation

$$\nabla^2 u = \frac{1}{c^2}\left(\frac{\partial^2 u}{\partial t^2} + \alpha\frac{\partial u}{\partial t}\right),$$

find damped inhomogeneous plane wave solutions of prescribed complex frequency $\omega = \Omega e^{\mathrm{i}\delta}$: determine T and ϕ. For which value of δ is the angle between the planes of equal phase and the planes of equal amplitude the most acute?

2. For the equation

$$\alpha_{ij}\frac{\partial^2 u}{\partial x_i \partial x_j} = \beta \frac{\partial u}{\partial t},$$

find damped inhomogeneous plane wave solutions with prescribed complex angular frequency $\omega = \Omega e^{i\delta}$.

7

Description of elliptical polarization

In this chapter, we consider elliptically polarized transverse homogeneous waves. Such waves are described by a vector field $\boldsymbol{u}(\boldsymbol{x}, t)$ whose polarization ellipse lies in a plane orthogonal to the propagation direction.

For instance, for electromagnetic waves in isotropic media, the vector field $\boldsymbol{u}(\boldsymbol{x}, t)$ may be the electric or magnetic field, the electric displacement, or the magnetic induction. For transverse elastic waves in isotropic media, the vector field $\boldsymbol{u}(\boldsymbol{x}, t)$ is the displacement field.

As the waves are transverse, we only consider amplitude bivectors $\boldsymbol{A}$ belonging to the same plane, a plane orthogonal to the propagation direction. We are interested in the different ways of characterizing the position and shape of the polarization ellipse, and in their links with the bivector $\boldsymbol{A}$. In particular, the Stokes parameters and the Poincaré sphere are introduced.

We also consider linear transformations of the amplitude bivector $\boldsymbol{A}$, and introduce the Jones and Mueller matrices associated with such transformations.

7.1 Transverse homogeneous waves

Using rectangular cartesian coordinate axes x, y, z, with the z-axis along the propagation direction, transverse homogeneous waves are described by a vector field $\boldsymbol{u}(\boldsymbol{x}, t)$ of the form

$$\boldsymbol{u}(z, t) = \{\boldsymbol{A}e^{i\tau}\}^+, \quad \tau = kz - \omega t. \tag{7.1.1}$$

Here the amplitude bivector $\boldsymbol{A}$ lies in the xy-plane, and thus, introducing the orthonormal triad $\mathbf{i}, \mathbf{j}, \mathbf{k}$ along the coordinate axes, we have

$$\boldsymbol{A} = A_1\mathbf{i} + A_2\mathbf{j}. \tag{7.1.2}$$

In general, the wave (7.1.1) is elliptically polarized. It is linearly polarized when $\boldsymbol{A}$ has a real direction, and it is circularly polarized when $\boldsymbol{A}$ is isotropic. The polarization ellipse is the directional ellipse of the bivector $\boldsymbol{A}$.

Because plane waves are obtained as solutions of linear equations, the amplitude bivector $\boldsymbol{A}$ in (7.1.1) may be multiplied by an arbitrary complex factor, so that the polarization ellipse is changed into a similar and similarly situated ellipse. This suggests defining the 'polarization form' of the wave as the polarization ellipse, up to a scale factor as follows: the polarization form is characterized by the aspect ratio, the direction of the major axis and the orientation of the polarization ellipse. The size of the polarization ellipse is related to the 'intensity' I, defined by

$$I = \boldsymbol{A}\cdot\bar{\boldsymbol{A}} = A_1\bar{A}_1 + A_2\bar{A}_2. \tag{7.1.3}$$

Writing the complex numbers, A_1, A_2 in terms of their moduli and arguments,

$$A_1 = a_1 e^{i\delta_1}, \quad A_2 = a_2 e^{i\delta_2}, \tag{7.1.4}$$

we note that

$$\boldsymbol{A} = a_1 e^{i\delta_1}\mathbf{i} + a_2 e^{i\delta_2}\mathbf{j} = \boldsymbol{c} + i\boldsymbol{d}, \tag{7.1.5}$$

where

$$\begin{aligned} \boldsymbol{c} &= a_1 \cos\delta_1 \mathbf{i} + a_2 \cos\delta_2 \mathbf{j}, \\ \boldsymbol{d} &= a_1 \sin\delta_1 \mathbf{i} + a_2 \sin\delta_2 \mathbf{j}. \end{aligned} \tag{7.1.6}$$

The components u_1 and u_2 of the vector field $\boldsymbol{u}(z, t)$ along the x and y-axes, respectively, are given by

$$u_1 = a_1 \cos(\tau + \delta_1), \quad u_2 = a_2 \cos(\tau + \delta_2). \tag{7.1.7}$$

Thus, at fixed z, u_1 and u_2 are oscillatory, the (real) amplitude of the oscillations being a_1 and a_2, respectively, and the phases being δ_1 and δ_2, respectively.

From (7.1.7), we note that the polarization ellipse is contained within a rectangular box whose sides are parallel to the x and y-axes, have lengths $2a_1$ and $2a_2$, respectively, and are tangential to the ellipse (see Figure 7.1). The angle α which the diagonal of the box in the first quadrant makes with the x-axis is given by

$$\tan\alpha = \frac{a_2}{a_1}. \tag{7.1.8}$$

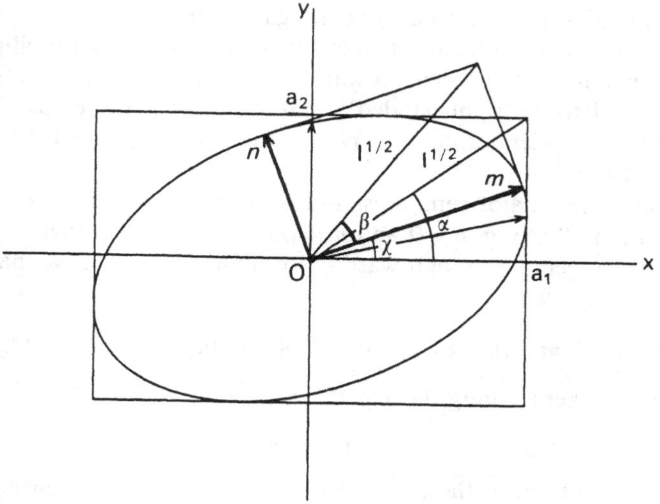

Figure 7.1 *Ellipticity β and azimuth χ of a polarization ellipse.*

Exercise 7.1

1. Derive the conditions that $a_1, a_2, \delta_1, \delta_2$ have to satisfy so that the wave be (a) linearly polarized; (b) circularly polarized.

7.2 Determination of the polarization ellipse from orthogonal components

For elliptically polarized transverse homogeneous waves, it is frequently the case that the amplitudes a_1, a_2, and the phase difference $\delta = \delta_2 - \delta_1$ of the orthogonal field components (7.1.7) may be obtained from observations. From this information the polarization ellipse is to be constructed. In particular, the polarization form and thus the aspect ratio, the direction of the major axis and the orientation of this ellipse are to be determined.

The polarization ellipse is the ellipse of the bivector $\boldsymbol{A}$ given by (7.1.5), or, equivalently, the ellipse of the bivector $\boldsymbol{A}'$ given by

$$\boldsymbol{A}' = \mathrm{e}^{-\mathrm{i}\delta_1}\boldsymbol{A} = a_1\mathbf{i} + a_2\mathrm{e}^{\mathrm{i}\delta}\mathbf{j} = \boldsymbol{c}' + \mathrm{i}\boldsymbol{d}', \tag{7.2.1}$$

where

$$\boldsymbol{c}' = a_1\mathbf{i} + a_2\cos\delta\,\mathbf{j}, \quad \boldsymbol{d}' = a_2\sin\delta\,\mathbf{j}. \tag{7.2.2}$$

We note that $\boldsymbol{c}'$, $\boldsymbol{d}'$ is a pair of conjugate radii of this ellipse, and that the orientation is from $\boldsymbol{c}'$ toward $\boldsymbol{d}'$. Thus, if $\sin\delta > 0$, the ellipse is oriented from the x-axis toward the y-axis (the polarization form is said to be 'right-handed'), and if $\sin\delta < 0$, the ellipse has the opposite orientation (the polarization form is said to be 'left-handed').

Let $\boldsymbol{m}$ be the major semi-axis, and $\boldsymbol{n}$ the minor semi-axis of the polarization ellipse: $\boldsymbol{m}\cdot\boldsymbol{n} = 0$ and $m = |\boldsymbol{m}| > n = |\boldsymbol{n}|$, the orientation of the ellipse being from $\boldsymbol{m}$ toward $\boldsymbol{n}$. Recalling section 2.2, we have, for some angle θ,

$$\boldsymbol{A}'' \equiv \boldsymbol{m} + \mathrm{i}\boldsymbol{n} = \mathrm{e}^{\mathrm{i}\theta}\boldsymbol{A}' = \mathrm{e}^{\mathrm{i}(\theta-\delta_1)}\boldsymbol{A} = \mathrm{e}^{\mathrm{i}\theta}(a_1\mathbf{i} + a_2\mathrm{e}^{\mathrm{i}\delta}\mathbf{j}), \tag{7.2.3}$$

or, in components along the x and y-axes,

$$m_1 + \mathrm{i}n_1 = \mathrm{e}^{\mathrm{i}\theta}a_1, \quad m_2 + \mathrm{i}n_2 = \mathrm{e}^{\mathrm{i}(\theta+\delta)}a_2. \tag{7.2.4}$$

A characterization of the polarization form and of the intensity is easily obtained from $\boldsymbol{m}$ and $\boldsymbol{n}$.

The polarization form is characterized by two angles β and χ (Figure 7.1). The angle β, called the 'ellipticity', is defined by

$$\tan\beta = \pm\frac{n}{m}, \quad -\frac{\pi}{4} \leqslant \beta \leqslant +\frac{\pi}{4}, \tag{7.2.5}$$

where the + sign or the − sign has to be chosen according to whether the polarization form is right-handed or left-handed. Thus β determines the aspect ratio and orientation of the polarization ellipse. The angle χ, called the 'azimuth of the major axis', or for short 'azimuth', is the angle that the major semi-axis $\boldsymbol{m}$ makes with the x-axis:

$$m_1 = m\cos\chi, \quad m_2 = m\sin\chi, \quad 0 \leqslant \chi < 2\pi. \tag{7.2.6}$$

The intensity I, defined by (7.1.3), is given in terms of m and n by

$$I = \boldsymbol{A}\cdot\bar{\boldsymbol{A}} = \boldsymbol{A}''\cdot\bar{\boldsymbol{A}}'' = m^2 + n^2, \tag{7.2.7}$$

so that it is the square of the diagonal of the rectangle constructed upon the major and minor semi-axes of the ellipse.

Now, we have to determine β and χ (polarization form), and I (intensity) from the knowledge of a_1, a_2, and δ. First, we note that

$$m = \sqrt{I}\cos\beta, \quad n = \pm\sqrt{I}\sin\beta, \tag{7.2.8}$$

and thus

$$m_1 = \sqrt{I}\cos\beta\cos\chi, \qquad m_2 = \sqrt{I}\cos\beta\sin\chi,$$
$$n_1 = -\sqrt{I}\sin\beta\sin\chi, \quad n_2 = \sqrt{I}\sin\beta\cos\chi. \tag{7.2.9}$$

Hence, forming the basic invariants of the ellipse (section 1.1), we have

$$m^2 + n^2 = I = a_1^2 + a_2^2, \tag{7.2.10a}$$

$$\pm mn = I\sin\beta\cos\beta = a_1 a_2 \sin\delta. \tag{7.2.10b}$$

Thus (7.2.10a) gives the intensity in terms of a_1 and a_2. From (7.2.10) we also have

$$\sin\beta\cos\beta = \frac{a_1 a_2}{a_1^2 + a_2^2}\sin\delta, \tag{7.2.11}$$

and hence, using the angle α, defined by (7.1.8),

$$\sin 2\beta = \sin 2\alpha \sin\delta. \tag{7.2.12}$$

Thus (7.2.12) gives the ellipticity β in terms of a_1, a_2 and δ.

The angle χ has still to be determined in terms of a_1, a_2 and δ. From (7.2.4), we note that

$$m_1^2 + n_1^2 = a_1^2, \quad m_2^2 + n_2^2 = a_2^2, \tag{7.2.13}$$

and

$$(m_2 + in_2)(m_1 - in_1) = a_1 a_2 e^{i\delta}. \tag{7.2.14}$$

Thus, taking the real and imaginary parts of (7.2.14), we have

$$a_1 a_2 \cos\delta = m_1 m_2 + n_1 n_2, \tag{7.2.15}$$

$$a_1 a_2 \sin\delta = m_1 n_2 - m_2 n_1. \tag{7.2.16}$$

Inserting (7.2.9) into (7.2.13), we obtain

$$a_1^2 = I(\cos^2\beta\cos^2\chi + \sin^2\beta\sin^2\chi),$$
$$a_2^2 = I(\cos^2\beta\sin^2\chi + \sin^2\beta\cos^2\chi), \tag{7.2.17}$$

and hence,

$$a_1^2 - a_2^2 = I\cos 2\beta\cos 2\chi. \tag{7.2.18}$$

Also, inserting (7.2.9) into (7.2.15) and (7.2.16) yields

$$2a_1a_2 \cos\delta = I \cos 2\beta \sin 2\chi, \tag{7.2.19}$$

$$2a_1a_2 \sin\delta = I \sin 2\beta. \tag{7.2.20}$$

Hence, from (7.2.18) and (7.2.19), we obtain

$$\tan 2\chi = \frac{2a_1a_2}{a_1^2 - a_2^2} \cos\delta, \tag{7.2.21}$$

and so, using the angle α defined by (7.1.8),

$$\tan 2\chi = \tan 2\alpha \cos\delta. \tag{7.2.22}$$

Thus (7.2.22) gives the azimuth χ of the major axis in terms of a_1, a_2 and δ.

Remark From (7.2.18) and (7.2.10a), we have

$$\cos 2\beta \cos 2\chi = \frac{a_1^2 - a_2^2}{a_1^2 + a_2^2}, \tag{7.2.23}$$

and so, using the angle α defined by (7.1.8),

$$\cos 2\beta \cos 2\chi = \cos 2\alpha. \tag{7.2.24}$$

Also, from (7.2.19) and (7.2.20), we note that

$$\sin 2\chi \tan\delta = \tan 2\beta. \tag{7.2.25}$$

Examining (7.2.12), (7.2.22), (7.2.24) and (7.2.25), it is seen that 2α, 2β, 2χ may be interpreted as the sides of a right-angled spherical triangle (Appendix, (A.10), (A.13), (A.9) and (A.15), and Figure A.2, with $a = 2\alpha$, $b = 2\chi$, $c = 2\beta$, $\hat{C} = \delta$).

Exercises 7.2

1. Find the ellipticity β, azimuth χ and intensity I for the waves

$$\text{(a)} \begin{cases} u_1 = \sqrt{3}a\cos\tau, \\ u_2 = -a\sin\tau. \end{cases} \qquad \text{(b)} \begin{cases} u_1 = a\cos\tau, \\ u_2 = a\cos(\tau + \tfrac{1}{3}\pi). \end{cases}$$

2. Characterize in terms of β and χ the following polarization forms:
   (a) linear polarization along **i**,
   (b) linear polarization along **j**,
   (c) circular polarization (right-handed or left-handed).
3. Let β and χ be the ellipticity and azimuth of the ellipse of the

bivector $\boldsymbol{A}$. Find the ellipticity β_* and azimuth χ_* of the ellipse of the bivector $\boldsymbol{A}_* = \bar{\boldsymbol{A}}_\perp$ (see section 2.5 for the definition of the reciprocal bivector $\boldsymbol{A}_\perp$). Write $\boldsymbol{A}''$ and $\boldsymbol{A}''_*$ in terms of β and χ.

7.3 Stokes parameters

The amplitude bivector $\boldsymbol{A}$ of the wave (7.1.1) may always be written as a linear combination of two linearly independent bivectors in its plane. In section 2.12, different decompositions of this type have been obtained. Thus, the vector field (7.1.1) may always be decomposed into the sum of two vector fields of waves with given distinct polarization forms (the ellipses of polarization of the two waves may not be similar and similarly situated). Here we are interested in the intensity of the wave (7.1.1), and of the two waves of the decomposition.

Let $\boldsymbol{P}$ and $\boldsymbol{Q}$ be two linearly independent bivectors in the xy-plane. Then, $\boldsymbol{A}$ may be decomposed in the form

$$\boldsymbol{A} = \lambda \boldsymbol{P} + \mu \boldsymbol{Q}, \qquad (7.3.1)$$

where the complex coefficients λ, μ may be easily obtained using the reciprocal bivectors, $\boldsymbol{P}_\perp$, $\boldsymbol{Q}_\perp$ of the bivectors $\boldsymbol{P}$, $\boldsymbol{Q}$ (section 2.5):

$$\lambda = \frac{\boldsymbol{A}\cdot\boldsymbol{Q}_\perp}{\boldsymbol{P}\cdot\boldsymbol{Q}_\perp}, \quad \mu = \frac{\boldsymbol{A}\cdot\boldsymbol{P}_\perp}{\boldsymbol{Q}\cdot\boldsymbol{P}_\perp}. \qquad (7.3.2)$$

The intensity of the wave with amplitude bivector $\boldsymbol{A}$ may thus be written

$$I = \boldsymbol{A}\cdot\bar{\boldsymbol{A}} = \lambda\bar{\lambda}\boldsymbol{P}\cdot\bar{\boldsymbol{P}} + \mu\bar{\mu}\boldsymbol{Q}\cdot\bar{\boldsymbol{Q}} + \lambda\bar{\mu}\boldsymbol{P}\cdot\bar{\boldsymbol{Q}} + \bar{\lambda}\mu\bar{\boldsymbol{P}}\cdot\boldsymbol{Q}. \qquad (7.3.3)$$

We note that this is not the sum of the intensities of the waves with amplitudes $\boldsymbol{A}^{(1)} = \lambda\boldsymbol{P}$ and $\boldsymbol{A}^{(2)} = \mu\boldsymbol{Q}$ into which the wave is decomposed. However, it will reduce to the sum of the intensities, whatever the bivector $\boldsymbol{A}$ may be, if and only if we choose

$$\boldsymbol{Q} = \bar{\boldsymbol{P}}_\perp. \qquad (7.3.4)$$

The decomposition (7.3.1) is then the decomposition (2.12.18).

From now on, choosing $\boldsymbol{Q}$ given by (7.3.4), we have

$$\boldsymbol{A} = \lambda\boldsymbol{P} + \mu\boldsymbol{Q} \quad \text{with} \quad \lambda = \frac{\boldsymbol{A}\cdot\bar{\boldsymbol{P}}}{\boldsymbol{P}\cdot\bar{\boldsymbol{P}}}, \quad \mu = \frac{\boldsymbol{A}\cdot\bar{\boldsymbol{Q}}}{\boldsymbol{Q}\cdot\bar{\boldsymbol{Q}}}, \qquad (7.3.5)$$

and the intensity is decomposed into the sum

$$I = \boldsymbol{A}\cdot\bar{\boldsymbol{A}} = \lambda\bar{\lambda}\boldsymbol{P}\cdot\bar{\boldsymbol{P}} + \mu\bar{\mu}\boldsymbol{Q}\cdot\bar{\boldsymbol{Q}}. \tag{7.3.6}$$

The ellipses of polarization of the waves with amplitude bivectors $\boldsymbol{A}^{(1)} = \lambda\boldsymbol{P}$ and $\boldsymbol{A}^{(2)} = \mu\boldsymbol{Q} = \mu\bar{\boldsymbol{P}}_{\perp}$ have the same aspect ratio, are described in opposite senses, and their major axes are perpendicular. The polarization forms of these waves are said to be 'opposite' and the waves are said to be 'oppositely polarized'.

Let

$$\boldsymbol{P}' = \frac{\boldsymbol{P}}{(\boldsymbol{P}\cdot\bar{\boldsymbol{P}})^{1/2}}, \quad \boldsymbol{Q}' = \frac{\boldsymbol{Q}}{(\boldsymbol{Q}\cdot\bar{\boldsymbol{Q}})^{1/2}}, \quad \boldsymbol{P}'\cdot\bar{\boldsymbol{P}}' = \boldsymbol{Q}\cdot\bar{\boldsymbol{Q}}' = 1, \tag{7.3.7}$$

so that (7.3.5) may be written

$$\boldsymbol{A} = \lambda'\boldsymbol{P}' + \mu'\boldsymbol{Q}' \quad \text{with} \quad \lambda' = \boldsymbol{A}\cdot\bar{\boldsymbol{P}}', \quad \mu' = \boldsymbol{A}\cdot\bar{\boldsymbol{Q}}'. \tag{7.3.8}$$

Also, the intensity (7.3.6) becomes

$$I = \boldsymbol{A}\cdot\bar{\boldsymbol{A}} = \boldsymbol{A}^{(1)}\cdot\bar{\boldsymbol{A}}^{(1)} + \boldsymbol{A}^{(2)}\cdot\bar{\boldsymbol{A}}^{(2)} = \lambda'\bar{\lambda}' + \mu\bar{\mu}'. \tag{7.3.9}$$

Now, introducing the ellipticity β' and the azimuth χ' of the ellipse of the bivector $\boldsymbol{P}$, we may write

$$\boldsymbol{P}' = \cos\beta'\hat{\mathbf{r}} + \mathrm{i}\sin\beta'\hat{\mathbf{s}}, \quad \boldsymbol{Q}' = \sin\beta'\hat{\mathbf{r}} - \mathrm{i}\cos\beta'\hat{\mathbf{s}}, \tag{7.3.10}$$

where $\hat{\mathbf{r}}$ and $\hat{\mathbf{s}}$ are two orthogonal unit vectors, χ' being the angle between $\hat{\mathbf{r}}$ and $\mathbf{i}$:

$$\hat{\mathbf{r}} = \cos\chi'\mathbf{i} + \sin\chi'\mathbf{j}, \quad \hat{\mathbf{s}} = -\sin\chi'\mathbf{i} + \cos\chi'\mathbf{j}. \tag{7.3.11}$$

Then, if the intensities of the waves with amplitude bivectors $\boldsymbol{A}^{(1)} = \lambda'\boldsymbol{P}'$ and $\boldsymbol{A}^{(2)} = \mu'\boldsymbol{Q}'$, are denoted by $I^{(1)}$ and $I^{(2)}$, respectively, we have, on using (7.1.2),

$$\begin{aligned} I^{(1)} = \lambda'\bar{\lambda}' &= (\boldsymbol{A}\cdot\bar{\boldsymbol{P}}')(\bar{\boldsymbol{A}}\cdot\boldsymbol{P}') \\ &= |A_1(\cos\beta'\cos\chi' + \mathrm{i}\sin\beta'\sin\chi') \\ &\quad + A_2(\cos\beta'\sin\chi' - \mathrm{i}\sin\beta'\cos\chi')|^2 \\ &= A_1\bar{A}_1(\cos^2\beta'\cos^2\chi' + \sin^2\beta'\sin^2\chi') \\ &\quad + A_2\bar{A}_2(\cos^2\beta'\sin^2\chi' + \sin^2\beta'\cos^2\chi') \\ &\quad + \tfrac{1}{2}A_1\bar{A}_2(\cos 2\beta'\sin 2\chi' + \mathrm{i}\sin 2\beta') \\ &\quad + \tfrac{1}{2}\bar{A}_1A_2(\cos 2\beta'\sin 2\chi' - \mathrm{i}\sin 2\beta'), \end{aligned} \tag{7.3.12}$$

$$\begin{aligned}
I^{(2)} &= \mu'\bar{\mu}' = (\boldsymbol{A}\cdot\bar{\boldsymbol{Q}}')(\bar{\boldsymbol{A}}\cdot\boldsymbol{Q}') \\
&= |A_1(\sin\beta'\cos\chi' - \mathrm{i}\cos\beta'\sin\chi') \\
&\quad + A_2(\sin\beta'\sin\chi' + \mathrm{i}\cos\beta'\cos\chi')|^2 \\
&= A_1\bar{A}_1(\sin^2\beta'\cos^2\chi' + \cos^2\beta'\sin^2\chi') \\
&\quad + A_2\bar{A}_2(\sin^2\beta'\sin^2\chi' + \cos^2\beta'\cos^2\chi') \\
&\quad - \tfrac{1}{2}A_1\bar{A}_2(\cos 2\beta'\sin 2\chi' + \mathrm{i}\sin 2\beta') \\
&\quad - \tfrac{1}{2}\bar{A}_1 A_2(\cos 2\beta'\sin 2\chi' - \mathrm{i}\sin 2\beta').
\end{aligned} \tag{7.3.13}$$

Hence,

$$I^{(1)} = \lambda'\bar{\lambda}' = \tfrac{1}{2}(I + Q\cos 2\beta'\cos 2\chi' + U\cos 2\beta'\sin 2\chi' + V\sin 2\beta'), \tag{7.3.14}$$

$$I^{(2)} = \mu'\bar{\mu}' = \tfrac{1}{2}(I - Q\cos 2\beta'\cos 2\chi' - U\cos 2\beta'\sin 2\chi' - V\sin 2\beta'), \tag{7.3.15}$$

where

$$\begin{aligned}
I &= A_1\bar{A}_1 + A_2\bar{A}_2 = \boldsymbol{A}\cdot\bar{\boldsymbol{A}}, \\
Q &= A_1\bar{A}_1 - A_2\bar{A}_2, \\
U &= A_1\bar{A}_2 + \bar{A}_1 A_2, \\
V &= \mathrm{i}(A_1\bar{A}_2 - \bar{A}_1 A_2) = \mathrm{i}\mathbf{k}\cdot(\boldsymbol{A}\times\bar{\boldsymbol{A}}).
\end{aligned} \tag{7.3.16}$$

Thus, for every decomposition of the wave with amplitude bivector $\boldsymbol{A}$ into two oppositely polarized waves, the intensities of these two waves are given as linear combinations of the four parameters I, Q, U, V. These are called the 'Stokes parameters' of the wave and are given by (7.3.16) in terms of the components of the bivector $\boldsymbol{A}$ along the x and y-axes. There is an infinity of such decompositions: a particular decomposition corresponds to a choice of the polarization form of the bivector $\boldsymbol{P}$, that is to a choice of the angles β' and χ'.

The definition (7.3.16) of the Stokes parameters may also be written

$$I = \bar{A}^{\mathrm{T}}\boldsymbol{\Pi}_0 A, \quad Q = \bar{A}^{\mathrm{T}}\boldsymbol{\Pi}_1 A, \quad U = \bar{A}^{\mathrm{T}}\boldsymbol{\Pi}_2 A, \quad V = \bar{A}^{\mathrm{T}}\boldsymbol{\Pi}_3 A, \tag{7.3.17}$$

where $\boldsymbol{\Pi}_0$, $\boldsymbol{\Pi}_1$, $\boldsymbol{\Pi}_2$, $\boldsymbol{\Pi}_3$ are the following hermitian 2×2 matrices:

$$\boldsymbol{\Pi}_0 = \begin{bmatrix} 1 & 0 \\ 0 & 1 \end{bmatrix}, \quad \boldsymbol{\Pi}_1 = \begin{bmatrix} 1 & 0 \\ 0 & -1 \end{bmatrix},$$

$$\boldsymbol{\Pi}_2 = \begin{bmatrix} 0 & 1 \\ 1 & 0 \end{bmatrix}, \quad \boldsymbol{\Pi}_3 = \begin{bmatrix} 0 & -\mathrm{i} \\ \mathrm{i} & 0 \end{bmatrix}. \tag{7.3.18}$$

Using the notation $s_0 = I$, $s_1 = Q$, $s_2 = U$, $s_3 = V$ for the Stokes parameters, (7.3.17) also reads

$$s_B = \bar{A}^{\mathrm{T}}\boldsymbol{\Pi}_B A, \quad (B = 0, 1, 2, 3). \tag{7.3.19}$$

The matrices $\boldsymbol{\Pi}_1, \boldsymbol{\Pi}_2, \boldsymbol{\Pi}_3$ are known as 'Pauli matrices'. All the Stokes parameters have the same dimension, that of intensity, and from their definition (7.3.16) it follows that

$$Q^2 + U^2 + V^2 = I^2. \tag{7.3.20}$$

We note that if the bivector $\boldsymbol{A}$ is replaced by $e^{i\phi}\boldsymbol{A}$, where $e^{i\phi}$ is an arbitrary phase factor, the Stokes parameters s_B remain unchanged.

Using (7.1.4), it is easy to write the Stokes parameters in terms of the amplitudes a_1, a_2, and the phase difference $\delta = \delta_2 - \delta_1$ of the orthogonal field components (7.1.7). Then, using (7.2.18), (7.2.19), (7.2.20) and (7.2.8), we obtain

$$\begin{aligned} s_0 &= I = a_1^2 + a_2^2 = m^2 + n^2, \\ s_1 &= Q = a_1^2 - a_2^2 = I\cos 2\beta\cos 2\chi = (m^2 - n^2)\cos 2\chi, \\ s_2 &= U = 2a_1a_2\cos\delta = I\cos 2\beta\sin 2\chi = (m^2 - n^2)\sin 2\chi, \\ s_3 &= V = 2a_1a_2\sin\delta = I\sin 2\beta = \pm 2mn. \end{aligned} \tag{7.3.21}$$

The intensity ($I = \boldsymbol{A}\cdot\bar{\boldsymbol{A}}$) and the polarization form (ellipticity β and azimuth χ) of the wave (7.1.1) are completely determined by the four Stokes parameters. Thus the polarization ellipse of the wave may be characterized by giving the Stokes parameters.

In (7.3.21), β and χ are the ellipticity and azimuth of the ellipse of the amplitude bivector $\boldsymbol{A}$. In the decomposition (7.3.5), the bivector $\boldsymbol{P}$ may be chosen arbitrarily and β' and χ' denote the ellipticity and azimuth of its ellipse. If $\boldsymbol{P}$ is chosen to be $\boldsymbol{A}$ itself, $\boldsymbol{P} = \boldsymbol{A}$, then $\lambda = 1$ and $\mu = 0$ in (7.3.5), and $\beta' = \beta, \chi' = \chi$. Then, using (7.3.21), we note that (7.3.14) and (7.3.15) yield $I^{(1)} = I$ and $I^{(2)} = 0$.

Remark Let $\boldsymbol{H}$ be the 2×2 complex matrix defined by

$$H_{\alpha\beta} = A_\alpha \bar{A}_\beta, \quad (\alpha, \beta = 1, 2). \tag{7.3.22}$$

This matrix is hermitian, $\bar{\boldsymbol{H}}^{\mathrm{T}} = \boldsymbol{H}$, and its trace is the intensity, $\operatorname{tr}\boldsymbol{H} = \boldsymbol{A}\cdot\bar{\boldsymbol{A}} = I$. From (7.3.16), it is easily seen that

$$\boldsymbol{H} = \frac{1}{2}\begin{bmatrix} I + Q & U - iV \\ U + iV & I - Q \end{bmatrix}, \tag{7.3.23}$$

which also reads

$$2H = I\Pi_0 + Q\Pi_1 + U\Pi_2 + V\Pi_3 = s_A\Pi_A, \quad (A = 0, 1, 2, 3). \quad (7.3.24)$$

Thus the knowledge of the matrix H is equivalent to the knowledge of the four Stokes parameters (I, Q, U, V). In terms of the matrix H, the relation (7.3.20) that the Stokes parameters have to satisfy, reads

$$\det H = 0. \quad (7.3.25)$$

From (7.3.21), we note that

$$H = \frac{I}{2}\begin{bmatrix} 1 + \cos 2\beta \cos 2\chi & \cos 2\beta \sin 2\chi - \mathrm{i} \sin 2\beta \\ \cos 2\beta \sin 2\chi + \mathrm{i} \sin 2\beta & 1 - \cos 2\beta \cos 2\chi \end{bmatrix}. \quad (7.3.26)$$

Finally, we also note that the intensity I is the trace norm of the hermitian matrix H, thus $I = \sqrt{\mathrm{tr}(H^2)}$.

Exercises 7.3

1. For the waves of Exercise 7.2.1, find the Stokes parameters.
2. Write (7.3.14) and (7.3.15) using the Stokes parameters of the waves with amplitude bivectors P', Q'.
3. Write (7.3.14) and (7.3.15) using the hermitian matrix H and the hermitian matrices $H'^{(1)}$ and $H'^{(2)}$ of the waves with amplitude bivectors P' and Q', respectively.
4. Show that

$$s_A = \mathrm{tr}(H\Pi_A), \quad (A = 0, 1, 2, 3). \quad (7.3.27)$$

5. Find the intensities and polarization forms corresponding to the following 2×2 hermitian matrices with zero determinant:

$$H_q = \begin{bmatrix} 1 & 0 \\ 0 & 0 \end{bmatrix}, \quad H_u = \frac{1}{2}\begin{bmatrix} 1 & 1 \\ 1 & 1 \end{bmatrix}, \quad H_v = \frac{1}{2}\begin{bmatrix} 1 & -\mathrm{i} \\ \mathrm{i} & 1 \end{bmatrix}. \quad (7.3.28)$$

6. Up to a phase factor $(e^{i\phi})$, write the bivector A in terms of the Stokes parameters.

7.4 The Poincaré sphere

Equation (7.3.20) and the representation of the Stokes parameters by the expressions in (7.3.21) immediately suggest consideration of a sphere of radius I (the wave intensity) in a space referred to rectangular cartesian coordinates Q, U, V. This sphere is called the 'Poincaré sphere' (see Figure 7.2). From (7.3.21), it is clear that a point P on this sphere, of coordinates Q, U, V, has latitude $2\beta(-\pi/2 \leqslant 2\beta \leqslant +\pi/2)$ and longitude $2\chi(0 \leqslant 2\chi < 2\pi)$.

Thus, to a point P on the Poincaré sphere corresponds a polarization form of ellipticity β and azimuth χ, and an intensity I. If we are interested only in the polarization form, we may take the radius $I = 1$ without loss in generality.

Right-handed polarization forms correspond to points of the 'northern' hemisphere ($0 < 2\beta \leqslant \frac{1}{2}\pi$, $V > 0$), and left-handed polarization forms to points of the 'southern' hemisphere ($-\frac{1}{2}\pi \leqslant 2\beta < 0$, $V < 0$).

Thus, to a point P of latitude 2β and longitude 2χ, we associate the amplitude bivector A'' given by (7.2.3), which, using (7.2.9) with $I = 1$, may be written (see Exercise 7.2.3)

$$A'' = (\cos\beta\cos\chi - \mathrm{i}\sin\beta\sin\chi)\mathbf{i} + (\cos\beta\sin\chi + \mathrm{i}\sin\beta\cos\chi)\mathbf{j}, \quad (7.4.1)$$

or, equivalently,

$$A'' = \cos\beta\hat{\mathbf{m}} + \mathrm{i}\sin\beta\hat{\mathbf{n}}, \quad (7.4.2)$$

where $\hat{\mathbf{m}}$ and $\hat{\mathbf{n}}$ are unit vectors along $\boldsymbol{m}$ and $\boldsymbol{n}$, and are given by

$$\hat{\mathbf{m}} = \cos\chi\mathbf{i} + \sin\chi\mathbf{j}, \quad \hat{\mathbf{n}} = -\sin\chi\mathbf{i} + \cos\chi\mathbf{j}. \quad (7.4.3)$$

For points on the 'equator', $\beta = 0$, and then

$$A'' = \hat{\mathbf{m}} = \cos\chi\mathbf{i} + \sin\chi\mathbf{j}, \quad (7.4.4)$$

and the wave is linearly polarized (the polarization ellipse degenerates into a segment of a straight line). For $\chi = 0$, the point P is on the positive Q-axis and $A'' = \mathbf{i}$ so that the polarization is along the x-axis. For $\chi = \frac{1}{2}\pi$, then $2\chi = \pi$, the point P is on the negative Q-axis and $A'' = \mathbf{j}$ so that the polarization is along the y-axis. When χ takes the intermediate value $\chi = \pi/4$, then $2\chi = \frac{1}{2}\pi$, the point P is on the positive U-axis and $A'' = 2^{-1/2}(\mathbf{i} + \mathbf{j})$, so that the direction of polarization is tilted at $\frac{1}{4}\pi$ to the positive x-axis. If $\chi = \frac{3}{4}\pi$, then $2\chi = \frac{3}{2}\pi$, P is on the negative V-axis and $A'' = 2^{-1/2}(\mathbf{j} - \mathbf{i})$, so that the direction of polarization is tilted at $\frac{1}{4}\pi$ to the negative x-axis.

For $\beta = \frac{1}{4}\pi$, the point P is at the 'north pole' of the Poincaré sphere ($Q = U = 0$, $V = 1$) and from (7.4.2) and (7.4.3),

$$A'' = \frac{2^{1/2}}{2}(\hat{\mathbf{m}} + \mathrm{i}\hat{\mathbf{n}}) = \frac{\mathrm{e}^{-\mathrm{i}\chi}}{2^{1/2}}(\mathbf{i} + \mathrm{i}\mathbf{j}), \quad (7.4.5)$$

so that the wave is right-handed circularly polarized. Similarly, for $\beta = -\frac{1}{4}\pi$, the point P is at the 'south pole' of the Poincaré sphere ($Q = U = 0$, $V = -1$) and in this case

$$A'' = \frac{2^{1/2}}{2}(\hat{\mathbf{m}} - \mathrm{i}\hat{\mathbf{n}}) = \frac{\mathrm{e}^{\mathrm{i}\chi}}{2^{1/2}}(\mathbf{i} - \mathrm{i}\mathbf{j}), \tag{7.4.6}$$

so that the wave is left-handed circularly polarized.

Points on the same 'meridian' ($2\chi =$ constant) represent all possible polarization forms with a given azimuth, that is with a given direction of the major axis. Points on the same 'parallel' ($2\beta =$ constant) represent all polarization forms with a given ellipticity, that is with a given aspect ratio and handedness.

To the point P_* of latitude $2\beta_* = -2\beta$ and longitude $2\chi_* = 2\chi + \pi$, which is antipodal to the point P of latitude 2β and longitude 2χ, corresponds the polarization form opposite to that of the point P (see Exercise 7.2.3): the corresponding amplitude bivector A''_* is given by

$$A''_* = \cos\beta\hat{\mathbf{n}} + \mathrm{i}\sin\beta\hat{\mathbf{m}}, \tag{7.4.7}$$

and we have $\bar{A}''\cdot A''_* = 0$. In particular the polarization forms corresponding to the north and south poles (right-handed and left-handed circular polarization) are opposite.

Now, recalling the results of section 7.2, the angles α and δ may be interpreted on the Poincaré sphere. Let P_0 be the point of the sphere on the positive Q-axis, let P be the point of latitude 2β and longitude 2χ, and let E be the point on the equator of longitude 2χ (Figure 7.2). From the remark of section 7.2, it follows that 2α is the side $\widehat{PP_0}$ of the spherical triangle PP_0E, and that δ is the angle $\hat{P}_0$ of this triangle.

Exercises 7.4

1. Let $A_R = 2^{-1/2}(\mathbf{i} + \mathrm{i}\mathbf{j})$ and $A_L = 2^{-1/2}(\mathbf{i} - \mathrm{i}\mathbf{j})$. Find the locus on the Poincaré sphere of the points representing the polarization forms of the waves with amplitude bivectors

$$A = a_R A_R + a_L A_L,$$

where a_R and a_L are **real** parameters.

2. Let P of coordinates Q, U, V, be a point on the Poincaré sphere of unit radius. Let P_0^* be the point of coordinates $(-1, 0, 0)$ on

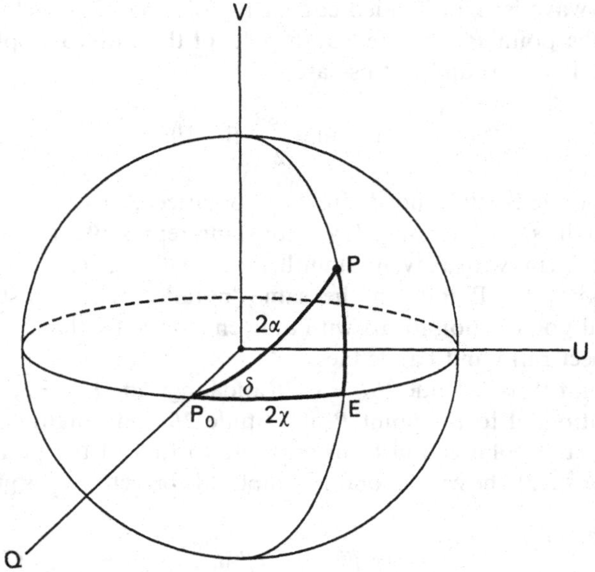

Figure 7.2 *The Poincaré sphere.*

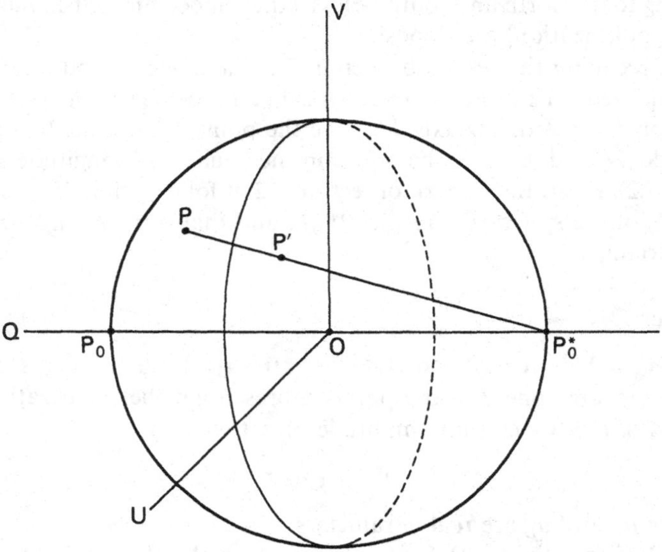

Figure 7.3 *Stereographic projection and the UV-plane.*

the sphere and consider the stereographic projection of center P_0^* onto the UV-plane. Let P' be the stereographic projection of P, and let u, v be the coordinates of P' (Figure 7.3). Show that

$$u + \mathrm{i}v = \frac{A_2}{A_1}.$$

Note: $u + \mathrm{i}v = A_2 A_1^{-1} = \zeta$ (say) is sometimes called the 'complex polarization variable'.

7.5 Linear transformations of amplitude bivectors. Jones and Mueller matrices

Devices transforming the amplitude bivector $\boldsymbol{A}$ of a wave of the type (7.1.1) into another amplitude bivector $\tilde{\boldsymbol{A}}$, also in the xy-plane, are frequently used in optics. Polarizers and retarders are examples of such devices.

Here we only consider devices such that the transformation is a linear transformation of $\boldsymbol{A}$ into $\tilde{\boldsymbol{A}}$. Then, using the orthonormal basis $\mathbf{i}, \mathbf{j}$, the transformation may be written

$$\tilde{\boldsymbol{A}} = \boldsymbol{J}\boldsymbol{A}, \quad \tilde{A}_\alpha = J_{\alpha\beta} A_\beta, \quad (\alpha, \beta = 1, 2). \qquad (7.5.1)$$

The 2×2 complex matrix $\boldsymbol{J} = (J_{\alpha\beta})$ is called the 'Jones matrix' of the device.

We now wish to obtain the Stokes parameters $\tilde{I}, \tilde{Q}, \tilde{U}, \tilde{V}$ of the transformed wave with amplitude bivector $\tilde{\boldsymbol{A}}$ in terms of the Stokes parameters I, Q, U, V of the wave with amplitude bivector $\boldsymbol{A}$. To achieve this, we first consider the hermitian matrices $H_{\alpha\beta} = A_\alpha \bar{A}_\beta$ and $\tilde{H}_{\alpha\beta} = \tilde{A}_\alpha \bar{\tilde{A}}_\beta$ (see remark of section 7.3). From (7.5.1), we obtain

$$\tilde{H}_{\alpha\beta} = J_{\alpha\gamma} \bar{J}_{\beta\delta} H_{\gamma\delta}, \quad \tilde{\mathbf{H}} = \boldsymbol{J}\boldsymbol{H}\bar{\boldsymbol{J}}^{\mathrm{T}}, \qquad (7.5.2)$$

which shows in particular that the transformation of the marix $\boldsymbol{H}$ into the matrix $\tilde{\boldsymbol{H}}$ is linear. Also, recalling (7.3.22) and (7.3.23), it is clear that the Stokes parameters $\tilde{s}_A = (\tilde{I}, \tilde{Q}, \tilde{U}, \tilde{V})$ depend linearly on the Stokes parameters $s_A = (I, Q, U, V)$. We thus write

$$\tilde{s}_A = \mu_{AB} s_B, \quad (A, B = 0, 1, 2, 3), \qquad (7.5.3)$$

where $\boldsymbol{\mu} = (\mu_{AB})$ is a real 4×4 matrix. This matrix is called the 'Mueller matrix' of the device.

The Mueller matrix may be easily obtained knowing the Jones matrix. Indeed, using the property of Exercise 7.3.4, we have, from

(7.5.2),

$$\tilde{s}_A = \mathrm{tr}(\tilde{\boldsymbol{H}}\boldsymbol{\Pi}_A) = \mathrm{tr}(\boldsymbol{J}\boldsymbol{H}\bar{\boldsymbol{J}}^{\mathrm{T}}\boldsymbol{\Pi}_A), \quad (A = 0, 1, 2, 3), \tag{7.5.4}$$

and, using (7.3.24), this becomes

$$\tilde{s}_A = \tfrac{1}{2}\mathrm{tr}(\boldsymbol{J}\boldsymbol{\Pi}_B\bar{\boldsymbol{J}}^{\mathrm{T}}\boldsymbol{\Pi}_A)s_B, \tag{7.5.5}$$

where $\boldsymbol{\Pi}_A$ $(A = 0, 1, 2, 3)$, are the matrices defined by (7.3.19) (unit matrix and Pauli matrices). Thus, the Mueller matrix is given in terms of the Jones matrix by

$$\mu_{AB} = \tfrac{1}{2}\mathrm{tr}(\boldsymbol{J}\boldsymbol{\Pi}_B\bar{\boldsymbol{J}}^{\mathrm{T}}\boldsymbol{\Pi}_A) = \tfrac{1}{2}\mathrm{tr}(\boldsymbol{\Pi}_A\boldsymbol{J}\boldsymbol{\Pi}_B\bar{\boldsymbol{J}}^{\mathrm{T}}). \tag{7.5.6}$$

We note that the elements of such a Mueller matrix are not independent. Indeed, the Stokes parameters have to satisfy (7.3.20). Thus, if γ denotes the 4×4 matrix

$$\gamma = \begin{bmatrix} +1 & 0 & 0 & 0 \\ 0 & -1 & 0 & 0 \\ 0 & 0 & -1 & 0 \\ 0 & 0 & 0 & -1 \end{bmatrix}, \tag{7.5.7}$$

the linear transformation (7.5.3) must be such that $s_A\gamma_{AB}s_B = 0$ implies $\tilde{s}_A\gamma_{AB}\tilde{s}_B = 0$. From this, it may be proved (Exercise 7.5.1) that the Mueller matrix $\boldsymbol{\mu} = (\mu_{AB})$ satisfies

$$\boldsymbol{\mu}^{\mathrm{T}}\gamma\boldsymbol{\mu} = \lambda\gamma, \tag{7.5.8}$$

for some real λ (depending on $\boldsymbol{\mu}$).

Because the Stokes parameters s_A and $\tilde{s}_A$ characterize the amplitude bivectors $\boldsymbol{A}$ and $\tilde{\boldsymbol{A}}$, each up to a phase factor, it is possible to obtain, up to a phase factor $(e^{i\phi})$, the Jones matrix $\boldsymbol{J}$ of a device when its Mueller matrix $\boldsymbol{\mu}$ is known. From the identity

$$(\boldsymbol{N}^{\mathrm{T}}\boldsymbol{\Pi}_C\boldsymbol{K})(\boldsymbol{L}^{\mathrm{T}}\boldsymbol{\Pi}_C\boldsymbol{M}) = 2(\boldsymbol{K}\cdot\boldsymbol{L})(\boldsymbol{M}\cdot\boldsymbol{N}), \quad (C = 0, 1, 2, 3), \tag{7.5.9}$$

valid for any bivectors $\boldsymbol{K}$, $\boldsymbol{L}$, $\boldsymbol{M}$, $\boldsymbol{N}$ and which may be checked explicitly, and using the definition (7.3.18) of the matrices $\boldsymbol{\Pi}_c$, we have

$$\Pi_C^{\alpha\beta}\Pi_C^{\gamma\nu} = 2\delta^{\alpha\nu}\delta^{\beta\gamma}, \quad (\alpha, \beta, \gamma, \nu = 1, 2). \tag{7.5.10}$$

Hence, from (7.5.6), we obtain

$$\tfrac{1}{2}\Pi_A^{\alpha\beta}\mu_{AB}\Pi_B^{\gamma\nu} = J_{\alpha\nu}\bar{J}_{\beta\gamma}. \tag{7.5.11}$$

Taking first $(\beta, \gamma) = (\alpha, \nu)$, this formula gives the squared moduli of the complex elements J_{11}, J_{12}, J_{21}, J_{22} of the Jones matrix in terms of the known elements of the Mueller matrix $\boldsymbol{\mu}$. Then, assuming these moduli known and considering the various possible $(\beta, \gamma) \neq (\alpha, \nu)$, we obain the differences between the arguments of the complex numbers $J_{11}, J_{12}, J_{21}, J_{22}$. Thus, (7.5.11) gives, up to a phase factor, the elements of the Jones matrix in terms of the elements of the Mueller matrix.

Now, if we are interested only in the polarization **forms** (and not in the intensities), the transformation (7.5.3) defines a transformation on the Poincaré sphere of unit radius. Using the notation

$$\sigma_i = \left(\frac{Q}{I}, \frac{U}{I}, \frac{V}{I}\right), \quad \tilde{\sigma}_i = \left(\frac{\tilde{Q}}{\tilde{I}}, \frac{\tilde{U}}{\tilde{I}}, \frac{\tilde{V}}{\tilde{I}}\right), \quad (\mathrm{i} = 1, 2, 3), \qquad (7.5.12)$$

for the coordinates on this sphere before and after transformation, we obtain, from (7.5.3),

$$\tilde{\sigma}_i = \frac{\mu_{i0} + \mu_{ij}\sigma_j}{\mu_{00} + \mu_{0k}\sigma_k}, \quad (i, j, k = 1, 2, 3). \qquad (7.5.13)$$

Also, the transformation of the polarization forms may be described in terms of the ratios $A_2 A_1^{-1} = \zeta$ (say) and $\tilde{A}_2 \tilde{A}_1^{-1} = \tilde{\zeta}$ (say). Indeed from (7.5.1), we obtain

$$\tilde{\zeta} = \frac{J_{22}\zeta + J_{21}}{J_{12}\zeta + J_{11}}. \qquad (7.5.14)$$

If $\det J \neq 0$, this is a 'homographic transformation', also called a 'conformal transformation', of the complex polarization variable ζ into the complex polarization variable $\tilde{\zeta}$. From Exercise 7.4.2, it follows that, in the complex UV-plane, the points $\zeta = u + iv$ and $\tilde{\zeta} = \tilde{u} + i\tilde{v}$ are the stereographic projections (of centre P_0^* of coordinates $(-1, 0, 0)$) of the points of the Poincaré sphere with coordinates σ_i and $\tilde{\sigma}_i$, respectively. Hence, the transformation (7.5.13) of the Poincaré sphere of unit radius, is described in stereographic projection as the transformation (7.5.14) of the complex UV-plane. The properties of such transformations are well known (e.g. Ahlfors, 1953).

If $\det J = 0$, we have $J_{22} J_{12}^{-1} = J_{21} J_{11}^{-1}$ and (7.5.14) reduces to

$$\tilde{\zeta} = \frac{J_{22}}{J_{12}} = \frac{J_{21}}{J_{11}}, \qquad (7.5.15)$$

and $\tilde{\zeta}$ is thus independent of ζ. All the polarization forms are transformed into the same polarization form. Thus all the points of the Poincaré sphere are mapped onto a single point of this sphere.

The transformation (7.5.14) with $\det J \neq 0$ constitute the 'group of homographic transformations', also called the 'conformal group'. This group is generated by the translations, the rotations, the homothetic transformations, and the inversion ($\tilde{\zeta} = \zeta^{-1}$). The corresponding transformations on the Poincaré sphere form the group of automorphisms of this sphere. These transformations are one-to-one, carry circles into circles, (but, in general, great circles are not carried into great circles), and preserve the angles between oriented circles. Further properties may be found in Ahlfors (1953) and Azzam and Bashara (1977).

The eigenbivectors of the Jones matrix are particularly important. Indeed if $\boldsymbol{A}$ is along an eigenbivector of $\boldsymbol{J}$, then $\tilde{\boldsymbol{A}}$ is parallel to $\boldsymbol{A}$, which means that the polarization form of the wave with amplitude bivector $\boldsymbol{A}$ is not modified by the device. Thus, to eigenbivectors of the Jones matrix correspond fixed points of the transformation (7.5.13) of the Poincaré sphere, or, equivalently, fixed points of the transformation (7.5.14) of the stereographic projection onto the complex UV-plane.

Here we consider only the general case when the Jones matrix $\boldsymbol{J}$ has two different eigenvalues α, β. Let $\boldsymbol{P}$ and $\boldsymbol{Q}$ be the corresponding eigenbivectors (each defined up to a complex scalar factor). Using the decomposition (7.3.1) and (7.3.2) of $\boldsymbol{A}$, we have

$$\boldsymbol{J}\boldsymbol{A} = \alpha \frac{\boldsymbol{A}\cdot\boldsymbol{Q}_\perp}{\boldsymbol{P}\cdot\boldsymbol{Q}_\perp}\boldsymbol{P} + \beta \frac{\boldsymbol{A}\cdot\boldsymbol{P}_\perp}{\boldsymbol{Q}\cdot\boldsymbol{P}_\perp}\boldsymbol{Q} \tag{7.5.16}$$

for every bivector $\boldsymbol{A}$, and hence

$$\boldsymbol{J} = \alpha \frac{\boldsymbol{P}\otimes\boldsymbol{Q}_\perp}{\boldsymbol{P}\cdot\boldsymbol{Q}_\perp} + \beta \frac{\boldsymbol{Q}\otimes\boldsymbol{P}_\perp}{\boldsymbol{Q}\cdot\boldsymbol{P}_\perp}. \tag{7.5.17}$$

This gives the Jones matrix of a device knowing its eigenbivectors (each up to a factor) and its eigenvalues. Because (7.5.17) is not modified when $\boldsymbol{P}_\perp$ and $\boldsymbol{Q}_\perp$ are multiplied by arbitrary factors, it may also be written in the form

$$\boldsymbol{J} = \alpha \frac{\boldsymbol{P}\otimes\boldsymbol{Q}'}{\boldsymbol{P}\cdot\boldsymbol{Q}'} + \beta \frac{\boldsymbol{Q}\otimes\boldsymbol{P}'}{\boldsymbol{Q}\cdot\boldsymbol{P}'}, \tag{7.5.18}$$

with

$$\boldsymbol{P} = P_1\mathbf{i} + P_2\mathbf{j}, \quad \mathbf{Q} = Q_1\mathbf{i} + Q_2\mathbf{j}, \tag{7.5.19}$$

and

$$P' = -P_2\mathbf{i} + P_1\mathbf{j}, \quad Q' = -Q_2\mathbf{i} + Q_1\mathbf{j}. \tag{7.5.20}$$

Now corresponding to the spectral decomposition (7.5.18) of the Jones matrix, we may obtain a decomposition of the Mueller matrix. Introducing (7.5.18) into (7.5.6), we obtain

$$2\mu_{AB} = \alpha\bar{\alpha}\frac{(\bar{P}^{\mathrm{T}}\Pi_A P)(Q'^{\mathrm{T}}\Pi_B\bar{Q}')}{(P\cdot Q')(\bar{P}\cdot\bar{Q}')} + \beta\bar{\beta}\frac{(\bar{Q}^{\mathrm{T}}\Pi_A Q)(P'^{\mathrm{T}}\Pi_B\bar{P}')}{(Q\cdot P')(\bar{Q}\cdot\bar{P}')}$$
$$+ 2\left\{\alpha\bar{\beta}\frac{(\bar{Q}^{\mathrm{T}}\Pi_A P)(Q'^{\mathrm{T}}\Pi_B\bar{P}')}{(P\cdot Q')(\bar{P}'\cdot\bar{Q})}\right\}^{+}. \tag{7.5.21}$$

Using the identity (7.5.9), this may also be written as

$$\mu_{AB} = \alpha\bar{\alpha}\frac{(\bar{P}^{\mathrm{T}}\Pi_A P)(Q'^{\mathrm{T}}\Pi_B\bar{Q}')}{(\bar{P}^{\mathrm{T}}\Pi_C P)(Q'^{\mathrm{T}}\Pi_C\bar{Q}')} + \beta\bar{\beta}\frac{(\bar{Q}^{\mathrm{T}}\Pi_A Q)(P'^{\mathrm{T}}\Pi_B\bar{P}')}{(\bar{Q}^{\mathrm{T}}\Pi_C Q)(P'^{\mathrm{T}}\Pi_C\bar{P}')}$$
$$+ 2\left\{\alpha\bar{\beta}\frac{(\bar{Q}^{\mathrm{T}}\Pi_A P)(Q'^{\mathrm{T}}\Pi_B\bar{P}')}{(\bar{Q}^{\mathrm{T}}\Pi_C P)(Q'^{\mathrm{T}}\Pi_C\bar{P}')}\right\}^{+}. \tag{7.5.22}$$

But it is easily seen that

$$P'^{\mathrm{T}}\Pi_B\bar{P}' = \gamma_{BC}\bar{P}^{\mathrm{T}}\Pi_C P, \quad Q'^{\mathrm{T}}\Pi_B\bar{P}' = \gamma_{BC}\bar{P}^{\mathrm{T}}\Pi_B Q, \text{ etc.} \tag{7.5.23}$$

Hence, recalling (7.3.19), the Stokes parameters $s_A^{(1)}$ and $s_A^{(2)}$, corresponding respectively to the eigenbivectors P and Q, are given by $s_A^{(1)} = \bar{P}^{\mathrm{T}}\Pi_A P$ and $s_A^{(2)} = \bar{Q}^{\mathrm{T}}\Pi_A Q$, and thus we obtain

$$\mu_{AB} = \alpha\bar{\alpha}\frac{s_A^{(1)}\gamma_{BC}s_C^{(2)}}{s_D^{(1)}\gamma_{DE}s_E^{(2)}} + \beta\bar{\beta}\frac{s_A^{(2)}\gamma_{BC}s_C^{(1)}}{s_D^{(1)}\gamma_{DE}s_E^{(2)}} + 2\left\{\alpha\bar{\beta}\frac{(\bar{Q}^{\mathrm{T}}\Pi_A P)\gamma_{BC}(\bar{P}^{\mathrm{T}}\Pi_C Q)}{(\bar{Q}^{\mathrm{T}}\Pi_D P)\gamma_{DE}(\bar{P}^{\mathrm{T}}\Pi_E Q)}\right\}^{+}. \tag{7.5.24}$$

Now, the last term in (7.5.24) may also be expressed in terms of the Stokes parameters $s_A^{(1)}, s_A^{(2)}$. Indeed, introducing the hermitian matrices $H_{\alpha\beta}^{(1)} = P_\alpha\bar{P}_\beta$ and $H_{\alpha\beta}^{(2)} = Q_\alpha\bar{Q}_\beta$, and using $2\boldsymbol{H}^{(1)} = s_A^{(1)}\Pi_A$ and $2\boldsymbol{H}^{(2)} = s_A^{(2)}\Pi_A$, we note that

$$(\bar{Q}^{\mathrm{T}}\Pi_A P)(\bar{P}^{\mathrm{T}}\Pi_C Q) = \mathrm{tr}(\Pi_A \boldsymbol{H}^{(1)}\Pi_C \boldsymbol{H}^{(2)}) = \tfrac{1}{4}s_B^{(1)}s_D^{(2)}\,\mathrm{tr}(\Pi_A\Pi_B\Pi_C\Pi_D). \tag{7.5.25}$$

But, using the identity (see Exercise 7.5.2)

$$\tfrac{1}{2}\mathrm{tr}(\Pi_A\Pi_B\Pi_C\Pi_D) = \delta_{AB}\delta_{CD} + \delta_{AD}\delta_{BC} - \gamma_{AC}\gamma_{BD} - \mathrm{i}\gamma_{AE}\gamma_{CF}e_{EBFD}, \tag{7.5.26}$$

where ε_{ABCD} denotes the alternating symbol ($\varepsilon_{0123} = +1$), (7.5.25) becomes

$$2(\bar{Q}^{\mathrm{T}}\Pi_A P)(\bar{P}^{\mathrm{T}}\Pi_C Q) = s_A^{(1)}s_C^{(2)} + s_A^{(2)}s_C^{(1)} - \gamma_{AC}(s_B^{(1)}\gamma_{BD}s_D^{(2)}) \\ - \mathrm{i}\gamma_{AE}\gamma_{CF}\mathrm{e}_{EBFD}s_B^{(1)}s_D^{(2)}. \qquad (7.5.27)$$

Using this in (7.5.24), we finally obtain

$$(s_D^{(1)}\gamma_{DE}s_E^{(2)})\mu_{AB} = \{\alpha\bar{\alpha} - (\alpha\bar{\beta})^+\}s_A^{(1)}\gamma_{BC}s_C^{(2)} + \{\beta\bar{\beta} - (\alpha\bar{\beta})^+\}s_A^{(2)}\gamma_{BC}s_C^{(1)} \\ + (\alpha\bar{\beta})^+(s_D^{(1)}\gamma_{DE}s_E^{(2)})\delta_{AB} + (\alpha\bar{\beta})^-\gamma_{AC}\mathrm{e}_{CBDE}s_D^{(1)}s_E^{(2)}. \qquad (7.5.28)$$

This formula gives the Mueller matrix μ_{AB} of a device in terms of the eigenvalues α, β of the Jones matrix, and of the Stokes parameters $s_A^{(1)}, s_A^{(2)}$ associated with the corresponding eigenbivectors of this Jones matrix. The Stokes parameters $s_A^{(1)}$ and $s_A^{(2)}$ characterize the polarization forms that are not modified by the device. The complex numbers α and β are the factors multiplying the complex amplitudes of the waves with polarization forms $s_A^{(1)}$ and $s_A^{(2)}$, respectively. The squared moduli $\alpha\bar{\alpha}$ and $\beta\bar{\beta}$ characterize the action of the device on the intensities of these two waves: they are the factors multiplying these intensities. The argument of $\alpha\bar{\beta}$ characterizes the action of the device on the relative phase of these two waves: it is the phase retardation of one with respect to the other.

Exercises 7.5

1. Show that if a linear transformation $\tilde{s}_A = \mu_{AB}s_A$ is such that $s_A\gamma_{AB}s_B = 0$ implies $\tilde{s}_A\gamma_{AB}\tilde{s}_B = 0$, then $\boldsymbol{\mu}^{\mathrm{T}}\boldsymbol{\gamma}\boldsymbol{\mu} = \lambda\boldsymbol{\gamma}$.
2. Prove the identity (7.5.26). **Hint**: separate the cases when 0, 1, 2, 3 or all of the indices A, B, C, D are zero.
3. Check that the Mueller matrix given by (7.5.28) has $s_B^{(1)}$ and $s_B^{(2)}$ as eigenvectors. What are the corresponding eigenvalues?
4. Find the coordinates σ_i $(i = 1, 2, 3)$ of a point on the Poincaré sphere of unit radius in terms of the complex polarization variable $\zeta = A_2A_1^{-1}$.
5. Find an expression for the Jones matrix of a device given the complex polarization variables $\zeta^{(1)}, \zeta^{(2)}$ of its eigenbivectors and its eigenvalues α, β (assume $\alpha \neq \beta$).
6. A device is said to be 'conservative' when, for every amplitude bivector, the intensity is the same before and after the

transformation. Show that a device is conservative if and only if its Jones matrix is unitary ($\boldsymbol{J}\bar{\boldsymbol{J}}^{\mathrm{T}} = \mathbf{1}$). What can be said about the corresponding Mueller matrix in this case? What is the corresponding transformation on the Poincaré sphere?

7.6 Polarizers and retarders

Here we consider two special devices acting linearly on the amplitude bivector: the ideal polarizer and the retarder. These are used in optics (Shurcliff, 1962; Azzam and Bashara 1977).

7.6.1 Ideal polarizers

An 'ideal polarizer' is a device characterized by a Jones matrix having two eigenbivectors $\boldsymbol{P}$ and $\boldsymbol{Q} = \bar{\boldsymbol{P}}'$ defining opposite polarization forms, the corresponding eigenvalues being $\alpha = 1$ and $\beta = 0$. Thus when a wave with the polarization form of $\mathbf{P}$ impinges upon the polarizer, it emerges unchanged; when a wave with the opposite polarization form (polarization form of $\bar{\boldsymbol{P}}'$) impinges upon it, no wave emerges.

From (7.5.18), we immediately see that the Jones matrix of an ideal polarizer is

$$\boldsymbol{J} = \frac{\boldsymbol{P} \otimes \bar{\boldsymbol{P}}}{\boldsymbol{P} \cdot \bar{\boldsymbol{P}}}. \tag{7.6.1}$$

From (7.5.28), we obtain its Mueller matrix, on noting that here $s_A^{(2)} = \gamma_{AB} s_B^{(1)}$, because opposite polarization forms are represented by antipodal points on the Poincaré sphere. We have

$$\mu_{AB} = \frac{s_A^{(1)} s_B^{(1)}}{s_C^{(1)} s_C^{(1)}}. \tag{7.6.2}$$

The Jones matrix of an ideal polarizer is singular, $\det \boldsymbol{J} = 0$, and thus the transformation (7.5.14) of the complex polarization variable reduces to (7.5.15), which reads, on using (7.6.1),

$$\zeta = \frac{P_2}{P_1}. \tag{7.6.3}$$

All the polarization forms are transformed into the polarization form of $\boldsymbol{P}$. All the points of the Poincaré sphere are mapped onto the single point associated with this polarization form.

When the polarization form of the eigenbivector $\boldsymbol{P}$ is elliptic, the polarizer is said to be 'elliptic'. The polarizer is said to be 'linear' when the polarization form of $\boldsymbol{P}$ is linear, so that, in this case, without loss in generality, the eigenbivector $\boldsymbol{P}$ corresponding to the eigenvalue $\alpha = 1$ may be taken to be a unit real vector $\hat{\mathbf{p}}$.

Consider now an ideal linear polarizer with $\hat{\mathbf{p}}$ inclined at an angle θ with the x-axis:

$$\boldsymbol{P} = \hat{\mathbf{p}} = \cos\theta\mathbf{i} + \sin\theta\mathbf{j}. \tag{7.6.4}$$

The corresponding Stokes parameters are

$$s_A^{(1)} = (1, \cos 2\theta, \sin 2\theta, 0). \tag{7.6.5}$$

Consider now an elliptically polarized wave of unit intensity ($I = 1$) which is incident on this ideal linear polarizer. Let

$$s_A = (1, \cos 2\beta \cos 2\chi, \cos 2\beta \sin 2\chi, \sin 2\beta) \tag{7.6.6}$$

be its Stokes parameters. The intensity, $\tilde{I}$, of the emergent wave may be obtained using the Mueller matrix (7.6.2). Indeed, writing $\tilde{I} = \tilde{s}_0 = \mu_{0B} s_B$, we obtain

$$\tilde{I} = \tfrac{1}{2} s_B^{(1)} s_B = \tfrac{1}{2} + \tfrac{1}{2}\cos 2\beta \cos 2(\chi - \theta). \tag{7.6.7}$$

On the Poincaré sphere of unit radius, the polarization form of the eigenvector $\hat{\mathbf{p}}$ is represented by the point P_1 on the equator with longitude 2θ. The polarization form of the incident wave is represented by the point P with latitude 2β and longitude 2χ (Figure 7.4).

Let N be the north pole of the sphere. Then, from the spherical triangle PNP_1, we have, using formula (A.1) of the Appendix,

$$\cos(\widehat{P_1 P}) = \cos(\widehat{NP_1})\cos(\widehat{NP}) + \sin(\widehat{NP_1})\sin(\widehat{NP})\cos 2(\chi - \theta). \tag{7.6.8}$$

But $\widehat{NP_1} = \tfrac{1}{2}\pi$ and $\widehat{NP} = \tfrac{1}{2}\pi - 2\beta$, so that

$$\cos(\widehat{P_1 P}) = \cos 2\beta \cos 2(\chi - \theta). \tag{7.6.9}$$

Hence (7.6.7) becomes

$$\tilde{I} = \cos^2\left(\frac{\widehat{P_1 P}}{2}\right). \tag{7.6.10}$$

Thus for the ideal linear polarizer that transmits unchanged the polarization form represented by the point P_1 of the Poincaré sphere,

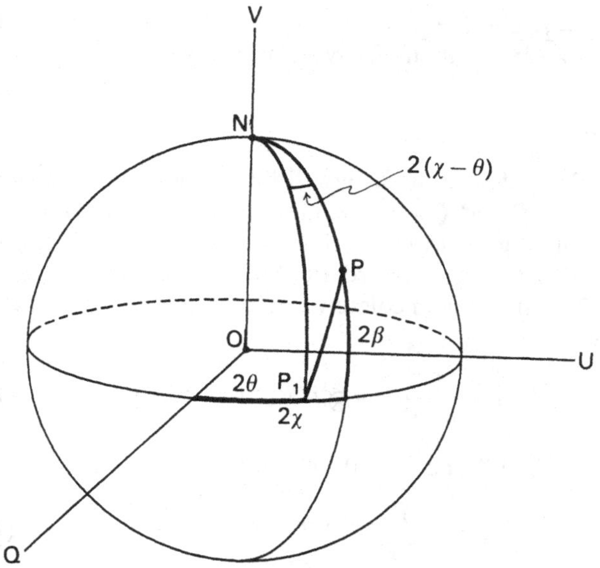

Figure 7.4 *Action of a linear polarizer interpreted on the Poincaré sphere.*

the emerging wave has intensity $\cos^2 \frac{1}{2}\widehat{P_1 P}$, where P is the point of the Poincaré sphere representing the polarization form of the incident wave (taken to be of unit intensity). It may be shown that this result remains valid for the more general case when the ideal polarizer is elliptical ($\boldsymbol{P}$ is a bivector) and the corresponding point P_1 on the Poincaré sphere need not lie on the equator (see Exercise 7.6.1).

Suppose now that the point P_1 representing the polarizer is given and that the transmitted intensity $\tilde{I}$ is also given. Then, from (7.6.10), the point P representing the incident wave must lie on a small circle on the sphere, the pole of which is at P_1.

Note that if the polarization form of the incident wave is opposite to that of the polarizer, P and P_1 are antipodal, $\widehat{P_1 P} = \pi$, and $\tilde{I} = 0$: we retrieve the fact that there is no emerging wave. Also, if the polarization form of the incident wave is the same as that of the polarizer, P and P_1 coincide, $\widehat{P_1 P} = 0$, and $\tilde{I} = 1$, in accordance with the fact that the incident wave emerges unchanged. Finally, note that if P is at a pole (north or south) and if P_1 is on the equator, then $P_1(\hat{P}) = \frac{1}{2}\pi$ and $\tilde{I} = \frac{1}{2}$: an ideal linear polarizer transmits half the intensity of an incident circularly polarized wave.

7.6.2 Retarders

A 'retarder' is a device characterized by a Jones matrix having two eigenbivectors $\boldsymbol{P}$ and $\boldsymbol{Q} = \bar{\boldsymbol{P}}'$ defining opposite polarization forms, the corresponding eigenvalues being $\alpha = e^{i\varphi}$, $\beta = e^{-i\varphi}$, where φ is real (2φ is called the 'relative retardation' of the retarder).

From (7.5.18), we immediately see that the Jones matrix of a retarder is

$$\boldsymbol{J} = \frac{1}{\boldsymbol{P}\cdot\bar{\boldsymbol{P}}}(e^{i\varphi}\boldsymbol{P}\otimes\bar{\boldsymbol{P}} + e^{-i\varphi}\bar{\boldsymbol{P}}'\otimes\boldsymbol{P}'), \tag{7.6.11}$$

on recalling (7.5.19) and (7.5.20). We note that

$$\boldsymbol{J}\bar{\boldsymbol{J}}^{\mathrm{T}} = \frac{1}{\boldsymbol{P}\cdot\bar{\boldsymbol{P}}}(\boldsymbol{P}\otimes\bar{\boldsymbol{P}} + \bar{\boldsymbol{P}}'\otimes\boldsymbol{P}') = \mathbf{1}, \tag{7.6.12}$$

so that $\boldsymbol{J}$ is unitary. It follows that, whatever the polarization of the incident wave may be, the intensity of the emerging wave is the same as the intensity of the incident wave (see Exercise 7.5.6).

Also (see Exercise 7.5.6), the corresponding Mueller matrix is such that $\mu_{00} = 1$, $\mu_{0i} = \mu_{i0} = 0$, and μ_{ij} is orthogonal ($i, j = 1, 2, 3$). Hence, from (7.5.13), $\tilde{\sigma}_i = \mu_{ij}\sigma_j$, and the corresponding transformation of the Poincaré sphere of unit radius is a rotation. Taking $\boldsymbol{P}\cdot\bar{\boldsymbol{P}} = 1$ without loss in generality, so that $s_0^{(1)} = 1$, we obtain from (7.5.28)

$$\mu_{ij} = (1 - \cos 2\varphi)s_i^{(1)}s_j^{(1)} + \cos 2\varphi\,\delta_{ij} + (\sin 2\varphi)e_{ijk}s_k^{(1)}. \tag{7.6.13}$$

This is the matrix of the rotation of angle -2φ (2φ being the retardation) about the axis $s_i^{(1)}$. Thus the points of coordinates $s_i^{(1)}$ and $-s_i^{(1)}$ associated with the polarization forms of the eigenbivectors are fixed, while all the other points of the Poincaré sphere rotate through the angle -2φ about the axis passing through these points.

When the polarization form of the eigenbivector $\boldsymbol{P}$ is elliptic, the retarder is said to be 'elliptic'. When this polarization form is linear the retarder is said to be 'linear'.

As an example, consider a linear retarder with $\boldsymbol{P} = \mathbf{i}$ along the x-axis. Then, $s_i^{(1)} = (1, 0, 0)$, and from (7.6.13),

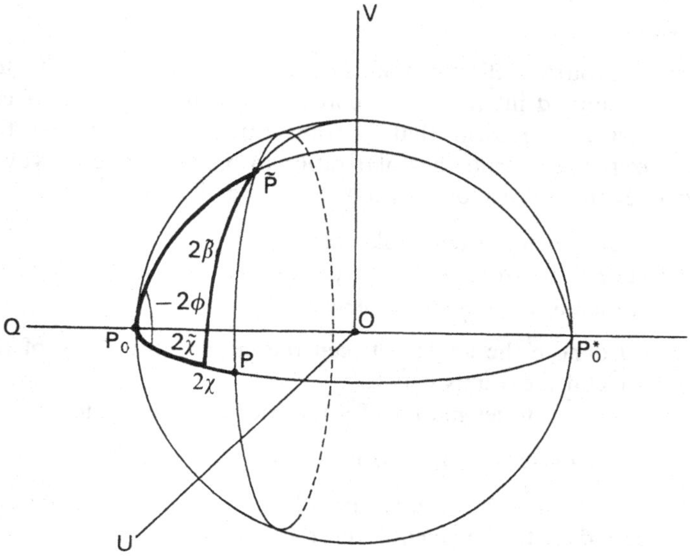

Figure 7.5 *Action of a linear retarder interpreted on the Poincaré sphere (rotation about the axis OQ).*

$$\mu_{ij} = \begin{bmatrix} 1 & 0 & 0 \\ 0 & \cos 2\varphi & \sin 2\varphi \\ 0 & -\sin 2\varphi & \cos 2\varphi \end{bmatrix}. \tag{7.6.14}$$

Now assume that the incident wave is linearly polarized in a direction tilted at an angle χ with respect to the x-axis. Then $\sigma_i = (\cos 2\chi, \sin 2\chi, 0)$. The coordinates $\tilde{\sigma}_i = (\cos 2\tilde{\beta} \cos 2\tilde{\chi}, \cos 2\tilde{\beta} \sin 2\tilde{\chi}, \sin 2\tilde{\beta})$ of the point of the Poincaré sphere associated with the polarization form of the emerging wave are then given by

$$\cos 2\tilde{\beta} \cos 2\tilde{\chi} = \cos 2\chi, \quad \cos 2\tilde{\beta} \sin 2\tilde{\chi} = \cos 2\varphi \sin 2\chi,$$
$$\sin 2\tilde{\beta} = -\sin 2\varphi \sin 2\chi. \tag{7.6.15}$$

The situation is represented in Figure 7.5: the point P represents the polarization form of the incident wave, and the point P_0 represents the polarization form of the eigenbivector $\boldsymbol{P} = \mathbf{i}$ of the retarder. The polarization form of the emerging wave is represented by the point $\tilde{P}$, which is obtained from P by a rotation of angle -2φ about the axis OP_0.

Exercises 7.6

1. For an arbitrary elliptic polarizer, derive the result (7.6.10) for the transmitted intensity (P_1 represents on the Poincaré sphere the polarization form that is transmitted unchanged by the polarizer; P represents the polarization form of the incident wave).
2. Write explicitly the Jones and Mueller matrices of

   (a) a 'right-circular' ideal polarizer;
   (b) a 'linear' retarder with retardation $2\varphi = \frac{1}{2}\pi$ (such a retarder is called a 'quarter-wave plate').

   Let θ denote the angle between the eigenvector $\boldsymbol{P} = \hat{\boldsymbol{p}}$ of the retarder and the x-axis.
3. Show that the Jones matrix of a retarder may be written

$$\boldsymbol{J} = (\cos\varphi)\boldsymbol{\Pi}_0 + \mathrm{i}(\sin\varphi)s_i^{(1)}\boldsymbol{\Pi}_i, \quad (i = 1, 2, 3),$$

   where 2φ is the retardation, and $s_A^{(1)}$ are the Stokes parameters corresponding to the eigenbivector $\boldsymbol{P}(\boldsymbol{P}\cdot\bar{\boldsymbol{P}} = 1)$.
4. A 'partial polarizer' is a device characterized by a Jones matrix having two eigenbivectors $\boldsymbol{P}$, $\boldsymbol{Q} = \bar{\boldsymbol{P}}'$ defining opposite polarization forms, with corresponding eigenvalues $\alpha = \mathrm{e}^{+k}$, $\beta = \mathrm{e}^{-k}$ (k real).

   Write the expressions for the Jones matrix and the Mueller matrix of a partial polarizer.

8

Energy flux

Waves carry energy. For that reason it is useful to have expressions for the energy flux across a surface element for trains of propagating waves whether these are homogeneous or inhomogeneous. We consider systems, such as linearized elasticity, for which there is conservation of energy. We also deal exclusively with linear systems in which the sum of solutions of the field equations is also a solution of the field equations. Generally, the most relevant quantities are not the specific values of energy flux and energy density, but rather, for time harmonic waves, the mean values of these quantities, the means being taken over a cycle at a point. The reason for this is that in many systems, for example in dealing with the propagation of light, the frequencies are so high that there are very many fluctuations in a second.

Here we try to take as general a standpoint as possible, abstracting general features from the types of systems considered. Results are derived without making explicit use of the constitutive equations describing the systems. What we obtain are general results for all the systems which satisfy the criteria of being linear, conservative and in which the energy flux and energy density involve only products or sums of products of pairs of field quantities.

8.1 Conservation of energy

The energy flux vector is denoted by $\boldsymbol{r}(\boldsymbol{x}, t)$, and the energy density by $e(\boldsymbol{x}, t)$. If V is a fixed volume then the energy contained within V at any time t is $\int\int\int_V e(\boldsymbol{x}, t)\,\mathrm{d}V$. If the closed surface S surrounding V has unit outward normal $\mathbf{n}$ then the flux of energy out across S per unit time is $\int\int_S \boldsymbol{r}\cdot\mathbf{n}\,\mathrm{d}S$. This is equal to the time rate of decrease of energy within the surface, assuming, of course, that there are no sources or sinks of energy. Thus we have the integral form of energy conservation:

$$\iint_S \boldsymbol{r}\cdot\mathbf{n}\,\mathrm{d}S = -\frac{\partial}{\partial t}\iiint_V e\,\mathrm{d}V. \tag{8.1.1}$$

Using the divergence theorem and assuming the integrand is continuous we obtain the point form of the conservation of energy

$$\nabla\cdot\boldsymbol{r} + \frac{\partial e}{\partial t} = 0. \tag{8.1.2}$$

For example, for nondissipative and nondispersive dielectrics with symmetric permittivity and permeability tensors,

$$\boldsymbol{r} = \mathscr{E} \times \mathscr{H}, \quad e = \tfrac{1}{2}(\mathscr{D}\cdot\mathscr{E} + \mathscr{B}\cdot\mathscr{H}), \tag{8.1.3}$$

where $\mathscr{E}$ and $\mathscr{H}$ are the electric and magnetic field, $\mathscr{D}$ the electric displacement and $\mathscr{B}$ is the magnetic induction (Chapter 9). For a purely elastic system

$$r_i = -t_{ij}v_j, \quad e = \tfrac{1}{2}\rho v_i v_i + \tfrac{1}{2}t_{ij}e_{ij}, \tag{8.1.4}$$

where $\boldsymbol{t}$ is the Cauchy stress tensor, $\boldsymbol{v}$ the particle velocity, ρ the material density and $\boldsymbol{e}$ the infinitesimal strain tensor (Chapter 10).

Note in both these examples that the energy flux is a product of a pair of field quantities: $\mathscr{E}$ and $\mathscr{H}$ in the electromagnetic case; $\boldsymbol{t}$ and $\boldsymbol{v}$ in the elastic case. Also note that the energy density is a sum of products of pairs of field quantities.

For waves of period $\tau = 2\pi\omega^{-1}$, the mean energy flux $\tilde{\boldsymbol{r}}$, and the mean energy density $\tilde{e}$, are given by

$$\tilde{\boldsymbol{r}} = \frac{1}{\tau}\int_0^\tau \boldsymbol{r}(\boldsymbol{x},t)\,\mathrm{d}t, \quad \tilde{e} = \frac{1}{\tau}\int_0^\tau e(\boldsymbol{x},t)\,\mathrm{d}t. \tag{8.1.5}$$

8.2 Homogeneous plane waves

We consider the propagation of a single infinite train of elliptically polarized time-harmonic homogeneous plane waves propagating in a homogeneous linear conservative dispersive system.

We make three assumptions.

(a) The energy flux vector $\boldsymbol{r}$, and the energy density e, involve products of pairs of field quantities.
(b) There is neither internal energy supply nor dissipation.

(c) The system is linear in the sense that if one field quantity such as displacement, velocity, stress or magnetic field, for example, is of the form $\boldsymbol{A}\exp\mathrm{i}(\boldsymbol{k}\cdot\boldsymbol{x}-\omega t)$, then every other field quantity is of the similar form $\boldsymbol{B}\exp\mathrm{i}(\boldsymbol{k}\cdot\boldsymbol{x}-\omega t)$. Here $\boldsymbol{A}$ and $\boldsymbol{B}$ may be (complex) scalars, vectors or tensors.

The first assumption is motivated by the examples of the electromagnetic and elastic systems, whilst the third assumption is essentially a statement about the types of systems under consideration. They are linear systems in which a linear combination of solutions is also a solution. The second assumption means that the systems under consideration are such that the conservation of energy equation (8.1.2) holds.

No assumption need be made about whether or not the system is anisotropic, nor about whether or not it is subject to internal constraints, such as incompressibility or inextensibility in a certain direction, as occurs in elasticity theory.

Suppose now that for the system under consideration the dispersion relation is

$$\omega = f(\boldsymbol{k}), \tag{8.2.1}$$

where f is a known function. If ω and $\boldsymbol{k}$ satisfy (8.2.1), then a wave train represented by

$$\boldsymbol{C}\exp\mathrm{i}(\boldsymbol{k}\cdot\boldsymbol{x}-\omega t), \tag{8.2.2}$$

may propagate in the system. Here $\boldsymbol{C}$ is a constant scalar, vector or tensor, in general complex. Due to assumption (a) the energy flux vector associated with (8.2.2) must have the form

$$\boldsymbol{r} = \boldsymbol{\Lambda}\exp 2\mathrm{i}(\boldsymbol{k}\cdot\boldsymbol{x}-\omega t) + \bar{\boldsymbol{\Lambda}}\exp -2\mathrm{i}(\boldsymbol{k}\cdot\boldsymbol{x}-\omega t) + \boldsymbol{\mu}, \tag{8.2.3}$$

where $\boldsymbol{\mu}$ is a real constant vector and $\boldsymbol{\Lambda}$ is a constant bivector. These vectors are functions of ω and $\boldsymbol{k}$, and of the parameters specifying the system. The reason it must have this form is that it is a product of the field quantities both of the form (8.2.2) and in determining the real vector $\boldsymbol{r}$, the real parts must be taken in each of the expressions in the product. Clearly, using (8.1.5), $\tilde{\boldsymbol{r}} = \boldsymbol{\mu}$.

The energy density e associated with the wave train, has a form similar to that of $\boldsymbol{r}$. Thus

$$e = A\exp\{2\mathrm{i}(\boldsymbol{k}\cdot\boldsymbol{x}-\omega t)\} + \bar{A}\exp\{-2\mathrm{i}(\boldsymbol{k}\cdot\boldsymbol{x}-\omega t)\} + \gamma, \tag{8.2.4}$$

where γ is a real constant and A is a complex constant. The constants are functions of ω and $\boldsymbol{k}$ and of the parameters specifying the system. Clearly $\tilde{e} = \gamma$.

The energy flux velocity $\boldsymbol{g}$ may be defined as

$$\boldsymbol{g} = \frac{\text{mean energy flux}}{\text{mean energy density}} = \frac{\tilde{\boldsymbol{r}}}{\tilde{e}}. \tag{8.2.5}$$

For the train (8.2.2),

$$\boldsymbol{g} = \frac{\boldsymbol{\mu}}{\gamma}. \tag{8.2.6}$$

Suppose now that $\boldsymbol{k}$ is replaced by $\boldsymbol{k} + \mathrm{i}\varepsilon\boldsymbol{k}'$ where ε is infinitesimal and $\boldsymbol{k}'$ is arbitrary and real. In order that (8.2.1) be satisfied, ω must be replaced by $\omega + \mathrm{i}\varepsilon\omega'$, where

$$\omega' = \boldsymbol{k}' \cdot \frac{\partial f}{\partial \boldsymbol{k}} = k_j' \frac{\partial f}{\partial k_j}. \tag{8.2.7}$$

Here the Taylor expansion of $f(\boldsymbol{k} + \mathrm{i}\varepsilon\boldsymbol{k}')$ has been used and terms higher than first order in ε have been neglected.

The wave train is now of the form

$$(\boldsymbol{C} + \varepsilon\boldsymbol{C}')\exp\mathrm{i}(\boldsymbol{k}\cdot\boldsymbol{x} - \omega t)\exp(-\varepsilon\boldsymbol{k}'\cdot\boldsymbol{x} + \varepsilon\omega' t), \tag{8.2.8}$$

where $\boldsymbol{C}'$ is some constant scalar, vector or tensor, in general complex.

The corresponding energy flux is a function of ε, say $\boldsymbol{r}(\varepsilon)$. Using assumption (a) we infer that it has the form

$$\begin{aligned}\boldsymbol{r}(\varepsilon) = [(\boldsymbol{\Lambda} + \varepsilon\boldsymbol{\Phi})\exp 2\mathrm{i}(\boldsymbol{k}\cdot\boldsymbol{x} - \omega t) + (\bar{\boldsymbol{\Lambda}} + \varepsilon\bar{\boldsymbol{\Phi}})\exp - 2\mathrm{i}(\boldsymbol{k}\cdot\boldsymbol{x} - \omega t)\\ + (\boldsymbol{\mu} + \varepsilon\boldsymbol{\psi})]\exp[-2\varepsilon\boldsymbol{k}'\cdot\boldsymbol{x} + 2\varepsilon\omega' t],\end{aligned} \tag{8.2.9}$$

where $\boldsymbol{\Phi}$ is some bivector and $\boldsymbol{\psi}$ is some real vector. Similarly, e is also a function of ε, say $e(\varepsilon)$, and has the form

$$\begin{aligned}e(\varepsilon) = [(A + \varepsilon B)\exp 2\mathrm{i}(\boldsymbol{k}\cdot\boldsymbol{x} - \omega t) + (\bar{A} + \varepsilon\bar{B})\exp - 2\mathrm{i}(\boldsymbol{k}\cdot\boldsymbol{x} - \omega t)\\ + (\gamma + \varepsilon\delta)]\exp[-2\varepsilon\boldsymbol{k}'\cdot\boldsymbol{x} + 2\varepsilon\omega' t],\end{aligned} \tag{8.2.10}$$

where B is some complex scalar and δ some real scalar.

On inserting (8.2.9) and (8.2.10) into (8.1.2), the equation of conservation of energy, we find

$$(\boldsymbol{\mu} + \varepsilon\boldsymbol{\psi})\cdot\boldsymbol{k}' = (\gamma + \varepsilon\delta)\omega'. \tag{8.2.11}$$

Hence

$$\boldsymbol{\mu}\cdot\boldsymbol{k}' = \gamma\omega', \tag{8.2.12}$$

and using (8.2.7) this gives

$$\left(\boldsymbol{\mu}-\gamma\frac{\partial f}{\partial \boldsymbol{k}}\right)\cdot\boldsymbol{k}'=0. \tag{8.2.13}$$

This holds for arbitrary $\boldsymbol{k}'$, because no restrictions have been placed on $\boldsymbol{k}'$. Hence

$$\boldsymbol{\mu}=\gamma\frac{\partial f}{\partial \boldsymbol{k}}. \tag{8.2.14}$$

Thus, by (8.2.6)

$$\frac{\partial f}{\partial \boldsymbol{k}}=\frac{\boldsymbol{\mu}}{\gamma}=\boldsymbol{g}. \tag{8.2.15}$$

Thus for the train of elliptically polarized plane waves, the energy flux velocity is equal to $\partial\omega(\boldsymbol{k})/\partial\boldsymbol{k}$, which is called the **group velocity**.

8.3 Inhomogeneous plane waves

We consider the propagation of a single infinite train of elliptically polarized time-harmonic small amplitude inhomogeneous plane waves in a homogeneous conservative system.

Thus the propagating field quantity is assumed to have the form

$$\boldsymbol{A}\exp i\omega(\boldsymbol{S}\cdot\boldsymbol{x}-t). \tag{8.3.1}$$

Here $\boldsymbol{A}$ is the amplitude bivector, ω the real angular frequency and $\boldsymbol{S}$ the slowness bivector. The partial differential equations with constant coefficients governing the system will lead to a propagation condition of the form

$$\boldsymbol{L}\boldsymbol{A}=0, \tag{8.3.2}$$

where $\boldsymbol{L}$ is a second order tensor, which, in general, may depend on $\boldsymbol{S}$ and on ω.

Now to the three assumptions of section 8.2, we add a fourth, namely that $\boldsymbol{L}$ depends only upon $\boldsymbol{S}$ and thus not on the angular frequency ω:

$$\boldsymbol{L}=\boldsymbol{L}(\boldsymbol{S}). \tag{8.3.3}$$

It follows that the secular equation corresponding to the propagation condition (8.3.2) reads

$$Q(\boldsymbol{S})=\det\boldsymbol{L}(\boldsymbol{S})=0, \tag{8.3.4}$$

where Q is independent of ω. Examples of systems in which this is the case include classical linear homogeneous anisotropic elasticity theory (Chapter 10), flow of ideal incompressible fluids, and Maxwell's equations for homogeneous nondissipative and non-dispersive anisotropic dielectrics (Chapter 9).

When an $\boldsymbol{S}$ is obtained which satisfies the secular equation $Q(\boldsymbol{S}) = 0$, the corresponding $\boldsymbol{A}$ is determined from the propagation condition (8.3.2). Note that the amplitude eigenbivector $\boldsymbol{A}$ so obtained is independent of ω, because $\boldsymbol{L}$ is independent of ω. The propagation condition (8.3.2) is homogeneous and hence any scalar multiple of $\boldsymbol{A}$ is also a suitable amplitude bivector. To accommodate boundary or initial conditions which may involve the frequency ω, the field is assumed to have the form

$$f(\omega)\boldsymbol{A}\exp \mathrm{i}\omega(\boldsymbol{S}\cdot\boldsymbol{x} - t). \tag{8.3.5}$$

Here $f(\omega)$ is an arbitrary function of ω, $\boldsymbol{S}$ is a solution of the secular equation $Q(\boldsymbol{S}) = 0$, and $\boldsymbol{A}$ is the corresponding eigenbivector of (8.3.2).

If one field quantity has the form (8.3.5) then every other field quantity entering the expressions for the energy flux and energy density has the similar form

$$f(\omega)\boldsymbol{B}\exp \mathrm{i}\omega(\boldsymbol{S}\cdot\boldsymbol{x} - t), \tag{8.3.6}$$

where the same function $f(\omega)$ occurs in both expressions and $\boldsymbol{B}$ may be a (complex) scalar, vector or tensor (see the examples of classical linearized elasticity (Chapter 10) and electromagnetism (Chapter 9)).

The energy flux vector has the form given by

$$\begin{aligned} \boldsymbol{r} = &\{f(\omega)\boldsymbol{A}\exp \mathrm{i}\omega(\boldsymbol{S}\cdot\boldsymbol{x} - t) + \text{c.c.}\} \\ &\times \{f(\omega)\boldsymbol{B}\exp \mathrm{i}\omega(\boldsymbol{S}\cdot\boldsymbol{x} - t) + \text{c.c.}\}, \end{aligned} \tag{8.3.7}$$

so that

$$\begin{aligned} \boldsymbol{r} = &[\{f(\omega)\}^2\boldsymbol{\Gamma}\exp[2\mathrm{i}\omega(\boldsymbol{S}\cdot\boldsymbol{x} - t)] + \text{c.c.}] \\ &+ f(\omega)\overline{f(\omega)}\,\boldsymbol{v}\exp[\mathrm{i}\omega(\boldsymbol{S} - \bar{\boldsymbol{S}})\cdot\boldsymbol{x}], \end{aligned} \tag{8.3.8}$$

where $\boldsymbol{\Gamma}$ is a bivector and $\boldsymbol{v}$ is a real vector, both independent of ω.

Similarly, the energy density, e, has the form

$$\begin{aligned} e = &[\{f(\omega)\}^2 B\exp[2\mathrm{i}\omega(\boldsymbol{S}\cdot\boldsymbol{x} - t)] + \text{c.c.}] \\ &+ f(\omega)\overline{f(\omega)}\beta\exp[\mathrm{i}\omega(\boldsymbol{S} - \bar{\boldsymbol{S}})\cdot\boldsymbol{x}]. \end{aligned} \tag{8.3.9}$$

where B is a complex constant and β is a real constant, both independent of ω.

Taking the mean over a period, we note that

$$\tilde{\boldsymbol{r}} = f\bar{f}\boldsymbol{\nu}\exp[\mathrm{i}\omega(\boldsymbol{S}-\bar{\boldsymbol{S}})\cdot\boldsymbol{x}] = f\bar{f}\boldsymbol{\nu}\exp(-2\omega\boldsymbol{S}^{-}\cdot\boldsymbol{x}),$$
$$\tilde{e} = f\bar{f}\beta\exp[\mathrm{i}\omega(\boldsymbol{S}-\bar{\boldsymbol{S}})\cdot\boldsymbol{x}] = f\bar{f}\beta\exp(-2\omega\boldsymbol{S}^{-}\cdot\boldsymbol{x}), \qquad (8.3.10)$$

In contrast with the case of homogeneous waves, the mean energy flux and mean energy density depend upon position. This is due to the spatial attenuation of the inhomogeneous waves.

For further reference we introduce here the 'weighted mean' energy flux vector $\hat{\mathbf{r}}$ and the 'weighted mean' energy density $\hat{e}$. These are defined by

$$\hat{\mathbf{r}} = \tilde{\boldsymbol{r}}\exp(2\omega\boldsymbol{S}^{-}\cdot\boldsymbol{x}) = f\bar{f}\boldsymbol{\nu},$$
$$\hat{e} = \tilde{e}\exp(2\omega\boldsymbol{S}^{-}\cdot\boldsymbol{x}) = f\bar{f}\beta, \qquad (8.3.11)$$

and are constant for a given inhomogeneous wave solution.

Now inserting (8.3.8) and (8.3.9) into the conservation of energy equation (8.1.2) which has to be satisfied for all $\boldsymbol{x}$ and t, we conclude that

$$\boldsymbol{\nu}\cdot(\boldsymbol{S}-\bar{\boldsymbol{S}}) = 0, \qquad (8.3.12)$$

or, equivalently,

$$\tilde{\boldsymbol{r}}\cdot\boldsymbol{S}^{-} = 0. \qquad (8.3.13)$$

Thus the mean energy flux vector is parallel to the planes of constant amplitude, $\boldsymbol{S}^{-}\cdot\boldsymbol{x} = \text{constant}$.

Because the solutions are valid for arbitrary ω, they are equally valid when the real ω is replaced by a complex ω^*, whose real part is ω, the real frequency of the waves under consideration. Then $\boldsymbol{r}$, given by (8.3.8), is replaced by $\boldsymbol{r}(\omega^*)$, say, given by

$$\boldsymbol{r}(\omega^*) = \{[f(\omega^*)]^2\boldsymbol{\Gamma}\exp[2\mathrm{i}\omega^*(\boldsymbol{S}\cdot\boldsymbol{x}-t)] + \text{c.c.}\}$$
$$+ f(\omega^*)\overline{f(\omega^*)}\boldsymbol{\nu}\exp[\mathrm{i}\omega^*(\boldsymbol{S}\cdot\boldsymbol{x}-t) - \mathrm{i}\overline{\omega^*}(\bar{\boldsymbol{S}}\cdot\boldsymbol{x}-t)]. \qquad (8.3.14)$$

Here $\boldsymbol{\Gamma}$ and $\boldsymbol{\nu}$ are unchanged because they are independent of ω. Similarly, the energy density, e, is replaced by $e(\omega^*)$, say, given by

$$e(\omega^*) = \{[f(\omega^*)]^2 B\exp[2\mathrm{i}\omega^*(\boldsymbol{S}\cdot\boldsymbol{x}-t)] + \text{c.c.}\}$$
$$+ f(\omega^*)\overline{f(\omega^*)}\beta\exp[\mathrm{i}\omega^*(\boldsymbol{S}\cdot\boldsymbol{x}-t) - \mathrm{i}\overline{\omega^*}(\bar{\boldsymbol{S}}\cdot\boldsymbol{x}-t)]. \qquad (8.3.15)$$

The energy conservation equation (8.1.2) now gives

$$\boldsymbol{v}\cdot(\omega^*\boldsymbol{S}-\overline{\omega^*}\overline{\boldsymbol{S}})=\beta(\omega^*-\overline{\omega^*}). \tag{8.3.16}$$

Using (8.3.12) it follows that

$$\boldsymbol{v}\cdot\boldsymbol{S}=\beta, \tag{8.3.17}$$

or equivalently

$$\tilde{\boldsymbol{r}}\cdot\boldsymbol{S}=\tilde{e}. \tag{8.3.18}$$

Hence, taking real and imaginary parts,

$$\tilde{\boldsymbol{r}}\cdot\boldsymbol{S}^+=\tilde{e}, \tag{8.3.19a}$$

$$\tilde{\boldsymbol{r}}\cdot\boldsymbol{S}^-=0. \tag{8.3.19b}$$

Thus the component of the mean energy flux vector along the normal to the planes of constant phase is equal to the phase speed ($|\boldsymbol{S}^+|^{-1}$) times the mean energy density. Using (8.2.5) we have, for the energy flux velocity $\boldsymbol{g}$,

$$\boldsymbol{g}\cdot\boldsymbol{S}^+=1,\quad \boldsymbol{g}\cdot\boldsymbol{S}^-=0. \tag{8.3.20}$$

Of course if $\boldsymbol{S}^-=\boldsymbol{0}$, so that the waves are not attenuated, then the result $\boldsymbol{g}\cdot\boldsymbol{S}^+=1$ is still valid.

Exercise 8.1

Examine the case when ω in (8.3.1) is complex. How does this affect the results (8.3.12) and (8.3.17)?

8.4 Two or more wave trains

Because the field components combine in pairs in the formation of the energy flux vector and the energy density, it is sufficient to consider just two wave trains to obtain results which are valid for two or more wave trains.

Suppose the trains have the same angular frequency ω, and have slownesses $\boldsymbol{S}_1$ and $\boldsymbol{S}_2$. Then the energy flux vector for the superposition of the two trains has the form

$$\begin{aligned}\boldsymbol{r}=&\{f_1(\omega)\boldsymbol{A}_1\exp i\omega(\boldsymbol{S}_1\cdot\boldsymbol{x}-t)+f_2(\omega)\boldsymbol{A}_2\exp i\omega(\boldsymbol{S}_2\cdot\boldsymbol{x}-t)+\text{c.c.}\}\\&\times\{f_1(\omega)\boldsymbol{B}_1\exp i\omega(\boldsymbol{S}_1\cdot\boldsymbol{x}-t)+f_2(\omega)\boldsymbol{B}_2\exp i\omega(\boldsymbol{S}_2\cdot\boldsymbol{x}-t)+\text{c.c.}\}\end{aligned}$$

$$= \sum_{\alpha=1}^{2} \{f_\alpha^2(\omega)\mathbf{\Lambda}_\alpha \exp 2i\omega(\mathbf{S}_\alpha\cdot\mathbf{x} - t) + \text{c.c.}\}$$

$$+ \{f_1 f_2 \mathbf{\Lambda}_{12} \exp i\omega[(\mathbf{S}_1 + \mathbf{S}_2)\cdot\mathbf{x} - 2t] + \text{c.c.}\} + \tilde{\mathbf{r}}, \qquad (8.4.1)$$

where $\mathbf{A}_\alpha$, $\mathbf{B}_\alpha$ may be complex scalars, vectors or tensors, independent of ω. Also $\mathbf{\Lambda}_\alpha$, $\mathbf{\Lambda}_{12}$ are bivectors independent of ω; $f_1(\omega)$ and $f_2(\omega)$ are arbitrary functions of ω, and $\tilde{\mathbf{r}}$ is the mean of $\mathbf{r}$, given by

$$\tilde{\mathbf{r}} = \sum_{\alpha=1}^{2} f_\alpha \bar{f}_\alpha \boldsymbol{\mu}_\alpha \exp i\omega(\mathbf{S}_\alpha - \bar{\mathbf{S}}_\alpha)\cdot\mathbf{x}$$

$$+ [f_1 \bar{f}_2 \mathbf{\Phi}_{12} \exp i\omega(\mathbf{S}_1 - \bar{\mathbf{S}}_2)\cdot\mathbf{x} + \text{c.c.}]. \qquad (8.4.2)$$

Here $\boldsymbol{\mu}_\alpha$ are real vectors and $\mathbf{\Phi}_{12}$ is a bivector, all independent of ω.

Similarly, for the energy density of the combined wave trains, we have

$$e = \sum_{\alpha=1}^{2} \{f_\alpha^2 B_\alpha \exp 2i\omega(\mathbf{S}_\alpha\cdot\mathbf{x} - t) + \text{c.c.}\}$$

$$+ [f_1 \bar{f}_2 B_{12} \exp\{i\omega(\mathbf{S}_1 + \bar{\mathbf{S}}_2)\cdot\mathbf{x} - 2t\} + \text{c.c.}] + \tilde{e}, \qquad (8.4.3)$$

where B_α, B_{12} are complex scalars independent of ω, and $\tilde{e}$ is the mean of e, given by

$$\tilde{e} = \sum_{\alpha=1}^{2} f_\alpha \bar{f}_\alpha \gamma_\alpha \exp i\omega(\mathbf{S}_\alpha - \bar{\mathbf{S}}_\alpha)\cdot\mathbf{x} + [f_1 \bar{f}_2 \Psi_{12} \exp i\omega(\mathbf{S}_1 - \bar{\mathbf{S}}_2)\cdot\mathbf{x} + \text{c.c.}]. \qquad (8.4.4)$$

Here γ_α are real scalars, and Ψ_{12} is a complex scalar, all independent of ω.

Insertion of (8.4.1) and (8.4.3) into the energy conservation equation (8.1.2) gives

$$\boldsymbol{\mu}_1\cdot(\mathbf{S}_1 - \bar{\mathbf{S}}_1) = \boldsymbol{\mu}_2\cdot(\mathbf{S}_2 - \bar{\mathbf{S}}_2) = \mathbf{\Phi}_{12}\cdot(\mathbf{S}_1 - \bar{\mathbf{S}}_2) = 0. \qquad (8.4.5)$$

As before, if ω is replaced by the complex ω^* whose real part is ω, the real frequency of the wave trains under consideration, then $\tilde{\mathbf{r}}$, given by (8.4.2), is replaced by $\tilde{\mathbf{r}}(\omega^*)$ say, given by

$$\tilde{\boldsymbol{r}}(\omega^*) = \sum_{\alpha=1}^{2} f_\alpha(\omega^*)\overline{f_\alpha(\omega^*)}\boldsymbol{\mu}_\alpha \exp[\mathrm{i}\omega^*(\boldsymbol{S}_\alpha\cdot\boldsymbol{x} - t) - \mathrm{i}\overline{\omega^*}(\bar{\boldsymbol{S}}_\alpha\cdot\boldsymbol{x} - t)]$$
$$+ \{f_1(\omega^*)\overline{f_2(\omega^*)}\boldsymbol{\Phi}_{12} \exp[\mathrm{i}\omega^*(\boldsymbol{S}_1\cdot\boldsymbol{x} - t) - \mathrm{i}\overline{\omega^*}(\bar{\boldsymbol{S}}_2\cdot\boldsymbol{x} - t)] + \text{c.c.}\}. \tag{8.4.6}$$

Also, $\tilde{e}$ is replaced by $\tilde{e}(\omega^*)$ say, given by

$$\tilde{e}(\omega^*) = \sum_{\alpha=1}^{2} f_\alpha(\omega^*)\overline{f_\alpha(\omega^*)}\gamma_\alpha \exp[\mathrm{i}\omega^*(\boldsymbol{S}_\alpha\cdot\boldsymbol{x} - t) - \mathrm{i}\overline{\omega^*}(\bar{\boldsymbol{S}}_\alpha\cdot\boldsymbol{x} - t)]$$
$$+ \{f_1(\omega^*)\overline{f_2(\omega^*)}\Psi_{12} \exp[\mathrm{i}\omega^*(\boldsymbol{S}_1\cdot\boldsymbol{x} - t) - \mathrm{i}\overline{\omega^*}(\bar{\boldsymbol{S}}_2\cdot\boldsymbol{x} - t)]$$
$$+ \text{c.c.}\}. \tag{8.4.7}$$

Now the energy conservation equation (8.1.2) gives

$$\boldsymbol{\mu}_1\cdot(\omega^*\boldsymbol{S}_1 - \overline{\omega^*}\bar{\boldsymbol{S}}_1) = \gamma_1(\omega^* - \bar{\omega}^*),$$
$$\boldsymbol{\mu}_2\cdot(\omega^*\boldsymbol{S}_2 - \overline{\omega^*}\bar{\boldsymbol{S}}_2) = \gamma_2(\omega^* - \bar{\omega}^*),$$
$$\boldsymbol{\Phi}_{12}\cdot(\omega^*\boldsymbol{S}_1 - \overline{\omega^*}\bar{\boldsymbol{S}}_2) = \Psi_{12}(\omega^* - \bar{\omega}^*). \tag{8.4.8}$$

Thus, by using (8.4.5), it follows that

$$\boldsymbol{\mu}_1\cdot\boldsymbol{S}_1 = \gamma_1, \quad \boldsymbol{\mu}_2\cdot\boldsymbol{S}_2 = \gamma_2, \quad \boldsymbol{\Phi}_{12}\cdot\boldsymbol{S}_1 = \boldsymbol{\Phi}_{12}\cdot\bar{\boldsymbol{S}}_2 = \Psi_{12}. \tag{8.4.9}$$

These results are valid for two wave trains which have the same frequency but different slownesses $\boldsymbol{S}_1$ and $\boldsymbol{S}_2$. Similar results are valid for more than two trains.

These results are equally valid for two or more wave trains in which the frequency ω is no longer real but complex.

We present some examples.

Example 8.1 Surface waves

Suppose that two plane waves with slownesses $\boldsymbol{S}_1$ and $\boldsymbol{S}_2$ are propagating over the plane surface $\boldsymbol{m}\cdot\boldsymbol{x} = 0$. In order that the amplitude decay with distance from the surface, $\boldsymbol{S}_1^-$ and $\boldsymbol{S}_2^-$ are both parallel to $\boldsymbol{m}$. Some boundary condition is imposed on the surface and hence the waves must be in phase on $\boldsymbol{m}\cdot\boldsymbol{x} = 0$. Thus $\boldsymbol{S}_1^+$ and $\boldsymbol{S}_2^+$ must have the form

$$\boldsymbol{S}_1^+ = \frac{\boldsymbol{a}}{c} + p_1\boldsymbol{m}, \quad \boldsymbol{S}_2^+ = \frac{\boldsymbol{a}}{c} + p_2\boldsymbol{m}, \quad \boldsymbol{a}\cdot\boldsymbol{m} = 0, \tag{8.4.10}$$

where c^{-1} is the common in-surface component of the slownesses and p_1, p_2 are real constants. Thus $(\boldsymbol{S}_1 - \bar{\boldsymbol{S}}_1)$, $(\boldsymbol{S}_2 - \bar{\boldsymbol{S}}_2)$, and $(\boldsymbol{S}_1 - \bar{\boldsymbol{S}}_2)$ are all parallel to $\boldsymbol{m}$. Hence, using (8.4.5),

$$\boldsymbol{\mu}_1 \cdot \boldsymbol{m} = \boldsymbol{\mu}_2 \cdot \boldsymbol{m} = \boldsymbol{\Phi}_{12} \cdot \boldsymbol{m} = 0. \tag{8.4.11}$$

Then, using (8.4.9),

$$\boldsymbol{\mu}_1 \cdot \boldsymbol{a} = \gamma_1 c, \quad \boldsymbol{\mu}_2 \cdot \boldsymbol{a} = \gamma_2 c, \quad \boldsymbol{\Phi}_{12} \cdot \boldsymbol{a} = \Psi_{12} c. \tag{8.4.12}$$

Hence, using (8.4.2) and (8.4.4),

$$\tilde{\boldsymbol{r}} \cdot \boldsymbol{a} = c\tilde{e}, \quad \tilde{\boldsymbol{r}} \cdot \boldsymbol{m} = 0. \tag{8.4.13}$$

Here $\tilde{\boldsymbol{r}}$ and $\tilde{e}$ are the mean energy flux and energy density for the combined wave motion. Using (8.2.5), we have, for the energy flux velocity of the combined motion,

$$\boldsymbol{g} \cdot \boldsymbol{a} = c, \quad \boldsymbol{g} \cdot \boldsymbol{m} = 0. \tag{8.4.14}$$

Example 8.2 Homogeneous waves with a common propagation direction

Suppose q wave trains with real slownesses propagate in the same direction. Thus, $\boldsymbol{S}_\alpha = s_\alpha \boldsymbol{n}$, $\alpha = 1, \ldots, q$. By analogy with (8.4.2) and (8.4.4), the mean energy flux vector and mean energy density corresponding to the superposition of these waves are given by

$$\tilde{\boldsymbol{r}} = \sum_{\alpha=1}^{q} f_\alpha \bar{f}_\alpha \boldsymbol{\mu}_\alpha + \sum_{\alpha=1}^{q} \sum_{\beta=1}^{q} \{f_\alpha \bar{f}_\beta \boldsymbol{\Phi}_{\alpha\beta} \exp i\omega(s_\alpha - s_\beta)\boldsymbol{n}\cdot\boldsymbol{x} + \text{c.c.}\},$$

$$\tilde{e} = \sum_{\alpha=1}^{q} f_\alpha \bar{f}_\alpha \gamma_\alpha + \sum_{\alpha=1}^{q} \sum_{\beta=1}^{q} \{f_\alpha \bar{f}_\beta \Psi_{\alpha\beta} \exp i\omega(s_\alpha - s_\beta)\boldsymbol{n}\cdot\boldsymbol{x} + \text{c.c.}\}. \tag{8.4.15}$$

Using (8.4.5) and (8.4.9) we have

$$\boldsymbol{\Phi}_{\alpha\beta} \cdot \boldsymbol{n} = 0, \quad \Psi_{\alpha\beta} = 0, \quad s_\alpha \boldsymbol{\mu}_\alpha \cdot \boldsymbol{n} = \gamma_\alpha, \tag{8.4.16}$$

and hence,

$$\tilde{\boldsymbol{r}} \cdot \boldsymbol{n} = \sum_{\alpha=1}^{q} \frac{\tilde{e}_\alpha}{s_\alpha}. \tag{8.4.17}$$

Here $\tilde{\boldsymbol{r}}$ is the mean energy flux for the resultant motion and $\tilde{e}_\alpha$ is the mean energy density for the individual motion with slowness $s_\alpha \boldsymbol{n}$. We note that $\Psi_{\alpha\beta} = 0$, which means that the mean energy density of the resultant motion is equal to the sum of the mean energy densities of the individual motions.

Exercises 8.2

1. Determine the form (8.1.2) takes if there is energy dissipation.
2. What form does (8.3.13) take in the case of energy dissipation?

9

Electromagnetic plane waves

The propagation of time harmonic electromagnetic plane waves in nonabsorbing, nonoptically active, homogeneous, electrically anisotropic, but magnetically isotropic, dielectrics is considered here. Both homogeneous and inhomogeneous plane waves are considered. All such solutions are obtained.

The properties of the anisotropic dielectrics (crystals) considered here are described in terms of the permittivity tensor, $\boldsymbol{\kappa}$, assumed to be constant, real, symmetric positive definite. The crystals are classified according to the number of central circular sections of the 'index ellipsoid' $\boldsymbol{x}^{\mathrm{T}}\boldsymbol{\kappa}^{-1}\boldsymbol{x} = 1$, associated with the tensor $\boldsymbol{\kappa}^{-1}$. If there are two such central circular sections, the crystal is said to be 'biaxial', if there is one it is said to be 'uniaxial',.and if all sections are circular ($\boldsymbol{\kappa}^{-1} = \lambda\mathbf{1}$) it is said to be 'isotropic'. The normals to the planes of the central circular sections are called the 'optic axes'.

For biaxial and uniaxial crystals, it is seen that the results may be given simple explicit expressions using the unit vectors along the optic axes.

9.1 Maxwell's equations and constitutive equations

Maxwell's equations, in the absence of free charges and currents are

$$\boldsymbol{\nabla}\cdot\mathscr{D} = 0, \quad \boldsymbol{\nabla}\cdot\mathscr{B} = 0, \tag{9.1.1}$$

$$\boldsymbol{\nabla}\times\mathscr{E} + \partial_t\mathscr{B} = 0, \tag{9.1.2a}$$

$$\boldsymbol{\nabla}\times\mathscr{H} - \partial_t\mathscr{D} = 0, \tag{9.1.2b}$$

where $\mathscr{E}$ is the electric field, $\mathscr{B}$ the magnetic induction, $\mathscr{D}$ the electric displacement and $\mathscr{H}$ the magnetic field.

To these have to be added constitutive equations. Here we consider homogeneous anisotropic dielectrics (crystals) which are nonabsorbing and nonoptically active. For most of these crystals, the constitutive equations may be taken to be

$$\mathscr{D} = \boldsymbol{\kappa}\mathscr{E}, \qquad \mathscr{D}_i = \kappa_{ij}\mathscr{E}_j,$$
$$\mathscr{B} = \mu_0\mathscr{H}, \quad \mathscr{B}_i = \mu_0\mathscr{H}_i. \tag{9.1.3}$$

Here $\boldsymbol{\kappa}$ is the permittivity tensor of the dielectric, assumed constant, real, symmetric, positive definite, and μ_0 is a real constant, the magnetic permeability. In this chapter, we consider wave propagation in crystals described by the constitutive equations (9.1.3). It may be noted that it is sometimes necessary to consider media which are both electrically and magnetically anisotropic: the constant μ_0 has then to be replaced by a permeability tensor $\boldsymbol{\mu}$. The reader is referred to Boulanger and Hayes (1990) for a detailed study of this more general case.

Taking the dot product of (9.1.2a) with $\mathscr{H}$, of (9.1.2b) with $\mathscr{E}$, and subtracting, yields the identity

$$\mathscr{H}\cdot\partial_t\mathscr{B} + \mathscr{E}\cdot\partial_t\mathscr{D} + \nabla\cdot(\mathscr{E}\times\mathscr{H}) = 0. \tag{9.1.4}$$

Using the constitutive equations (9.1.3) we have, because $\boldsymbol{\kappa}$ is symmetric,

$$\mathscr{H}\cdot\partial_t\mathscr{B} + \mathscr{E}\cdot\partial_t\mathscr{D} = \tfrac{1}{2}\partial_t(\mu_0\mathscr{H}\cdot\mathscr{H} + \mathscr{E}^{\mathrm{T}}\boldsymbol{\kappa}\mathscr{E}) = \tfrac{1}{2}\partial_t(\mathscr{B}\cdot\mathscr{H} + \mathscr{D}\cdot\mathscr{E}). \tag{9.1.5}$$

Hence an energy conservation equation

$$\partial_t e + \nabla\cdot\boldsymbol{r} = 0, \tag{8.1.2}$$

holds with an energy flux vector $\boldsymbol{r}$, and an energy density e, given by

$$\boldsymbol{r} = \mathscr{E}\times\mathscr{H}, \quad e = \tfrac{1}{2}(\mathscr{D}\cdot\mathscr{E} + \mathscr{B}\cdot\mathscr{H}). \tag{9.1.6}$$

9.2 The propagation condition

Here we consider time harmonic homogeneous and inhomogeneous plane waves. Thus, let (section 6.3)

$$(\mathscr{E}, \mathscr{H}, \mathscr{D}, \mathscr{B}) = \{(\boldsymbol{E}, \boldsymbol{H}, \boldsymbol{D}, \boldsymbol{B})\exp \mathrm{i}\omega(\boldsymbol{S}\cdot\boldsymbol{x} - t)\}^+, \tag{9.2.1}$$

where $\boldsymbol{S}$ is the slowness bivector and $\boldsymbol{E}, \boldsymbol{H}, \boldsymbol{D}, \boldsymbol{B}$ are amplitude bivectors of the fields $\mathscr{E}$, $\mathscr{H}$, $\mathscr{D}$, $\mathscr{B}$. We use the DE method, and thus, as explained in section 6.3, we write

$$\boldsymbol{S} = N\boldsymbol{C}, \quad \boldsymbol{C} = m\hat{\mathbf{m}} + \mathrm{i}\hat{\mathbf{n}}, \tag{9.2.2}$$

where N is a complex number, $\hat{\mathbf{m}}$ and $\hat{\mathbf{n}}$ are two orthogonal real unit vectors, and m is a real number. It is convenient here to introduce the complex skew-symmetric matrix $\mathbf{\Gamma}$ associated with the prescribed bivector $\boldsymbol{C}$ (section 4.4):

$$\Gamma_{ij} = -e_{ijk}C_k, \quad C_i = -\tfrac{1}{2}e_{ijk}\Gamma_{jk}. \qquad (9.2.3)$$

Also, let $\mathbf{\Pi}$ be the complex symmetric matrix defined by

$$\mathbf{\Pi} = \mathbf{\Gamma}\mathbf{\Gamma}^{\mathrm{T}} = -\mathbf{\Gamma}^2 = \mathbf{\Gamma}^{\mathrm{T}}\mathbf{\Gamma} = (\boldsymbol{C}\cdot\boldsymbol{C})\mathbf{1} - \boldsymbol{C}\otimes\boldsymbol{C}. \qquad (9.2.4)$$

Inserting the expressions (9.2.1) into Maxwell's equations (9.1.1) and (9.1.2), and using (9.2.2), yields

$$\boldsymbol{C}\cdot\boldsymbol{D} = 0, \qquad (9.2.5a)$$

$$\boldsymbol{C}\cdot\boldsymbol{B} = 0, \qquad (9.2.5b)$$

$$N\boldsymbol{C}\times\boldsymbol{E} = \boldsymbol{B} \text{ or } N\mathbf{\Gamma}\boldsymbol{E} = \boldsymbol{B}, \qquad (9.2.6)$$

$$N\boldsymbol{C}\times\boldsymbol{H} = -\boldsymbol{D} \text{ or } N\mathbf{\Gamma}\boldsymbol{H} = -\boldsymbol{D}. \qquad (9.2.7)$$

Also, the constitutive equations (9.1.3) yield for the amplitudes

$$\boldsymbol{D} = \boldsymbol{\kappa}\boldsymbol{E}, \qquad (9.2.8a)$$

$$\boldsymbol{B} = \mu_0\boldsymbol{H}. \qquad (9.2.8b)$$

From the equations (9.2.6), (9.2.7) and (9.2.8) we have

$$\mu_0\boldsymbol{H} = N\mathbf{\Gamma}\boldsymbol{E} = N\mathbf{\Gamma}\boldsymbol{\kappa}^{-1}\boldsymbol{D} = -N^2\mathbf{\Gamma}\boldsymbol{\kappa}^{-1}\mathbf{\Gamma}\boldsymbol{H}, \qquad (9.2.9)$$

and hence

$$\{\boldsymbol{Q}(\boldsymbol{C}) - \mu_0 N^{-2}\mathbf{1}\}\boldsymbol{H} = \mathbf{0}, \qquad (9.2.10)$$

where $\boldsymbol{Q}(\boldsymbol{C})$ is called 'the optical tensor' associated with the prescribed bivector $\boldsymbol{C}$, and is given by

$$\boldsymbol{Q}(\boldsymbol{C}) = \mathbf{\Gamma}^{\mathrm{T}}\boldsymbol{\kappa}^{-1}\mathbf{\Gamma}. \qquad (9.2.11)$$

It is a complex symmetric matrix depending on the choice of $\boldsymbol{C}$. Equation (9.2.10) is the 'propagation condition' written in terms of $\boldsymbol{H}$ (or $\boldsymbol{B} = \mu_0\boldsymbol{H}$). It is an eigenvalue problem for the complex symmetric matrix $\boldsymbol{Q}(\boldsymbol{C})$.

Similarly, from (9.2.6), (9.2.7) and (9.2.8) we also have

$$\boldsymbol{\kappa}\boldsymbol{E} = -N\mathbf{\Gamma}\boldsymbol{H} = \frac{-1}{\mu_0}N\mathbf{\Gamma}\boldsymbol{B} = \frac{1}{\mu_0}N^2\mathbf{\Pi}\boldsymbol{E}, \qquad (9.2.12)$$

and hence the propagation condition, in terms of $\boldsymbol{E}$, reads

$$(\mathbf{\Pi} - \mu_0 N^{-2}\boldsymbol{\kappa})\boldsymbol{E} = \mathbf{0}. \qquad (9.2.13)$$

Using $\boldsymbol{D} = \boldsymbol{\kappa}\boldsymbol{E}$ it becomes, in terms of $\boldsymbol{D}$ (Hayes, 1987),

$$(\boldsymbol{\Pi}\boldsymbol{\kappa}^{-1} - \mu_0 N^{-2}\mathbf{1})\boldsymbol{D} = \mathbf{0}, \tag{9.2.14}$$

which is an eigenvalue problem for the nonsymmetric matrix $\boldsymbol{\Pi}\boldsymbol{\kappa}^{-1}$. When $\boldsymbol{C}$ is not isotropic, $\boldsymbol{C}\cdot\boldsymbol{C} \neq 0$, the propagation condition (9.2.14) is equivalent to an eigenvalue problem for a complex symmetric matrix. Indeed, using (9.2.5a), it follows that $\boldsymbol{\Pi}\boldsymbol{D} = (\boldsymbol{C}\cdot\boldsymbol{C})\boldsymbol{D}$. Hence, multiplying (9.2.14) by $\boldsymbol{C}\cdot\boldsymbol{C}$ (assuming $\boldsymbol{C}\cdot\boldsymbol{C} \neq 0$), yields the eigenvalue problem

$$\{\boldsymbol{\Pi}\boldsymbol{\kappa}^{-1}\boldsymbol{\Pi} - \mu_0 N^{-2}(\boldsymbol{C}\cdot\boldsymbol{C})\mathbf{1}\}\boldsymbol{D} = \mathbf{0}, \tag{9.2.15}$$

for the complex symmetric matrix $\boldsymbol{\Pi}\boldsymbol{\kappa}^{-1}\boldsymbol{\Pi}$.

Finally, we note that (9.2.10) may also be written as

$$\boldsymbol{C} \times \{\boldsymbol{\kappa}^{-1}(\boldsymbol{C} \times \boldsymbol{H})\} + \mu_0 N^{-2}\boldsymbol{H} = \mathbf{0}, \tag{9.2.16}$$

or, using the identities (5.1.4) and (5.1.2),

$$(\det \boldsymbol{\kappa})^{-1}\boldsymbol{\kappa}\{\boldsymbol{C}(\boldsymbol{C}^{\mathrm{T}}\boldsymbol{\kappa}\boldsymbol{H}) - \boldsymbol{H}(\boldsymbol{C}^{\mathrm{T}}\boldsymbol{\kappa}\boldsymbol{C})\boldsymbol{\kappa}\} + \mu_0 N^{-2}\boldsymbol{H} = \mathbf{0}. \tag{9.2.17}$$

Because the left-hand sides of (9.2.17) and (9.2.10) are identical whatever the bivector $\boldsymbol{H}$ may be, it follows that the optical tensor $\boldsymbol{Q}(\boldsymbol{C})$, defined by (9.2.11), may also be written as

$$\boldsymbol{Q}(\boldsymbol{C}) = (\det \boldsymbol{\kappa})^{-1}\{(\boldsymbol{C}^{\mathrm{T}}\boldsymbol{\kappa}\boldsymbol{C})\boldsymbol{\kappa} - \boldsymbol{\kappa}(\boldsymbol{C}\otimes\boldsymbol{C})\boldsymbol{\kappa}\}. \tag{9.2.18}$$

9.3 The secular equation

The equation for the determination of the eigenvalues $\mu_0 N^{-2}$ is called the 'secular equation'. From (9.2.10), this equation reads

$$\det\{\boldsymbol{Q}(\boldsymbol{C}) - \mu_0 N^{-2}\mathbf{1}\} = 0. \tag{9.3.1}$$

Since $III_Q = \det(\boldsymbol{\Gamma}^{\mathrm{T}}\boldsymbol{\kappa}^{-1}\boldsymbol{\Gamma}) = 0$, it may be written explicitly as

$$-(\mu_0 N^{-2})^3 + I_Q(\mu_0 N^{-2})^2 - II_Q(\mu_0 N^{-2}) = 0, \tag{9.3.2}$$

where I_Q, II_Q, III_Q denote the principal invariants of the optical tensor $\boldsymbol{Q}(\boldsymbol{C}) = \boldsymbol{\Gamma}^{\mathrm{T}}\boldsymbol{\kappa}^{-1}\boldsymbol{\Gamma}$. We have

$$I_Q = \mathrm{tr}(\boldsymbol{\Gamma}^{\mathrm{T}}\boldsymbol{\kappa}^{-1}\boldsymbol{\Gamma}) = \mathrm{tr}(\boldsymbol{\Pi}\boldsymbol{\kappa}^{-1}) = (\boldsymbol{C}\cdot\boldsymbol{C})\,\mathrm{tr}(\boldsymbol{\kappa}^{-1}) - \boldsymbol{C}^{\mathrm{T}}\boldsymbol{\kappa}^{-1}\boldsymbol{C}, \tag{9.3.3}$$

or, equivalently, from (9.2.18),

$$I_Q = (\det \boldsymbol{\kappa})^{-1}\{(\boldsymbol{C}^{\mathrm{T}}\boldsymbol{\kappa}\boldsymbol{C})\,\mathrm{tr}\,\boldsymbol{\kappa} - \boldsymbol{C}^{\mathrm{T}}\boldsymbol{\kappa}^2\boldsymbol{C}\}. \tag{9.3.4}$$

That (9.3.4) is identical to (9.3.3) may be checked directly using the

Cayley–Hamilton theorem. Also, using the adjugate $\boldsymbol{\Gamma}_* = \boldsymbol{C} \otimes \boldsymbol{C}$ of $\boldsymbol{\Gamma}$, and the adjugate $\boldsymbol{\kappa}_*^{-1} = (\det \boldsymbol{\kappa})^{-1}\boldsymbol{\kappa}$ of $\boldsymbol{\kappa}^{-1}$ (section 5.1 and Exercises 5.1.2, 5.1.4, 5.1.5), we have

$$\begin{aligned} II_Q = I_{Q*} &= \operatorname{tr}(\boldsymbol{\Gamma}_*^{\mathrm{T}}\boldsymbol{\kappa}_*^{-1}\boldsymbol{\Gamma}_*) = (\det \boldsymbol{\kappa})^{-1} \operatorname{tr}\{(\boldsymbol{C}\otimes\boldsymbol{C})\boldsymbol{\kappa}(\boldsymbol{C}\otimes\boldsymbol{C})\} \\ &= (\det \boldsymbol{\kappa})^{-1}(\boldsymbol{C}\cdot\boldsymbol{C})(\boldsymbol{C}^{\mathrm{T}}\boldsymbol{\kappa}\boldsymbol{C}). \end{aligned} \tag{9.3.5}$$

One of the three roots of the secular equation (9.3.2) is zero, and the two others are solutions of the quadratic equation

$$\begin{aligned} &(\mu_0 N^{-2})^2 - \{(\boldsymbol{C}\cdot\boldsymbol{C})\operatorname{tr}(\boldsymbol{\kappa}^{-1}) - \boldsymbol{C}^{\mathrm{T}}\boldsymbol{\kappa}^{-1}\boldsymbol{C}\}\mu_0 N^{-2} \\ &\quad + (\det \boldsymbol{\kappa}^{-1})(\boldsymbol{C}\cdot\boldsymbol{C})(\boldsymbol{C}^{\mathrm{T}}\boldsymbol{\kappa}\boldsymbol{C}) = 0, \end{aligned} \tag{9.3.6}$$

or, equivalently,

$$(\det \boldsymbol{\kappa})(\mu_0 N^{-2})^2 - \{(\boldsymbol{C}^{\mathrm{T}}\boldsymbol{\kappa}\boldsymbol{C})\operatorname{tr}\boldsymbol{\kappa} - \boldsymbol{C}^{\mathrm{T}}\boldsymbol{\kappa}^2\boldsymbol{C}\}\mu_0 N^{-2} + (\boldsymbol{C}\cdot\boldsymbol{C})(\boldsymbol{C}^{\mathrm{T}}\boldsymbol{\kappa}\boldsymbol{C}) = 0. \tag{9.3.7}$$

Clearly, $\boldsymbol{C}$ is an eigenbivector of $\boldsymbol{Q}(\boldsymbol{C})$ corresponding to the eigenvalue zero:

$$\boldsymbol{Q}(\boldsymbol{C})\boldsymbol{C} = \boldsymbol{\Gamma}^{\mathrm{T}}\boldsymbol{\kappa}\boldsymbol{\Gamma}\boldsymbol{C} = \boldsymbol{\Gamma}^{\mathrm{T}}\boldsymbol{\kappa}(\boldsymbol{C}\times\boldsymbol{C}) = \mathbf{0}. \tag{9.3.8}$$

Exercise 9.1

Show that the secular equation $\det(\boldsymbol{\Pi}\boldsymbol{\kappa}^{-1} - \mu_0 N^{-2}\mathbf{1}) = 0$ obtained from (9.2.14) is identical with (9.3.1).

9.4 Orthogonality relations

Here we first derive orthogonality relations among the amplitude bivectors $\boldsymbol{E}, \boldsymbol{H}, \boldsymbol{D}, \boldsymbol{B}$ for a single homogeneous or inhomogeneous plane wave.

Then, when the secular equation has two different nonzero roots, we determine the orthogonality relations among the two sets of amplitude bivectors $\boldsymbol{E}_1, \boldsymbol{H}_1, \ldots$ and $\boldsymbol{E}_2, \boldsymbol{H}_2, \ldots$.

9.4.1 Orthogonality relations for one wave

From equations (9.2.6) and (9.2.7) we obtain, because $\boldsymbol{B}$ is parallel to $\boldsymbol{H}$,

$$\boldsymbol{E}\cdot\boldsymbol{H} = 0, \quad \boldsymbol{D}\cdot\boldsymbol{H} = 0, \tag{9.4.1}$$

and hence, using the constitutive equation (9.2.8a),

$$\boldsymbol{D}^{\mathrm{T}}\boldsymbol{\kappa}^{-1}\boldsymbol{H}=0, \quad \boldsymbol{E}^{\mathrm{T}}\boldsymbol{\kappa}\boldsymbol{H}=0 \tag{9.4.2}$$

The geometrical interpretation of the orthogonality of bivectors has been given in section 2.4. Thus, the orthogonal projections of the ellipses of $\boldsymbol{E}$ and of $\boldsymbol{D}$ upon the plane of the ellipse of $\boldsymbol{H}$ are both similar and similarly situated to the ellipse of $\boldsymbol{H}$ when rotated through a quadrant. The geometrical interpretation of the orthogonality of bivectors with respect to a metric has been given in section 5.8. Thus, the $\boldsymbol{\kappa}^{-1}$-projection of the ellipse of $\boldsymbol{D}$ onto the plane of the ellipse of $\boldsymbol{H}$ is similar and similarly situated to the polar reciprocal of the ellipse of $\boldsymbol{H}$ with respect to the section of the $\boldsymbol{\kappa}^{-1}$-ellipsoid by the plane of the bivector $\boldsymbol{H}$. Similarly, the $\boldsymbol{\kappa}$-projection of the ellipse of $\boldsymbol{E}$ onto the plane of the ellipse of $\boldsymbol{H}$ is similar and similarly situated to the polar reciprocal of the ellipse of $\boldsymbol{H}$ with respect to the section of the $\boldsymbol{\kappa}$-ellipsoid by the plane of the bivector $\boldsymbol{H}$.

Taking the dot product of (9.2.6) with $\boldsymbol{H}$, and the dot product of (9.2.7) with $\boldsymbol{E}$, we note that

$$\boldsymbol{D}\cdot\boldsymbol{E}=\boldsymbol{B}\cdot\boldsymbol{H}=\sigma \text{ (say)}, \tag{9.4.3}$$

and hence, using (9.2.8),

$$\sigma=\boldsymbol{E}^{\mathrm{T}}\boldsymbol{\kappa}\boldsymbol{E}=\boldsymbol{D}^{\mathrm{T}}\boldsymbol{\kappa}^{-1}\boldsymbol{D}=\mu_0\boldsymbol{H}\cdot\boldsymbol{H}=\frac{1}{\mu_0}\boldsymbol{B}\cdot\boldsymbol{B}. \tag{9.4.4}$$

Moreover, using (9.2.6) and (9.2.5a), we obtain

$$\boldsymbol{D}\times\boldsymbol{B}=N\boldsymbol{D}\times(\boldsymbol{C}\times\boldsymbol{E})=N(\boldsymbol{D}\cdot\boldsymbol{E})\boldsymbol{C}=N\sigma\boldsymbol{C}, \tag{9.4.5}$$

with σ given by (9.4.3).

9.4.2 *Orthogonality relations for two waves*

Let us assume that the quadratic equation (9.3.6), or equivalently (9.3.7), has two different nonzero roots $\mu_0 N_1^{-2} \neq \mu_0 N_2^{-2}$. Since these roots are eigenvalues of the optical tensor $\boldsymbol{Q}(\boldsymbol{C})$ which is complex and symmetric, it follows (section 3.1) that the corresponding eigenbivectors $\boldsymbol{H}_1$, $\boldsymbol{H}_2$ are orthogonal:

$$\boldsymbol{H}_1\cdot\boldsymbol{H}_2=0. \tag{9.4.6}$$

Then, using (9.2.7) and (9.2.5b), we have

$$\begin{aligned} \boldsymbol{D}_1 \times \boldsymbol{B}_2 &= -N_1(\boldsymbol{C} \times \boldsymbol{H}_1) \times \boldsymbol{B}_2 = N_1(\boldsymbol{H}_1 \cdot \boldsymbol{B}_2)\boldsymbol{C} = \mu_0 N_1(\boldsymbol{H}_1 \cdot \boldsymbol{H}_2)\boldsymbol{C} = \boldsymbol{0}, \\ \boldsymbol{D}_2 \times \boldsymbol{B}_1 &= -N_2(\boldsymbol{C} \times \boldsymbol{H}_2) \times \boldsymbol{B}_1 = N_2(\boldsymbol{H}_2 \cdot \boldsymbol{B}_1)\boldsymbol{C} = \mu_0 N_2(\boldsymbol{H}_2 \cdot \boldsymbol{H}_1)\boldsymbol{C} = \boldsymbol{0}. \end{aligned} \tag{9.4.7}$$

Hence, the bivector $\boldsymbol{D}_1$ is parallel to $\boldsymbol{B}_2$ which is parallel to $\boldsymbol{H}_2$, and the bivector $\boldsymbol{D}_2$ is parallel to $\boldsymbol{B}_1$ which is parallel to $\boldsymbol{H}_1$. Thus $\boldsymbol{D}_1$ and $\boldsymbol{D}_2$ are also orthogonal:

$$\boldsymbol{D}_1 \cdot \boldsymbol{D}_2 = 0. \tag{9.4.8}$$

Note that, assuming $\boldsymbol{C} \cdot \boldsymbol{C} \neq 0$, this also follows from the fact that $\boldsymbol{D}_1$ and $\boldsymbol{D}_2$ are eigenbivectors of the complex symmetric matrix $\boldsymbol{\Pi}\boldsymbol{\kappa}^{-1}\boldsymbol{\Pi}$ corresponding to different eigenvalues (9.2.15).

Because $\boldsymbol{D}_1$ and $\boldsymbol{H}_2$ are parallel, as are $\boldsymbol{D}_2$ and $\boldsymbol{H}_1$, and also $\boldsymbol{D}_1^{\mathrm{T}}\boldsymbol{\kappa}^{-1}\boldsymbol{H}_1 = \boldsymbol{D}_2^{\mathrm{T}}\boldsymbol{\kappa}^{-1}\boldsymbol{H}_2 = 0$, it follows that

$$\boldsymbol{D}_1^{\mathrm{T}}\boldsymbol{\kappa}^{-1}\boldsymbol{D}_2 = 0, \quad \boldsymbol{E}_1^{\mathrm{T}}\boldsymbol{\kappa}\boldsymbol{E}_2 = 0, \tag{9.4.9}$$

and

$$\boldsymbol{H}_1^{\mathrm{T}}\boldsymbol{\kappa}^{-1}\boldsymbol{H}_2 = 0. \tag{9.4.10}$$

9.5 Mean energy density and mean energy flux

Let $\tilde{e}$ and $\tilde{\boldsymbol{r}}$ be the mean energy density and mean energy flux for a single train of inhomogeneous waves, as defined in section 8.1. In dealing with inhomogeneous plane waves, we know that $\tilde{e} = \hat{e}\exp(-2\omega \boldsymbol{S}^- \cdot \boldsymbol{x})$ and $\tilde{\boldsymbol{r}} = \hat{\boldsymbol{r}}\exp(-2\omega \boldsymbol{S}^- \cdot \boldsymbol{x})$, where $\hat{e}$ and $\hat{\boldsymbol{r}}$ are called the weighted mean energy flux and the weighted mean energy density (section 8.3). Also, because the assumptions of sections 8.2 and 8.3 are valid here, we have (8.3.19)

$$\hat{\boldsymbol{r}} \cdot \boldsymbol{S}^+ = \hat{e}, \quad \hat{\boldsymbol{r}} \cdot \boldsymbol{S}^- = 0. \tag{9.5.1}$$

Here, from (9.1.5), we obtain

$$\hat{\boldsymbol{r}} = \tfrac{1}{2}(\boldsymbol{E} \times \bar{\boldsymbol{H}})^+ = \tfrac{1}{4}(\boldsymbol{E} \times \bar{\boldsymbol{H}} + \bar{\boldsymbol{E}} \times \boldsymbol{H}), \tag{9.5.2}$$

and

$$\hat{e} = \tfrac{1}{4}(\boldsymbol{D} \cdot \bar{\boldsymbol{E}} + \boldsymbol{B} \cdot \bar{\boldsymbol{H}}) = \tfrac{1}{4}(\boldsymbol{E}^{\mathrm{T}}\boldsymbol{\kappa}\bar{\boldsymbol{E}} + \mu_0 \boldsymbol{H} \cdot \bar{\boldsymbol{H}}) = \frac{1}{4}\left(\boldsymbol{D}^{\mathrm{T}}\boldsymbol{\kappa}^{-1}\bar{\boldsymbol{D}} + \frac{1}{\mu_0}\boldsymbol{B} \cdot \bar{\boldsymbol{B}}\right). \tag{9.5.3}$$

Also

$$\hat{r}\cdot(S^+ \times S^-) = \frac{i}{4}(E \times \bar{H} + \bar{E} \times H)\cdot(S \times \bar{S})$$

$$= \frac{i}{4}\{(S\cdot E)(\bar{S}\cdot\bar{H}) - (S\cdot\bar{H})(\bar{S}\cdot E) + (S\cdot\bar{E})(\bar{S}\cdot H) - (S\cdot H)(\bar{S}\cdot\bar{E})\}. \tag{9.5.4}$$

But, from (9.2.5b) and (9.2.8b), $S\cdot H = \mu_0^{-1} NC\cdot B = 0$, and thus (9.5.4) reduces to (Hayes, 1987)

$$\hat{r}\cdot(S^+ \times S^-) = \tfrac{1}{2}\{(S\cdot\bar{E})(\bar{S}\cdot H)\}^-. \tag{9.5.5}$$

From (9.5.1) and (9.5.5) it follows that, for inhomogeneous waves (S^- not parallel to S^+), the weighted mean energy flux $\hat{r}$ is given by

$$\{(S^+ \times S^-)\cdot(S^+ \times S^-)\}\hat{r} = \hat{e}S^- \times (S^+ \times S^-) + \tfrac{1}{2}\{(S\cdot\bar{E})(\bar{S}\cdot H)\}^- S^+ \times S^-. \tag{9.5.6}$$

In general it is not in the plane of the slowness bivector.

Also, using (9.2.8b) and (9.2.6), we note that

$$E \times \bar{H} = \frac{1}{\mu_0} E \times (\bar{S} \times \bar{E}) = \frac{1}{\mu_0}\{(E\cdot\bar{E})\bar{S} - (\bar{S}\cdot E)\bar{E}\}, \tag{9.5.7}$$

and hence

$$2\mu_0\hat{r} = (E\cdot\bar{E})S^+ - \{(S\cdot\bar{E})E\}^+. \tag{9.5.8}$$

Further,

$$4\hat{e} = E^{\mathrm{T}}\kappa\bar{E} + \frac{1}{\mu_0}(S \times E)\cdot(\bar{S} \times \bar{E})$$

$$= E^{\mathrm{T}}\kappa\bar{E} + \frac{1}{\mu_0}\{(S\cdot\bar{S})(E\cdot\bar{E}) - (S\cdot\bar{E})(\bar{S}\cdot E)\}. \tag{9.5.9}$$

Exercises 9.2

1. From (9.5.8) and (9.5.9), check that $\hat{r}\cdot S^+ = \hat{e}$ and $\hat{r}\cdot S^- = 0$.
2. Show that $S^+\cdot(E \times \bar{H})^- = 0$.

9.6 Homogeneous waves

To deal with homogeneous waves, we write $S = NC$ with $C = \hat{\mathbf{n}}$, where $\hat{\mathbf{n}}$ is a real unit vector (along the propagation direction). Thus,

$\mathbf{\Gamma}$, defined by (9.2.3), is now real, and $\mathbf{\Pi}$, defined by (9.2.4), reduces to

$$\mathbf{\Pi} = \mathbf{1} - \hat{\mathbf{n}} \otimes \hat{\mathbf{n}}. \tag{9.6.1}$$

It is the projection operator onto the plane $\hat{\mathbf{n}} \cdot \boldsymbol{x} = 0$, orthogonal to $\hat{\mathbf{n}}$.

Here we use the propagation condition (9.2.15), which reduces to

$$\{\mathbf{\Pi}\boldsymbol{\kappa}^{-1}\mathbf{\Pi} - \mu_0 N^{-2}\mathbf{1}\}\boldsymbol{D} = \mathbf{0}, \tag{9.6.2}$$

because $\hat{\mathbf{n}} \cdot \hat{\mathbf{n}} = 1$. This is an eigenvalue problem of the form considered in section 5.7 (take $\boldsymbol{\alpha} = \boldsymbol{\kappa}^{-1}$ in (5.7.3) and (5.7.21)). It follows that the eigenvectors $\boldsymbol{D}_1$ and $\boldsymbol{D}_2$ corresponding to the nonzero eigenvalues are along the principal axes of the elliptical section of the ellipsoid $\boldsymbol{x}^{\mathrm{T}}\boldsymbol{\kappa}^{-1}\boldsymbol{x} = 1$ by the plane $\hat{\mathbf{n}} \cdot \boldsymbol{x} = 0$. This ellipsoid is sometimes called the 'optical indicatrix' or the 'index ellipsoid'. Also, the corresponding eigenvalues $\mu_0 N_1^{-2}$ and $\mu_0 N_2^{-2}$ are the inverse of the squared lengths of the principal semi-axes of this elliptical section. Thus the slownesses N_1, N_2 of the waves are equal to $\mu_0{}^{1/2}$ times the lengths of these principal semi-axes (e.g. Born and Wolf, 1980).

When $\hat{\mathbf{n}}$ is orthogonal to a circular section of the ellipsoid $\boldsymbol{x}^{\mathrm{T}}\boldsymbol{\kappa}^{-1}\boldsymbol{x} = 1$, then $N_1 = N_2$, and there is a double nonzero eigenvalue for the eigenvalue problem (9.6.2). Such a direction $\hat{\mathbf{n}}$ is called an 'optic axis'. For propagation along an optic axis, $\boldsymbol{D}$ may be any vector or bivector in the plane orthogonal to $\hat{\mathbf{n}}$. In particular, $\boldsymbol{D}$ may be chosen to be isotropic: $\boldsymbol{D} \cdot \boldsymbol{D} = 0$, so that the $\mathscr{D}$-field is circularly polarized. Then, from (9.2.7), (9.2.8b) and (9.2.5b) with $\boldsymbol{C} = \hat{\mathbf{n}}$, it follows that $\boldsymbol{D} \cdot \boldsymbol{D} = N^2 \boldsymbol{H} \cdot \boldsymbol{H} = 0$. Hence, the $\mathscr{H}$- and $\mathscr{B}$-fields are also circularly polarized, but the $\mathscr{E}$-field is not circularly polarized. The optic axes are thus the only directions along which homogeneous waves with circularly polarized $\mathscr{D}$-, $\mathscr{B}$-, and $\mathscr{H}$-fields may propagate. For all other propagation directions there are two linearly polarized waves with different slownesses.

From section 5.6, we recall that the index ellipsoid $\boldsymbol{x}^{\mathrm{T}}\boldsymbol{\kappa}^{-1}\boldsymbol{x} = 1$ may have either two central circular sections, or one, or possibly an infinity. Thus, a crystal may have two optic axes ('biaxial crystal'), or one optic axis ('uniaxial crystal'), or an infinity ('isotropic crystal').

Now, we give analytic details for the propagation of homogeneous waves in the three types of crystals.

Let $\mathbf{s}, \mathbf{t}, \mathbf{u}$ denote unit vectors along the principal axes of the ellipsoid $\boldsymbol{x}^{\mathrm{T}}\boldsymbol{\kappa}^{-1}\boldsymbol{x} = 1$, and let $\lambda = \kappa_1^{-1}$, $\mu = \kappa_2^{-1}$, $\nu = \kappa_3^{-1}$ be the eigenvalues of $\boldsymbol{\kappa}^{-1}$, ordered $\lambda \geqslant \mu \geqslant \nu$. Thus $\kappa_1^{1/2}$, $\kappa_2^{1/2}$, $\kappa_3^{1/2}$ are the lengths of the semi-axes of this ellipsoid.

Case (a) Biaxial crystals

Adopting the same notation as in section 5.6 and taking $\boldsymbol{\alpha} = \boldsymbol{\kappa}^{-1}$, we see that the unit vectors $\mathbf{h}, \mathbf{k}$ along the optic axes are given by

$$\begin{aligned} \sqrt{\lambda - \nu}\,\mathbf{h} &= \sqrt{\lambda - \mu}\,\mathbf{s} + \sqrt{\mu - \nu}\,\mathbf{u}, \\ \sqrt{\lambda - \nu}\,\mathbf{k} &= \sqrt{\lambda - \mu}\,\mathbf{s} - \sqrt{\mu - \nu}\,\mathbf{u}. \end{aligned} \tag{5.6.6}$$

For propagation along an arbitrary direction $\hat{\mathbf{n}}$, we have (5.7.8)

$$\mu_0 N_{1,2}^{-2} = \tfrac{1}{2}(\lambda + \nu) - \tfrac{1}{2}(\lambda - \nu)\cos(\phi_1 \pm \phi_2), \tag{9.6.3}$$

where ϕ_1, ϕ_2 are the angles between $\hat{\mathbf{n}}$ and $\mathbf{h}$, $\mathbf{k}$, respectively. The corresponding $\boldsymbol{D}_1, \boldsymbol{D}_2$ are given, up to a scalar factor, by

$$\boldsymbol{D}_{1,2} = (\sin\phi_1)^{-1}\{\mathbf{h} - (\mathbf{h}\cdot\hat{\mathbf{n}})\hat{\mathbf{n}}\} \pm (\sin\phi_2)^{-1}\{\mathbf{k} - (\mathbf{k}\cdot\hat{\mathbf{n}})\hat{\mathbf{n}}\} = \hat{\mathbf{h}}^* \pm \hat{\mathbf{k}}^*. \tag{9.6.4}$$

(It is assumed here that $\hat{\mathbf{n}}$ does not lie in the plane containing both optic axes. See section 5.7.1, special case, when $\hat{\mathbf{n}}$ lies in this plane.)

For propagation along an optic axis $\mathbf{h}$, or $\mathbf{k}$, we have $\mu_0 N_1^{-2} = \mu_0 N_2^{-2} = \mu$, and $\mathbf{D}$ may be any bivector orthogonal to $\hat{\mathbf{n}}$.

Case (b) Uniaxial crystal

Let $\lambda = \mu$ ('negative crystal'). The index ellipsoid is now a spheroid, and the unit vector along the optic axis is $\mathbf{u}$ (section 5.7.2).

For propagation along an arbitrary direction $\hat{\mathbf{n}}$, we have

$$\begin{aligned} &\mu_0 N_1^{-2} = \lambda, && \text{('ordinary' wave)} \\ &\mu_0 N_2^{-2} = \tfrac{1}{2}(\lambda + \nu) - \tfrac{1}{2}(\lambda - \nu)\cos 2\phi, && \text{('extraordinary' wave)} \end{aligned} \tag{9.6.5}$$

where ϕ is the angle between $\hat{\mathbf{n}}$ and $\mathbf{u}$. The corresponding $\boldsymbol{D}_1$, $\boldsymbol{D}_2$ are given, up to a scalar factor, by

$$\begin{aligned} \boldsymbol{D}_1 &= \mathbf{u} \times \hat{\mathbf{n}}, \\ \boldsymbol{D}_2 &= \mathbf{u} - (\mathbf{u}\cdot\hat{\mathbf{n}})\hat{\mathbf{n}}. \end{aligned} \tag{9.6.6}$$

For propagation along the optic axis $\mathbf{u}$, we have $\mu_0 N_1^{-2} = \mu_0 N_2^{-2} = \lambda$, and $\boldsymbol{D}$ may be any bivector orthogonal to $\mathbf{u}$, that is, any bivector in the plane of $\boldsymbol{s}$ and $\boldsymbol{t}$.

The case when $\mu = \nu$ ('positive crystal') may be dealt with similarly.

Case (c) Isotropic crystal

Let $\lambda = \mu = \nu$. Hence $\boldsymbol{\kappa}^{-1} = \lambda \mathbf{1}$. The index ellipsoid is a sphere and every direction $\hat{\mathbf{n}}$ is an optic axis. For propagation along $\hat{\mathbf{n}}$, we have $\mu_0 N_1^{-2} = \mu_0 N_2^{-2} = \lambda$, and $\boldsymbol{D}$ may be any bivector orthogonal to $\hat{\mathbf{n}}$.

Exercises 9.3

1. For a homogeneous wave propagating along $\hat{\mathbf{n}}$, with slowness N, express $\boldsymbol{H}$ in terms of $\boldsymbol{D}$. Express the mean energy flux $\bar{\boldsymbol{r}}$ in terms of $\boldsymbol{D}$.
2. For the waves (9.6.3) and (9.6.4) propagating in biaxial crystals, obtain $\boldsymbol{H}_1, \boldsymbol{H}_2$ corresponding to $\boldsymbol{D}_1, \boldsymbol{D}_2$. Check that $\boldsymbol{H}_1 \parallel \boldsymbol{D}_2$ and $\boldsymbol{H}_2 \parallel \boldsymbol{D}_1$ (**Hint**: use the unit vectors $\hat{\mathbf{h}}^*, \hat{\mathbf{k}}^*$ along $\mathbf{h} - (\mathbf{h}\cdot\hat{\mathbf{n}})\hat{\mathbf{n}}$, $\mathbf{k} - (\mathbf{k}\cdot\hat{\mathbf{n}})\hat{\mathbf{n}}$).
3. For the biaxial case, draw a figure (in the plane orthogonal to $\hat{\mathbf{n}}$) of the elliptical section of $\mathbf{x}^{\mathrm{T}}\boldsymbol{\kappa}^{-1}\boldsymbol{x} = 1$ by the plane $\hat{\mathbf{n}}\cdot\boldsymbol{x} = 0$, with $\hat{\mathbf{h}}^*, \hat{\mathbf{k}}^*$ and $\boldsymbol{D}_1, \boldsymbol{D}_2, \boldsymbol{H}_1, \boldsymbol{H}_2$. Assume $\hat{\mathbf{n}}$ is not in the plane of the optic axes $\mathbf{h}$ and $\mathbf{k}$.
4. Draw the corresponding figure for the uniaxial case.

9.7 Inhomogeneous waves

9.7.1 Zero roots of the secular equation

The secular equation (9.3.1) has always a zero root, and the two other roots are the solutions $\mu_0 N_1^{-2}$, $\mu_0 N_2^{-2}$ of the quadratic equation (9.3.6), or equivalently (9.3.7). For homogeneous waves, these are always different from zero. However, for inhomogeneous waves, the possibility of having one of these roots, or both, equal to zero must be considered. When the prescribed bivector $\boldsymbol{C}$ is such that both roots are zero, no propagation is possible. Such a bivector $\boldsymbol{C}$ will be said to be 'critical'. When the prescribed bivector $\boldsymbol{C}$ is such that just one root is zero, only one propagation mode is possible. We examine in turn these two possibilities.

(a) Both roots zero. No propagation There are, in general, two cases when equation (9.3.6), or equivalently (9.3.7), has two zero roots, namely (i) when $\boldsymbol{C}\cdot\boldsymbol{C} = \boldsymbol{C}^{\mathrm{T}}\boldsymbol{\kappa}^{-1}\boldsymbol{C} = 0$, or (ii) when $\boldsymbol{C}^{\mathrm{T}}\boldsymbol{\kappa}\boldsymbol{C} = \boldsymbol{C}^{\mathrm{T}}\boldsymbol{\kappa}^2\boldsymbol{C} = 0$.

Case (i): $\boldsymbol{C}\cdot\boldsymbol{C} = \boldsymbol{C}^{\mathrm{T}}\boldsymbol{\kappa}^{-1}\boldsymbol{C} = 0$

The bivectors $\boldsymbol{C}$ whose ellipses are similar and similarly situated to a circular section of the index ellipsoid $\boldsymbol{x}^{\mathrm{T}}\boldsymbol{\kappa}^{-1}\boldsymbol{x} = 1$ are critical. Thus, when $\boldsymbol{C}$ is isotropic in a plane orthogonal to an optic axis, then no propagation is possible ($\boldsymbol{C}$ is critical).

Case (ii): $\boldsymbol{C}^{\mathrm{T}}\boldsymbol{\kappa}\boldsymbol{C} = \boldsymbol{C}^{\mathrm{T}}\boldsymbol{\kappa}^{2}\boldsymbol{C} = 0$

Here, we note that $\boldsymbol{C}^{\mathrm{T}}\boldsymbol{\kappa}\boldsymbol{C} = \boldsymbol{C}^{\mathrm{T}}\boldsymbol{\kappa}^{2}\boldsymbol{C} = 0$ may also be written as $(\boldsymbol{\kappa}\boldsymbol{C})^{\mathrm{T}}\boldsymbol{\kappa}^{-1}(\boldsymbol{\kappa}\boldsymbol{C}) = \boldsymbol{\kappa}\boldsymbol{C}\cdot\boldsymbol{\kappa}\boldsymbol{C} = 0$, which means that the ellipse of $\boldsymbol{\kappa}\boldsymbol{C}$ is similar and similar situated to a circular section of the index ellipsoid $\boldsymbol{x}^{\mathrm{T}}\boldsymbol{\kappa}^{-1}\boldsymbol{x} = 1$: $\boldsymbol{\kappa}\boldsymbol{C}$ is isotropic in a plane orthogonal to an optic axis. Thus, $\hat{\boldsymbol{n}}^{\mathrm{T}}\boldsymbol{\kappa}\boldsymbol{C} = 0$, where $\hat{\boldsymbol{n}}$ denotes a unit vector along an optic axis, and hence $\boldsymbol{C}$ is in a plane conjugate to the optic axis $\hat{\boldsymbol{n}}$ with respect to the ellipsoid $\boldsymbol{x}^{\mathrm{T}}\boldsymbol{\kappa}\boldsymbol{x} = 1$ ('Fresnel ellipsoid'). Hence, when the ellipse of $\boldsymbol{C}$ is similar and similarly situated to a section of the Fresnel ellipsoid $\boldsymbol{x}^{\mathrm{T}}\boldsymbol{\kappa}\boldsymbol{x} = 1$ by a plane conjugate to an optic axis with respect to this ellipsoid, then no propagation is possible ($\boldsymbol{C}$ is critical).

From the study of cases (i) and (ii), we draw the following conclusions about the critical bivectors $\boldsymbol{C}$ for biaxial crystals, uniaxial crystals, and isotropic media.

For biaxial crystals ($\lambda > \mu > \nu$), recall that $\boldsymbol{A}_h$ and $\boldsymbol{A}_k$ given by (5.6.15) are isotropic and orthogonal to the optic axes $\boldsymbol{h}$ and $\boldsymbol{k}$, respectively. Hence, the critical bivectors $\boldsymbol{C}$ are the bivectors parallel to $\boldsymbol{A}_h$, $\boldsymbol{A}_k$, $\boldsymbol{\kappa}^{-1}\boldsymbol{A}_h$, $\boldsymbol{\kappa}^{-1}\boldsymbol{A}_k$, or to their complex conjugates.

For uniaxial crystals ($\lambda = \mu > \nu$), recall that $\boldsymbol{A}_z$ given by (5.6.16) is isotropic and orthogonal to the optic axis $\boldsymbol{u}$. Here $\boldsymbol{\kappa}^{-1}\boldsymbol{A}_z = \lambda\boldsymbol{A}_z$, and cases (i) and (ii) coalesce. Hence, the critical bivectors $\boldsymbol{C}$ are the bivectors parallel to $\boldsymbol{A}_z$, or to its complex conjugate.

Finally, for isotropic media, every isotropic bivector $\boldsymbol{C}$ is critical.

(b) One root zero. One propagation mode There are in general two cases when equation (9.3.6), or equivalently (9.3.7), has just one zero root, namely when (i) $\boldsymbol{C}\cdot\boldsymbol{C} = 0$ and $\boldsymbol{C}^{\mathrm{T}}\boldsymbol{\kappa}^{-1}\boldsymbol{C} \neq 0$, or (ii) when $\boldsymbol{C}^{\mathrm{T}}\boldsymbol{\kappa}\boldsymbol{C} = 0$ and $\boldsymbol{C}^{\mathrm{T}}\boldsymbol{\kappa}^{2}\boldsymbol{C} \neq 0$. Of course, these are not feasible for isotropic media, so that isotropic media are not considered here.

Case (i): $\boldsymbol{C}\cdot\boldsymbol{C} = 0$, $\boldsymbol{C}^{\mathrm{T}}\boldsymbol{\kappa}^{-1}\boldsymbol{C} \neq 0$

This case occurs when $\boldsymbol{C}$ is isotropic ($\boldsymbol{C}\cdot\boldsymbol{C} = 0$), but not in the plane of a circular section of the $\boldsymbol{\kappa}^{-1}$-ellipsoid ($\boldsymbol{C}^{\mathrm{T}}\boldsymbol{\kappa}^{-1}\boldsymbol{C} \neq 0$,

$C \cdot C = 0$). Thus C is not critical, that is not in a plane orthogonal to an optic axis. Then, one root of (9.3.6) is zero, and the other one is given by

$$\mu_0 N^{-2} = -C^{\mathrm{T}} \kappa^{-1} C \neq 0. \tag{9.7.1}$$

For the amplitude D corresponding to this root, we have, using (9.2.7) and (9.2.8b),

$$D \cdot D = \mu_0^{-2} N^2 (C \times B) \cdot (C \times B) = \mu_0^{-2} N^2 \{(C \cdot C)(B \cdot B) - (C \cdot B)^2\} = 0. \tag{9.7.2}$$

Thus, the corresponding amplitude D is isotropic. But then, because $C \cdot C = C \cdot D = D \cdot D = 0$, it follows (section 2.6. Theorem 2.2) that D is parallel to C. Hence, up to an arbitrary complex factor, the amplitudes of the wave corresponding to (9.7.1) are given by

$$D = C, \quad E = \kappa^{-1} C, \quad B = N C \times \kappa^{-1} C. \tag{9.7.3}$$

We note, from (9.7.1), for this wave, that

$$B \cdot B = -N^2 (C^{\mathrm{T}} \kappa^{-1} C)^2 = \mu_0 C^{\mathrm{T}} \kappa^{-1} C \neq 0. \tag{9.7.4}$$

Thus, for the wave corresponding to (9.7.1), the $\mathscr{D}$-field is circularly polarized, but the $\mathscr{B}$-field is not.

Case (ii): $C^{\mathrm{T}} \kappa C = 0$, $C^{\mathrm{T}} \kappa^2 C \neq 0$

This case occurs when the ellipse of C is similar and similarly situated to any section of the Fresnel ellipsoid $x^{\mathrm{T}} \kappa x = 1$, except the sections by the planes conjugate to the optic axes with respect to this ellipsoid. (Recall the discussion for case a(ii) when both roots are zero.) Then, one root of (9.3.7) is zero, and the other one is given by

$$\mu_0 N^{-2} = -(\det \kappa)^{-1} C^{\mathrm{T}} \kappa^2 C \neq 0. \tag{9.7.5}$$

For the wave corresponding to this root, the propagation condition, written in the form (9.2.17), shows that $B = \mu_0 H$ is parallel to κC. Hence, up to an arbitrary complex factor, the amplitudes of the wave corresponding to (9.7.4) are given by

$$B = \kappa C, \quad D = \mu_0^{-1} N \kappa C \times C,$$
$$E = \mu_0^{-1} N \kappa^{-1} (\kappa C \times C) = \mu_0^{-1} N (\det \kappa)^{-1} (\kappa^2 C \times \kappa C). \tag{9.7.6}$$

We note that, for this wave,

$$B \cdot B = C^{\mathrm{T}} \kappa^2 C \neq 0, \quad D \cdot D = -\mu_0^{-1} (\det \kappa)(C \cdot C). \tag{9.7.7}$$

In general, neither the $\mathscr{B}$-field nor the $\mathscr{D}$-field is circularly polarized.

Exercise 9.4

Compute the weighted mean energy density $\hat{e}$ and the weighted mean energy flux $\hat{\mathbf{r}}$ for

(a) the wave given by (9.7.1), (9.7.3);
(b) the wave given by (9.7.5), (9.7.6).

9.7.2 *Results in terms of the optic axes*

Here we show that, for any prescribed bivector $\boldsymbol{C}$, the eigenvalues $\mu_0 N_1^{-2}$, $\mu_0 N_2^{-2}$, of the optical tensor $\boldsymbol{Q}(\boldsymbol{C})$, given by (9.2.11), and the corresponding amplitude bivectors $\boldsymbol{H}_1$, $\boldsymbol{H}_2$, may be expressed in terms of the unit vectors along the optic axes. We consider in turn the case of biaxial crystals, uniaxial crystals and isotropic crystals.

(a) Biaxial crystals ($\lambda > \mu > \nu$) Recalling that $\boldsymbol{\kappa}^{-1}$ has the Hamilton cyclic form ((5.6.5), with $\boldsymbol{\alpha} = \boldsymbol{\kappa}^{-1}$)

$$\boldsymbol{\kappa}^{-1} = \mu \mathbf{1} + \tfrac{1}{2}(\lambda - \nu)(\mathbf{h} \otimes \mathbf{k} + \mathbf{k} \otimes \mathbf{h}), \tag{9.7.8}$$

with $\lambda > \mu > \nu$, we have

$$\boldsymbol{Q}(\boldsymbol{C}) = \boldsymbol{\Gamma}^{\mathrm{T}} \boldsymbol{\kappa}^{-1} \boldsymbol{\Gamma} = \mu \boldsymbol{\Pi} + \tfrac{1}{2}(\lambda - \nu)(\boldsymbol{\Gamma}\mathbf{h} \otimes \boldsymbol{\Gamma}\mathbf{k} + \boldsymbol{\Gamma}\mathbf{k} \otimes \boldsymbol{\Gamma}\mathbf{h}). \tag{9.7.9}$$

Hence, the propagation condition (9.2.10) may be written as

$$\begin{aligned} &\{\mu(\boldsymbol{C}\cdot\boldsymbol{C}) - \mu_0 N^{-2}\}\boldsymbol{H} \\ &\quad - \tfrac{1}{2}(\lambda - \nu)\{\boldsymbol{\Gamma}\mathbf{h}(\mathbf{k}^{\mathrm{T}}\boldsymbol{\Gamma}\boldsymbol{H}) + \boldsymbol{\Gamma}\mathbf{k}(\mathbf{h}^{\mathrm{T}}\boldsymbol{\Gamma}\boldsymbol{H})\} = \mathbf{0}, \end{aligned} \tag{9.7.10}$$

because $\boldsymbol{\Pi}\boldsymbol{H} = (\boldsymbol{C}\cdot\boldsymbol{C})\boldsymbol{H}$.

(i) General case: $\boldsymbol{C}$ not in plane containing both optic axes

For the moment, we assume that the bivector $\boldsymbol{C}$ is not in the plane of $\mathbf{h}$ and $\mathbf{k}$, so that $\boldsymbol{C} \times \mathbf{h}$ and $\boldsymbol{C} \times \mathbf{k}$ are linearly independent. Hence, the bivector $\boldsymbol{H}$, which is orthogonal to $\boldsymbol{C}$, may be written as a linear combination of $\boldsymbol{C} \times \mathbf{h}$ and $\boldsymbol{C} \times \mathbf{k}$:

$$\boldsymbol{H} = \alpha \boldsymbol{C} \times \mathbf{h} + \beta \boldsymbol{C} \times \mathbf{k} = \alpha \boldsymbol{\Gamma}\mathbf{h} + \beta \boldsymbol{\Gamma}\mathbf{k}, \tag{9.7.11}$$

for some scalars α and β.

Introducing this in (9.7.10), and equating to zero the components along $\boldsymbol{\Gamma}\mathbf{h}$ and $\boldsymbol{\Gamma}\mathbf{k}$, we obtain the following homogeneous system for the unknowns α and β:

$$\{\mu(\boldsymbol{C}\cdot\boldsymbol{C})+\tfrac{1}{2}(\lambda-\nu)\mathbf{k}^{\mathrm{T}}\boldsymbol{\Gamma}^{\mathrm{T}}\boldsymbol{\Gamma}\mathbf{h}-\mu_0 N^{-2}\}\alpha+\tfrac{1}{2}(\lambda-\nu)(\mathbf{k}^{\mathrm{T}}\boldsymbol{\Gamma}^{\mathrm{T}}\boldsymbol{\Gamma}\mathbf{k})\beta=0,$$
$$\tfrac{1}{2}(\lambda-\nu)(\mathbf{h}^{\mathrm{T}}\boldsymbol{\Gamma}^{\mathrm{T}}\boldsymbol{\Gamma}\mathbf{h})\alpha+\{\mu(\boldsymbol{C}\cdot\boldsymbol{C})+\tfrac{1}{2}(\lambda-\nu)\mathbf{h}^{\mathrm{T}}\boldsymbol{\Gamma}^{\mathrm{T}}\boldsymbol{\Gamma}\mathbf{k}-\mu_0 N^{-2}\}\beta=0. \quad (9.7.12)$$

But

$$\mathbf{k}^{\mathrm{T}}\boldsymbol{\Gamma}^{\mathrm{T}}\boldsymbol{\Gamma}\mathbf{h}=\mathbf{h}^{\mathrm{T}}\boldsymbol{\Gamma}^{\mathrm{T}}\boldsymbol{\Gamma}\mathbf{k}=(\boldsymbol{C}\times\mathbf{h})\cdot(\boldsymbol{C}\times\mathbf{k})$$
$$=\frac{\lambda+\nu-2\mu}{\lambda-\nu}\boldsymbol{C}\cdot\boldsymbol{C}-(\boldsymbol{C}\cdot\mathbf{h})(\boldsymbol{C}\cdot\mathbf{k}), \quad (9.7.13)$$

and

$$\mathbf{h}^{\mathrm{T}}\boldsymbol{\Gamma}^{\mathrm{T}}\boldsymbol{\Gamma}\mathbf{h}=(\boldsymbol{C}\times\mathbf{h})\cdot(\boldsymbol{C}\times\mathbf{h}),\quad \mathbf{k}^{\mathrm{T}}\boldsymbol{\Gamma}^{\mathrm{T}}\boldsymbol{\Gamma}\mathbf{k}=(\boldsymbol{C}\times\mathbf{k})\cdot(\boldsymbol{C}\times\mathbf{k}). \quad (9.7.14)$$

Thus, the system (9.7.12) may be written

$$\{\tfrac{1}{2}(\lambda+\nu)\boldsymbol{C}\cdot\boldsymbol{C}-\tfrac{1}{2}(\lambda-\nu)(\boldsymbol{C}\cdot\mathbf{h})(\boldsymbol{C}\cdot\mathbf{k})-\mu_0 N^{-2}\}\alpha$$
$$+\tfrac{1}{2}(\lambda-\nu)(\boldsymbol{C}\times\mathbf{k})\cdot(\boldsymbol{C}\times\mathbf{k})\beta=0,$$
$$\tfrac{1}{2}(\lambda-\nu)(\boldsymbol{C}\times\mathbf{h})\cdot(\boldsymbol{C}\times\mathbf{h})\alpha$$
$$+\{\tfrac{1}{2}(\lambda+\nu)\boldsymbol{C}\cdot\boldsymbol{C}-\tfrac{1}{2}(\lambda-\nu)(\boldsymbol{C}\cdot\mathbf{h})(\boldsymbol{C}\cdot\mathbf{k})-\mu_0 N^{-2}\}\beta=0. \quad (9.7.15a)$$

The condition for nontrivial solutions of this system yields the eigenvalues

$$\mu_0 N_{1,2}^{-2}=\tfrac{1}{2}(\lambda\times\nu)\boldsymbol{C}\cdot\boldsymbol{C}$$
$$-\tfrac{1}{2}(\lambda-\nu)\{(\boldsymbol{C}\cdot\mathbf{h})(\boldsymbol{C}\cdot\mathbf{k})\pm\sqrt{(\boldsymbol{C}\times\mathbf{h})\cdot(\boldsymbol{C}\times\mathbf{h})}\sqrt{(\boldsymbol{C}\times\mathbf{k})\cdot(\boldsymbol{C}\times\mathbf{k})}\}. \quad (9.7.15b)$$

This result is a generalization of (9.6.3) to the case of inhomogeneous waves. For homogeneous waves, $\boldsymbol{C}=\hat{\mathbf{n}}$, and, $\boldsymbol{C}\cdot\mathbf{h}=\cos\phi_1$, $\boldsymbol{C}\cdot\mathbf{k}=\cos\phi_2$, $(\boldsymbol{C}\times\mathbf{h})\cdot(\boldsymbol{C}\times\mathbf{h})=\sin^2\phi_1$, $(\boldsymbol{C}\times\mathbf{k})\cdot(\boldsymbol{C}\times\mathbf{k})=\sin^2\phi_2$, and we retrieve (9.6.3).

For the moment, we assume that neither $\boldsymbol{C}\times\mathbf{h}$, nor $\boldsymbol{C}\times\mathbf{k}$ is isotropic, so that $N_1^{-2}\neq N_2^{-2}$ (the case of a double root will be considered in section 9.7.3). Then, introducing (9.7.15b) into the system (9.7.15a), we obtain the amplitudes $\boldsymbol{H}_1$, $\boldsymbol{H}_2$ corresponding to N_1^{-2}, N_2^{-2}. Up to an arbitrary scalar factor, we have

$$\boldsymbol{H}_{1,2}=\frac{\boldsymbol{C}\times\mathbf{h}}{\{(\boldsymbol{C}\times\mathbf{h})\cdot(\boldsymbol{C}\times\mathbf{h})\}^{1/2}}\mp\frac{\boldsymbol{C}\times\mathbf{k}}{\{(\boldsymbol{C}\times\mathbf{k})\cdot(\boldsymbol{C}\times\mathbf{k})\}^{1/2}}. \quad (9.7.16)$$

Using (9.2.7) and (9.2.4), we obtain the corresponding amplitudes

$\boldsymbol{D}_1$, $\boldsymbol{D}_2$:

$$\boldsymbol{D}_{1,2} = N_{1,2}\left\{\frac{\boldsymbol{\Pi}\mathbf{h}}{\{(\boldsymbol{C}\times\mathbf{h})\cdot(\boldsymbol{C}\times\mathbf{h})\}^{1/2}} \mp \frac{\boldsymbol{\Pi}\mathbf{k}}{\{(\boldsymbol{C}\times\mathbf{k})\cdot(\boldsymbol{C}\times\mathbf{k})\}^{1/2}}\right\}. \tag{9.7.17}$$

(ii) Special case: $\boldsymbol{C}$ in plane containing both optic axes

Here the bivector $\boldsymbol{C}$ is assumed to be in the plane of $\mathbf{h}$ and $\mathbf{k}$, that is the plane of the major and the minor axes of the index ellipsoid $\boldsymbol{x}^{\mathrm{T}}\boldsymbol{\kappa}^{-1}\boldsymbol{x} = 1$. Hence, $\boldsymbol{C}\times\mathbf{h}$ and $\boldsymbol{C}\times\mathbf{k}$ have both the same real direction, the direction of $\mathbf{h}\times\mathbf{k}$ (along the intermediate axis of the index ellipsoid). Thus, we now have

$$\frac{\boldsymbol{C}\times\mathbf{h}}{\{(\boldsymbol{C}\times\mathbf{h})\cdot(\boldsymbol{C}\times\mathbf{h})\}^{1/2}} = \varepsilon\frac{\boldsymbol{C}\times\mathbf{k}}{\{(\boldsymbol{C}\times\mathbf{k})\cdot(\boldsymbol{C}\times\mathbf{k})\}^{1/2}}, \tag{9.7.18}$$

with $\varepsilon = +1$ or -1, and from (9.7.13),

$$\begin{aligned}\varepsilon\sqrt{(\boldsymbol{C}\times\mathbf{h})\cdot(\boldsymbol{C}\times\mathbf{h})}\sqrt{(\boldsymbol{C}\times\mathbf{k})\cdot(\boldsymbol{C}\times\mathbf{k})}& \\ = \frac{\lambda+\nu-2\mu}{\lambda-\nu}\boldsymbol{C}\cdot\boldsymbol{C} - (\boldsymbol{C}\cdot\mathbf{h})(\boldsymbol{C}\cdot\mathbf{k}).&\end{aligned} \tag{9.7.19}$$

The eigenbivectors, $\boldsymbol{H}$, of the optical tensor (9.7.9), orthogonal to $\boldsymbol{C}$, are, up to a scalar factor,

$$\boldsymbol{H}' = \boldsymbol{\Pi}\mathbf{h}, \quad \boldsymbol{H}'' = \boldsymbol{C}\times\mathbf{h} = \boldsymbol{\Gamma}\mathbf{h}, \tag{9.7.20}$$

because $\boldsymbol{\Pi}^2\mathbf{h} = (\boldsymbol{C}\cdot\boldsymbol{C})\boldsymbol{\Pi}\mathbf{h}$, $(\boldsymbol{C}\times\mathbf{h})\cdot\boldsymbol{\Pi}\mathbf{h} = 0$, $\boldsymbol{\Pi}(\boldsymbol{C}\times\mathbf{h}) = (\boldsymbol{C}\cdot\boldsymbol{C})\boldsymbol{C}\times\mathbf{h}$. The corresponding eigenvalues are, respectively,

$$\begin{aligned}\mu_0 N'^{-2} &= \mu(\boldsymbol{C}\cdot\boldsymbol{C}),\\ \mu_0 N''^{-2} &= \mu(\boldsymbol{C}\cdot\boldsymbol{C}) + \varepsilon\sqrt{(\boldsymbol{C}\times\mathbf{h})\cdot(\boldsymbol{C}\times\mathbf{h})}\sqrt{(\boldsymbol{C}\times\mathbf{k})\cdot(\boldsymbol{C}\times\mathbf{k})}\\ &= (\lambda+\nu-\mu)\boldsymbol{C}\cdot\boldsymbol{C} - (\lambda-\nu)(\boldsymbol{C}\cdot\mathbf{h})(\boldsymbol{C}\cdot\mathbf{k}).\end{aligned} \tag{9.7.21}$$

Also, using (9.2.7) and (9.2.4), we obtain the amplitudes $\boldsymbol{D}'$ and $\boldsymbol{D}''$ corresponding to $\boldsymbol{H}'$ and $\boldsymbol{H}''$, respectively:

$$\boldsymbol{D}' = -N'(\boldsymbol{C}\cdot\boldsymbol{C})\boldsymbol{H}'', \quad \boldsymbol{D}'' = N''\boldsymbol{H}'. \tag{9.7.22}$$

We note that when (9.7.19) is substituted into (9.7.15), $\mu_0 N_1^{-2}$ and $\mu_0 N_2^{-2}$ reduce to $\mu_0 N'^{-2}$ and $\mu_0 N''^{-2}$ (for $\varepsilon = +1$, we have $N_1^{-2} = N'^{-2}$, $N_2^{-2} = N''^{-2}$, and for $\varepsilon = -1$, we have $N_1^{-2} = N''^{-2}$, $N_2^{-2} = N'^{-2}$). Hence, (9.7.15) remains valid even in the special case when $\boldsymbol{C}$ is in the plane containing both optic axes.

Remark From (9.7.8), it follows immediately that

$$(\boldsymbol{C}\times\mathbf{h})^{\mathrm{T}}\boldsymbol{\kappa}^{-1}(\boldsymbol{C}\times\mathbf{h})=\mu(\boldsymbol{C}\times\mathbf{h})\cdot(\boldsymbol{C}\times\mathbf{h}),$$
$$(\boldsymbol{C}\times\mathbf{k})^{\mathrm{T}}\boldsymbol{\kappa}^{-1}(\boldsymbol{C}\times\mathbf{k})=\mu(\boldsymbol{C}\times\mathbf{k})\cdot(\boldsymbol{C}\times\mathbf{k}). \qquad (9.7.23)$$

Also, using the identity of Exercise 5.2.5, with $\boldsymbol{\alpha}^{-1}=\boldsymbol{\kappa}$, we have

$$(\lambda+\nu)\boldsymbol{C}\cdot\boldsymbol{C}-(\lambda-\nu)(\boldsymbol{C}\cdot\mathbf{h})(\boldsymbol{C}\cdot\mathbf{k})$$
$$=(\lambda+\nu)\mu\boldsymbol{C}^{\mathrm{T}}\boldsymbol{\kappa}\boldsymbol{C}+\lambda\nu(\lambda-\nu)(\mathbf{h}^{\mathrm{T}}\boldsymbol{\kappa}\boldsymbol{C})(\mathbf{k}^{\mathrm{T}}\boldsymbol{\kappa}\boldsymbol{C}). \qquad (9.7.24)$$

Thus, the formula (9.7.15) for the eigenvalues of the optical tensor $\boldsymbol{Q}(\boldsymbol{C})$ may also be written as

$$\mu_0 N_{1,2}^{-2}=\tfrac{1}{2}(\lambda+\nu)\mu\boldsymbol{C}^{\mathrm{T}}\boldsymbol{\kappa}\boldsymbol{C}+\tfrac{1}{2}(\lambda-\nu)\{\lambda\nu(\mathbf{h}^{\mathrm{T}}\boldsymbol{\kappa}\boldsymbol{C})(\mathbf{k}^{\mathrm{T}}\boldsymbol{\kappa}\boldsymbol{C})$$
$$\mp\mu^{-1}\sqrt{(\boldsymbol{C}\times\mathbf{h})^{\mathrm{T}}\boldsymbol{\kappa}^{-1}(\boldsymbol{C}\times\mathbf{h})}\sqrt{(\boldsymbol{C}\times\mathbf{k})^{\mathrm{T}}\boldsymbol{\kappa}^{-1}(\boldsymbol{C}\times\mathbf{k})}\}. \qquad (9.7.25)$$

(b) Uniaxial crystals ($\lambda=\mu\neq\nu$) Let $\lambda=\mu\neq\nu$, so that

$$\boldsymbol{\kappa}^{-1}=\lambda\mathbf{1}+(\nu-\lambda)\mathbf{u}\otimes\mathbf{u}, \qquad (9.7.26)$$

and thus,

$$\boldsymbol{Q}(\boldsymbol{C})=\boldsymbol{\Gamma}^{\mathrm{T}}\boldsymbol{\kappa}^{-1}\boldsymbol{\Gamma}=\lambda\boldsymbol{\Pi}+(\nu-\lambda)\boldsymbol{\Gamma}\mathbf{u}\otimes\boldsymbol{\Gamma}\mathbf{u}. \qquad (9.7.27)$$

Clearly, the eigenbivectors, $\boldsymbol{H}$, of $\boldsymbol{Q}(\boldsymbol{C})$, orthogonal to $\boldsymbol{C}$, are, up to a scalar factor,

$$\boldsymbol{H}_1=\boldsymbol{\Pi}\mathbf{u}, \quad \boldsymbol{H}_2=\boldsymbol{C}\times\mathbf{u}=\boldsymbol{\Gamma}\mathbf{u}, \qquad (9.7.28)$$

and, from (9.2.7) and (9.2.4), we obtain the corresponding amplitudes $\boldsymbol{D}_1$, $\boldsymbol{D}_2$:

$$\boldsymbol{D}_1=-N_1(\boldsymbol{C}\cdot\boldsymbol{C})\boldsymbol{H}_2, \quad \boldsymbol{D}_2=N_2\boldsymbol{H}_1. \qquad (9.7.29)$$

The corresponding eigenvalues are, respectively,

$$\mu_0 N_1^{-2}=\lambda(\boldsymbol{C}\cdot\boldsymbol{C}), \quad \mu_0 N_2^{-2}=\nu(\boldsymbol{C}\cdot\boldsymbol{C})-(\nu-\lambda)(\boldsymbol{C}\cdot\mathbf{u})^2. \qquad (9.7.30)$$

Remark From (9.7.26), it follows that

$$\boldsymbol{\kappa}^{-1}\boldsymbol{C}=\lambda\boldsymbol{C}+(\nu-\lambda)(\boldsymbol{C}\cdot\mathbf{u})\mathbf{u}, \qquad (9.7.31)$$

and thus, because $\boldsymbol{\kappa}\mathbf{u}=\nu^{-1}\mathbf{u}$,

$$\boldsymbol{C}=\lambda\boldsymbol{\kappa}\boldsymbol{C}+\nu^{-1}(\nu-\lambda)(\boldsymbol{C}\cdot\mathbf{u})\mathbf{u}. \qquad (9.7.32)$$

Using this in (9.7.30), we note that the eigenvalues may also be written as

$$\mu_0 N_1^{-2} = \lambda^2 \boldsymbol{C}^{\mathrm{T}} \boldsymbol{\kappa} \boldsymbol{C} + \lambda \nu^{-1}(\nu - \lambda)(\boldsymbol{C}\cdot\mathbf{u})^2,$$
$$\mu_0 N_2^{-2} = \lambda \nu \boldsymbol{C}^{\mathrm{T}} \boldsymbol{\kappa} \boldsymbol{C}. \tag{9.7.33}$$

(c) Isotropic media Let $\lambda = \mu = \nu$, so that $\boldsymbol{\kappa}^{-1} = \lambda \mathbf{1}$, and thus

$$Q(\boldsymbol{C}) = \boldsymbol{\Gamma}^{\mathrm{T}} \boldsymbol{\kappa}^{-1} \boldsymbol{\Gamma} = \lambda \boldsymbol{\Pi}. \tag{9.7.34}$$

Clearly, any bivector $\boldsymbol{H}$ orthogonal to $\boldsymbol{C}$ is an eigenbivector of $Q(\boldsymbol{C})$, the corresponding eigenvalue being

$$\mu_0 N^{-2} = \lambda(\boldsymbol{C}\cdot\boldsymbol{C}). \tag{9.7.35}$$

Exercises 9.5

1. In the biaxial case, check that one of the roots (9.7.15), or equivalently (9.7.25) is zero when (a) $\boldsymbol{C}\cdot\boldsymbol{C} = 0$, or when (b) $\boldsymbol{C}^{\mathrm{T}}\boldsymbol{\kappa}\boldsymbol{C} = 0$. Obtain then an expression for the other root (propagating mode). From this find when both roots are zero (recall the discussion of section 9.7.1).
2. Same question for the uniaxial case.
3. Using (9.7.16), (9.7.17) and (9.7.28), (9.7.29) for the (a) biaxial and (b) uniaxial case, respectively, find the amplitudes $\boldsymbol{H}$ and $\boldsymbol{D}$ of the propagating mode corresponding to a prescribed isotropic $\boldsymbol{C}$: $\boldsymbol{C}\cdot\boldsymbol{C} = 0$ (not in a plane orthogonal to an optic axis). Check that $\boldsymbol{D}$ is isotropic and parallel to $\boldsymbol{C}$, but that $\boldsymbol{H}$ is not isotropic (in accordance with the results for when one root is zero, section 9.7.1).
4. For the waves with amplitudes $\boldsymbol{H}_{1,2}$ and $\boldsymbol{D}_{1,2}$ given by (9.7.16) and (9.7.17), check that $\boldsymbol{D}_1 \parallel \boldsymbol{H}_2$ and $\boldsymbol{D}_2 \parallel \boldsymbol{H}_1$.

9.7.3 Double nonzero root of the secular equation. $\mathscr{H}$, $\mathscr{B}$ and $\mathscr{D}$ circularly polarized

Here we are interested in waves with the $\mathscr{H}$-field, and thus also the $\mathscr{B}$-field, circularly polarized: $\boldsymbol{H}\cdot\boldsymbol{H} = \boldsymbol{B}\cdot\boldsymbol{B} = 0$. For such waves it follows from (9.4.4) that $\sigma = 0$, and then, from (9.4.5), that $\boldsymbol{D}$ is parallel to $\boldsymbol{B}$ and $\boldsymbol{H}$. Hence the $\mathscr{D}$-field is also circularly polarized.

Note that other waves, with the $\mathscr{D}$-field circularly polarized, but with $\mathscr{B} = \mu_0 \mathscr{H}$ elliptically polarized have been previously obtained for prescribed isotropic bivectors $\boldsymbol{C}$ which are not orthogonal to an optic axis (section 9.7.1(b) and Exercise 9.5.3).

Now, the propagation condition states that $\boldsymbol{H}$ must be an eigenbivector of the optical tensor $\boldsymbol{Q}(\boldsymbol{C})$ given by (9.2.11). In section 3.2 it has been shown that a necessary and sufficient condition that a complex symmetric matrix have an isotropic eigenbivector is that it has a multiple eigenvalue. Hence, waves with the $\mathscr{H}$-field circularly polarized may propagate if and only if $\boldsymbol{Q}(\boldsymbol{C})$ has a double nonzero eigenvalue, that is, if and only if the quadratic equation (9.3.6) (or equivalently (9.3.7)) has a double root. Here we seek the bivectors $\boldsymbol{C}$ for which this is the case. We consider in turn biaxial crystals, uniaxial crystals and isotropic media.

(a) Biaxial crystals $(\lambda > \mu > \nu)$ From (9.7.15), it is clear that $\mu_0 N_1^{-2} = \mu_0 N_2^{-2}$ when either

$$(\boldsymbol{C} \times \mathbf{h}) \cdot (\boldsymbol{C} \times \mathbf{h}) = 0 \quad \text{or} \quad (\boldsymbol{C} \times \mathbf{k}) \cdot (\boldsymbol{C} \times \mathbf{k}) = 0. \tag{9.7.36}$$

Thus the quadratic equation (9.3.6) has a double root when the projection of the ellipse of $\boldsymbol{C}$ onto a plane orthogonal to an optic axis ($\mathbf{h}$ or $\mathbf{k}$) is a circle. Because $\mathbf{h}$ and $\mathbf{k}$ are parallel to $\boldsymbol{A}_h \times \bar{\boldsymbol{A}}_h$ and $\boldsymbol{A}_k \times \bar{\boldsymbol{A}}_k$ respectively, where the isotropic bivectors $\boldsymbol{A}_h$ and $\boldsymbol{A}_k$ are defined by (5.6.15), the conditions (9.7.36) for a double root may equivalently be written as

$$(\boldsymbol{C} \cdot \boldsymbol{A}_h)(\boldsymbol{C} \cdot \bar{\boldsymbol{A}}_h) = 0, \tag{9.7.37a}$$

or

$$(\boldsymbol{C} \cdot \boldsymbol{A}_k)(\boldsymbol{C} \cdot \bar{\boldsymbol{A}}_k) = 0. \tag{9.7.37b}$$

From (9.7.15), it is clear that the double root is

$$\mu_0 N^{-2} = \tfrac{1}{2}(\lambda + \nu)\boldsymbol{C} \cdot \boldsymbol{C} - \tfrac{1}{2}(\lambda - \nu)(\boldsymbol{C} \cdot \mathbf{h})(\boldsymbol{C} \cdot \mathbf{k}), \tag{9.7.38}$$

and, using (9.7.36) and the result of Exercise 5.2.3, it is seen that this reduces to

$$\mu_0 N^{-2} = \lambda \nu (\boldsymbol{C} \cdot \mathbf{h})(\boldsymbol{C}^{\mathrm{T}} \boldsymbol{\kappa} \mathbf{h}), \tag{9.7.39a}$$

or

$$\mu_0 N^{-2} = \lambda \nu (\boldsymbol{C} \cdot \mathbf{k})(\boldsymbol{C}^{\mathrm{T}} \boldsymbol{\kappa} \mathbf{k}). \tag{9.7.39b}$$

Here we may assume that $\boldsymbol{C}$ is not in the plane of $\mathbf{h}$ and $\mathbf{k}$. Indeed, if $\boldsymbol{C}$ is in this plane and satisfies (9.7.36), then $\boldsymbol{C}$ is parallel to $\mathbf{h}$ or $\mathbf{k}$ (and hence the wave is homogeneous), a case which has been considered in section 9.6. Hence, the eigenbivectors $\boldsymbol{H}$ corresponding to the double root (9.7.38) may be obtained from the system (9.7.15).

We consider in turn the cases when $\boldsymbol{C} \times \mathbf{h}$ is isotropic, when $\boldsymbol{C} \times \mathbf{k}$ is isotropic, and when both are isotropic.

(i) $\boldsymbol{C} \times \mathbf{h}$ isotropic

If $(\boldsymbol{C} \times \mathbf{h})\cdot(\boldsymbol{C} \times \mathbf{h}) = 0$, with $(\boldsymbol{C} \times \mathbf{k})\cdot(\boldsymbol{C} \times \mathbf{k}) \neq 0$, we have $\alpha \neq 0$, $\beta = 0$ from (9.7.15) and (9.7.38), and thus, we obtain up to a scalar factor the eigenbivector $\boldsymbol{H} = \boldsymbol{C} \times \mathbf{h}$. Because $\mathbf{h}$ is parallel to $\boldsymbol{A}_h \times \bar{\boldsymbol{A}}_h$, we note, using (9.7.37a), that this isotropic eigenbivector $\boldsymbol{H}$ is parallel to $\boldsymbol{A}_h$ or $\bar{\boldsymbol{A}}_h$ according to whether $\boldsymbol{C}\cdot\boldsymbol{A}_h = 0$ or $\boldsymbol{C}\cdot\bar{\boldsymbol{A}}_h = 0$. Hence, when $\boldsymbol{C}\cdot\boldsymbol{A}_h = 0$, we have up to a scalar factor,

$$\boldsymbol{H} = \boldsymbol{A}_h, \tag{9.7.40a}$$

$$\boldsymbol{D} = \mathrm{i}N(\boldsymbol{C}\cdot\mathbf{h})\boldsymbol{A}_h, \tag{9.7.40b}$$

on using $\boldsymbol{C} \times \boldsymbol{A}_h = \mathrm{i}(\boldsymbol{C}\cdot\mathbf{h})\boldsymbol{A}_h$ (see Exercise 2.5.5). Similarly, when $\boldsymbol{C}\cdot\bar{\boldsymbol{A}}_h = 0$, we have, up to a scalar factor

$$\boldsymbol{H} = \bar{\boldsymbol{A}}_h, \tag{9.7.41a}$$

$$\boldsymbol{D} = -\mathrm{i}N(\boldsymbol{C}\cdot\mathbf{h})\bar{\boldsymbol{A}}_h. \tag{9.7.41b}$$

In (9.7.40b) and (9.7.41b), N is given by (9.7.39a).

Thus, when $\boldsymbol{C} \times \mathbf{h}$ is isotropic but when $\boldsymbol{C} \times \mathbf{k}$ is not, there is one propagation mode corresponding to the double root (9.7.39a), with $\mathscr{H}$, $\mathscr{B}$ and $\mathscr{D}$ circularly polarized in the plane $\mathbf{h}\cdot\boldsymbol{x} = 0$, orthogonal to the optic axis $\mathbf{h}$.

(ii) $\boldsymbol{C} \times \mathbf{k}$ isotropic

If $(\boldsymbol{C} \times \mathbf{k})\cdot(\boldsymbol{C} \times \mathbf{k}) = 0$, with $(\boldsymbol{C} \times \mathbf{h})\cdot(\boldsymbol{C} \times \mathbf{h}) \neq 0$, we have $\alpha = 0$, $\beta \neq 0$ from (9.7.15) and (9.7.38), and thus, we obtain, up to a scalar factor the eigenbivector $\boldsymbol{H} = \boldsymbol{C} \times \mathbf{k}$. The conclusions are analogous to those of case (i). Here, we have, up to a scalar factor,

$$\boldsymbol{H} = \boldsymbol{A}_k, \tag{9.7.42a}$$

$$\boldsymbol{D} = \mathrm{i}N(\boldsymbol{C}\cdot\mathbf{k})\boldsymbol{A}_k, \tag{9.7.42b}$$

when $\boldsymbol{C}\cdot\boldsymbol{A}_k = 0$, and

$$\boldsymbol{H} = \bar{\boldsymbol{A}}_k, \tag{9.7.43a}$$

$$\boldsymbol{D} = -\mathrm{i}N(\boldsymbol{C}\cdot\mathbf{k})\bar{\boldsymbol{A}}_k, \tag{9.7.43b}$$

when $\boldsymbol{C}\cdot\bar{\boldsymbol{A}}_k = 0$. In (9.7.42a) and (9.7.43b), N is given by (9.7.39b).

Thus, when $\boldsymbol{C} \times \mathbf{k}$ is isotropic but when $\boldsymbol{C} \times \mathbf{h}$ is not, there is one propagation mode corresponding to the double root (9.7.39b), with $\mathscr{H}$, $\mathscr{B}$ and $\mathscr{D}$ circularly polarized in the plane $\mathbf{k}\cdot\boldsymbol{x}=0$, orthogonal to the optic axis $\mathbf{k}$.

(iii) $\boldsymbol{C} \times \mathbf{h}$ *and* $\boldsymbol{C} \times \mathbf{k}$ *both isotropic*

We now consider the possibility that $\boldsymbol{C} \times \mathbf{h}$ and $\boldsymbol{C} \times \mathbf{k}$ are both isotropic:

$$(\boldsymbol{C} \times \mathbf{h})\cdot(\boldsymbol{C} \times \mathbf{h}) = \boldsymbol{C}\cdot\boldsymbol{C} - (\boldsymbol{C}\cdot\mathbf{h})^2 = \boldsymbol{C}\cdot\boldsymbol{C} - (\boldsymbol{C}\cdot\mathbf{k})^2$$
$$= (\boldsymbol{C} \times \mathbf{k})\cdot(\boldsymbol{C} \times \mathbf{k}) = 0. \tag{9.7.44}$$

For this to be the case, $\boldsymbol{C}$ must satisfy

$$(\boldsymbol{C}\cdot\mathbf{h})^2 - (\boldsymbol{C}\cdot\mathbf{k})^2 = \{\boldsymbol{C}\cdot(\mathbf{h}+\mathbf{k})\}\{\boldsymbol{C}\cdot(\mathbf{h}-\mathbf{k})\} = 0, \tag{9.7.45}$$

and thus

$$\boldsymbol{C}\cdot\mathbf{s} = 0 \quad \text{or} \quad \boldsymbol{C}\cdot\mathbf{u} = 0. \tag{9.7.46}$$

It follows that $\boldsymbol{C} \times \mathbf{h}$ and $\boldsymbol{C} \times \mathbf{k}$ are both isotropic when $\boldsymbol{C}$ is given by

$$\boldsymbol{C} = \left(\frac{\lambda - \nu}{\lambda - \mu}\right)^{1/2} \mathbf{u} \pm i\mathbf{t}, \tag{9.7.47a}$$

or

$$\boldsymbol{C} = \left(\frac{\lambda - \nu}{\mu - \nu}\right)^{1/2} \mathbf{s} \pm i\mathbf{t}. \tag{9.7.47b}$$

For these bivectors $\boldsymbol{C}$, the double root is

$$\mu_0 N^{-2} = \lambda \frac{\mu - \nu}{\lambda - \mu}, \tag{9.7.48a}$$

or

$$\mu_0 N^{-2} = \nu \frac{\lambda - \mu}{\mu - \nu}. \tag{9.7.48b}$$

We note that the slowness bivector $\boldsymbol{S} = N\boldsymbol{C}$ corresponding to (9.7.47) and (9.7.48) satisfies

$$\boldsymbol{S}\cdot\boldsymbol{S} = \frac{\mu_0}{\lambda} \quad \text{or} \quad \boldsymbol{S}\cdot\boldsymbol{S} = \frac{\mu_0}{\nu}, \tag{9.7.49}$$

so that for these waves the planes of constant amplitude are orthogonal to the planes of constant phase. Because now $\boldsymbol{C} \times \mathbf{h}$ and $\boldsymbol{C} \times \mathbf{k}$ are both isotropic, it follows from (9.7.15) and (9.7.38), that α

and β are both arbitrary, and so $\boldsymbol{H}$ may be any bivector orthogonal to $\boldsymbol{C}$. Thus, according to whether $\boldsymbol{C}$ is given by (9.7.47a) or (9.7.47b), we have

$$\boldsymbol{H}=\alpha_1\left(\left(\frac{\lambda-\mu}{\lambda-\nu}\right)^{1/2}\mathbf{u}\pm\mathrm{i}\mathbf{t}\right)+\alpha_2\left(\frac{\mu-\nu}{\lambda-\nu}\right)^{1/2}\mathbf{s}, \tag{9.7.50a}$$

$$\boldsymbol{D}=\pm\mathrm{i}N\left(\frac{\mu-\nu}{\lambda-\mu}\right)^{1/2}\left\{\alpha_1\left(\frac{\mu-\nu}{\lambda-\nu}\right)^{1/2}\mathbf{s}+\alpha_2\left(\left(\frac{\lambda-\mu}{\lambda-\nu}\right)^{1/2}\mathbf{u}\pm\mathrm{i}\mathbf{t}\right)\right\}, \tag{9.7.50b}$$

or

$$\boldsymbol{H}=\alpha_1\left(\left(\frac{\mu-\nu}{\lambda-\nu}\right)^{1/2}\mathbf{s}\pm\mathrm{i}\mathbf{t}\right)+\alpha_2\left(\frac{\lambda-\mu}{\lambda-\nu}\right)^{1/2}\mathbf{u}, \tag{9.7.51a}$$

$$\boldsymbol{D}=\mp\mathrm{i}N\left(\frac{\lambda-\mu}{\mu-\nu}\right)^{1/2}\left\{\alpha_1\left(\frac{\lambda-\mu}{\lambda-\nu}\right)^{1/2}\mathbf{u}+\alpha_2\left(\left(\frac{\mu-\nu}{\lambda-\nu}\right)^{1/2}\mathbf{s}\pm\mathrm{i}\mathbf{t}\right)\right\}, \tag{9.7.51b}$$

where α_1 and α_2 are arbitrary complex numbers. In (9.7.50b) and (9.7.51b), N is given by (9.7.48a) and (9.7.48b), respectively. For $\alpha_1=0$, the $\mathscr{H}$-field is linearly polarized while the $\mathscr{D}$-field is elliptically polarized in the plane of $\boldsymbol{C}$. For $\alpha_2=0$, the $\mathscr{H}$-field is elliptically polarized in the plane of $\boldsymbol{C}$, while the $\mathscr{D}$-field is linearly polarized. The $\mathscr{H}$-field and the $\mathscr{D}$-field are both circularly polarized when $(\alpha_2\alpha_1^{-1})^2=1$.

(b) Uniaxial crystals ($\lambda=\mu\neq\nu$) From (9.7.30), it is clear that $\mu_0 N_1^{-2}=\mu_0 N_2^{-2}$ when $\boldsymbol{C}\cdot\boldsymbol{C}-(\boldsymbol{C}\cdot\mathbf{u})^2=0$, that is, when

$$(\boldsymbol{C}\times\mathbf{u})\cdot(\boldsymbol{C}\times\mathbf{u})=0. \tag{9.7.52}$$

Thus the quadratic equation (9.3.6) has a double root when the projection of the ellipse of $\boldsymbol{C}$ onto the plane orthogonal to the optic axis (plane of $\mathbf{s}$ and $\mathbf{t}$) is a circle. As in the biaxial case, the condition (9.7.52) for a double root is equivalent to

$$(\boldsymbol{C}\cdot\boldsymbol{A}_z)(\boldsymbol{C}\cdot\bar{\boldsymbol{A}}_z)=0, \tag{9.7.53}$$

where $\boldsymbol{A}_z$ is defined by (5.6.16b). The double root is then

$$\mu_0 N^{-2}=\lambda(\boldsymbol{C}\cdot\mathbf{u})^2=\lambda(\boldsymbol{C}\cdot\boldsymbol{C}), \tag{9.7.54}$$

and corresponding to this double root, the optical tensor (9.7.27) has only one eigenbivector, $\boldsymbol{H}=\boldsymbol{C}\times\mathbf{u}$, defined up to a scalar factor.

Because $\mathbf{u}$ is parallel to $\boldsymbol{A}_z \times \bar{\boldsymbol{A}}_z$, we note, using (9.7.53), that this isotropic eigenbivector $\boldsymbol{H}$ is parallel to $\boldsymbol{A}_z$ or $\bar{\boldsymbol{A}}_z$ according to whether $\boldsymbol{C}\cdot\boldsymbol{A}_z = 0$ or $\boldsymbol{C}\cdot\bar{\boldsymbol{A}}_z = 0$. Hence, when $\boldsymbol{C}\cdot\boldsymbol{A}_z = 0$, we have, up to a scalar factor,

$$\boldsymbol{H} = \boldsymbol{A}_z, \tag{9.7.55a}$$

$$\boldsymbol{D} = \mathrm{i}N(\boldsymbol{C}\cdot\mathbf{u})\boldsymbol{A}_z, \tag{9.7.55b}$$

on using $\boldsymbol{C} \times \boldsymbol{A}_z = -\mathrm{i}(\boldsymbol{C}\cdot\mathbf{u})\boldsymbol{A}_z$. Similarly, when $\boldsymbol{C}\cdot\bar{\boldsymbol{A}}_z = 0$, we have, up to a scalar factor

$$\boldsymbol{H} = \bar{\boldsymbol{A}}_z, \tag{9.7.56a}$$

$$\boldsymbol{D} = -\mathrm{i}N(\boldsymbol{C}\cdot\mathbf{u})\bar{\boldsymbol{A}}_z. \tag{9.7.56b}$$

In (9.7.55b) and (9.7.56b), N is given by (9.7.54).

Thus, when $\boldsymbol{C} \times \mathbf{u}$ is isotropic, there is one propagation mode corresponding to the double root (9.7.54), with $\mathscr{H}$, $\mathscr{B}$ and $\mathscr{D}$ circularly polarized in the plane $\mathbf{u}\cdot\boldsymbol{r} = 0$, orthogonal to the optic axis.

(c) Isotropic media ($\lambda = \mu = \nu$) The root (9.7.35) is always a double root of the quadratic equation (9.3.6). Any bivector $\boldsymbol{H}$ orthogonal to $\boldsymbol{C}$ is a corresponding eigenbivector of the optical tensor (9.7.34). Writing $\boldsymbol{C} = m\hat{\mathbf{m}} + \mathrm{i}\hat{\mathbf{n}}$, we have

$$\boldsymbol{H} = \alpha_1 \boldsymbol{C}_\perp + \alpha_2 \hat{\mathbf{m}} \times \hat{\mathbf{n}}, \tag{9.7.57a}$$

$$\boldsymbol{D} = -\mathrm{i}mN\left\{\alpha_1\left(\frac{m^2-1}{m^2}\right)\hat{\mathbf{m}} \times \hat{\mathbf{n}} + \alpha_2 \boldsymbol{C}_\perp\right\}, \tag{9.7.57b}$$

where α_1 and α_2 are arbitrary complex numbers, and where $\boldsymbol{C}_\perp = m^{-1}\hat{\mathbf{m}} + \mathrm{i}\hat{\mathbf{n}}$ is the reciprocal of the bivector $\boldsymbol{C}$ (section 2.5). In (9.7.57b), N is given by (9.7.35), that is, $\mu_0 N^{-2} = \lambda(m^2 - 1)$. For $\alpha_1 = 0$, the $\mathscr{H}$-field is linearly polarized while the $\mathscr{D}$-field is elliptically polarized in the plane of $\boldsymbol{C}$. For $\alpha_2 = 0$, the $\mathscr{H}$-field is elliptically polarized in the plane of $\boldsymbol{C}$ while the $\mathscr{D}$-field is linearly polarized. The $\mathscr{H}$-field and the $\mathscr{D}$-field are both circularly polarized when $(\alpha_2\alpha_1^{-1})^2 = (m^2 - 1)m^{-2}$.

Exercises 9.6

1. In a biaxial crystal, for the wave (9.7.40) corresponding to a prescribed bivector $\boldsymbol{C}$ such that $\boldsymbol{C}\cdot\boldsymbol{A}_h = 0$, find the weighted mean energy density $\hat{e}$ and the weighted mean energy flux $\hat{\mathbf{r}}$.

2. In a biaxial crystal, for the wave (9.7.50) corresponding to $\boldsymbol{C}$ given by (9.7.47a), find the weighted mean energy density $\hat{e}$ and the weighted mean energy flux $\hat{\mathbf{f}}$. Analyse how the direction of $\hat{\mathbf{f}}$ varies when α_1 and α_2 are varied.
3. In a uniaxial crystal, for the wave (9.7.55) corresponding to a prescribed $\boldsymbol{C}$ such that $\boldsymbol{C}\cdot\boldsymbol{A}_z = 0$, find the weighted mean energy density $\hat{e}$ and the weighted mean energy flux $\hat{\mathbf{f}}$.
4. In an isotropic medium, for the wave (9.7.57) corresponding to an arbitrary given nonisotropic $\boldsymbol{C}$, find the weighted mean energy density $\hat{e}$ and the weighted mean energy flux $\hat{\mathbf{f}}$. Analyse how the direction of $\hat{\mathbf{f}}$ varies when α_1 and α_2 are varied.

10

Plane waves in linearized elasticity theory

In this chapter, the propagation of homogeneous and inhomogeneous time-harmonic plane waves within the context of linearized elasticity theory is considered.

First homogeneous waves in anisotropic elastic media are examined, and the acoustical tensor which is a function of the propagation direction, is introduced. The cases when the acoustical tensor has three simple eigenvalues, and when it has a double eigenvalue are examined. Results are presented concerning the mean energy flux and energy density. Two models of internally constrained materials, namely the incompressible material and the inextensible material, are considered and wave propagation in them briefly examined. The special cases of isotropic and transversely isotropic materials are studied in detail.

Next the propagation of inhomogeneous time-harmonic plane waves is investigated, using the DE-method introduced in Chapter 6. Here, the acoustical tensor is complex and symmetric. After dealing with the general theory of anisotropic media, the special cases of isotropic and transversely isotropic materials are studied in detail. As a particular case, for isotropic materials, we briefly develop the theory of elastic Rayleigh waves.

10.1 Constitutive equations and equations of motion

The constitutive equations describing a homogeneous anisotropic elastic body within the context of the infinitesimal strain theory are

$$t_{ij} = c_{ijkl}e_{kl}, \quad 2e_{kl} = u_{k,l} + u_{l,k}. \tag{10.1.1}$$

Here t_{ij} are the components (in a rectangular cartesian coordinate system) of the symmetric Cauchy stress tensor, and e_{kl} are the components of the infinitesimal strain tensor defined in terms of the derivatives of the displacement components u_i (the comma denotes the partial derivative with respect to the spatial coordinates: $u_{k,l} = \partial u_k/\partial x_l$). The elastic constants c_{ijkl} are assumed to have the symmetries

$$c_{ijkl} = c_{jikl} = c_{ijlk} = c_{klij}. \tag{10.1.2}$$

Owing to these symmetries, the number of possible independent elastic constants is reduced from 81 to at most 21. The traction vector, $\boldsymbol{t}_{(\mathbf{n})}$, across the material element of surface with unit outward normal $\mathbf{n}$, is given by

$$t_{(\mathbf{n})i} = t_{ij}n_j. \tag{10.1.3}$$

Introducing the constitutive equation (10.1.1) into the linearized balance of momentum

$$\rho\partial_t^2 u_i = t_{ij,j}, \tag{10.1.4}$$

in the absence of body forces, yields

$$\rho\partial_t^2 u_i = c_{ijkl}u_{k,lj}. \tag{10.1.5}$$

Here ρ denotes the density of the undeformed material. It is a constant.

Multiplying the balance of momentum (10.1.4) by the components $\dot{u}_i = \partial_t u_i$ of the particle velocity gives

$$\partial_t(\tfrac{1}{2}\rho\dot{\boldsymbol{u}}\cdot\dot{\boldsymbol{u}}) = (\dot{u}_i t_{ij})_{,j} \quad - t_{ij}\dot{u}_{i,j}. \tag{10.1.6}$$

But, using the symmetry of stress tensor t_{ij}, the constitutive equation (10.1.1), and (10.1.2), we have

$$t_{ij}\dot{u}_{i,j} = t_{ij}\dot{e}_{ij} = c_{ijkl}\dot{e}_{ij}e_{kl} = \tfrac{1}{2}\partial_t(t_{ij}e_{ij}). \tag{10.1.7}$$

Inserting this into (10.1.6), we obtain the energy conservation equation

$$\partial_t e + \boldsymbol{\nabla}\cdot\boldsymbol{r} = 0, \tag{8.1.2}$$

where the total energy $e = \kappa + \sigma$ is the sum of the kinetic energy density κ and the elastic stored energy density σ given by

$$\kappa = \tfrac{1}{2}\rho\dot{\boldsymbol{u}}\cdot\dot{\boldsymbol{u}}, \quad \sigma = \tfrac{1}{2}t_{ij}e_{ij}, \tag{10.1.8}$$

and where $\boldsymbol{r}$ is the energy flux whose components are given by

$$r_j = -\dot{u}_i t_{ij}. \tag{10.1.9}$$

10.2 The acoustical tensor for homogeneous waves

Now suppose that an infinite train of time-harmonic homogeneous plane waves propagates in the material. Thus, using the notation of sections 6.1 and 6.2, let the displacement $\boldsymbol{u}$ be given by

$$\boldsymbol{u} = \{\boldsymbol{A}\exp ik(\mathbf{n}\cdot\boldsymbol{x} - vt)\}^{+} = \{\boldsymbol{A}\exp i\omega(v^{-1}\mathbf{n}\cdot\boldsymbol{x} - t)\}^{+}. \quad (10.2.1)$$

Then, from the equations of motion (10.1.5), we obtain the propagation condition

$$Q_{ik}(\mathbf{n})A_k = \rho v^2 A_i, \quad (10.2.2)$$

where the tensor $\boldsymbol{Q}(\mathbf{n})$, depending on the propagation direction $\mathbf{n}$, has components

$$Q_{ik}(\mathbf{n}) = c_{ijkl}n_j n_l. \quad (10.2.3)$$

It is called the 'acoustical tensor' corresponding to the propagation direction $\mathbf{n}$. It is real and from (10.1.2) it follows that it is symmetric. The propagation condition (10.2.2) is the eigenvalue problem for this tensor. The equation

$$\det\{\boldsymbol{Q}(\mathbf{n}) - \rho v^2\mathbf{1}\} = 0, \quad (10.2.4)$$

for the determination of the eigenvalues ρv^2, and thus of the squared propagation speeds v^2, is called the 'secular equation'. It is a real cubic in v^2.

Because $\boldsymbol{Q}(\mathbf{n})$ is real and symmetric, it follows that all the roots of the cubic for v^2 are real for all directions $\mathbf{n}$. Let the roots be denoted by v_α^2, $(\alpha = 1, 2, 3)$. If there are no repeated roots, the corresponding eigenbivectors $\boldsymbol{A}^{(1)}, \boldsymbol{A}^{(2)}, \boldsymbol{A}^{(3)}$ (say) have real directions. Thus, $\boldsymbol{A}^{(1)} = \lambda^{(1)}\boldsymbol{a}^{(1)}$, $\boldsymbol{A}^{(2)} = \lambda^{(2)}\boldsymbol{a}^{(2)}$, $\boldsymbol{A}^{(3)} = \lambda^{(3)}\boldsymbol{a}^{(3)}$, where $\boldsymbol{a}^{(\alpha)}$, $(\alpha = 1, 2, 3)$ are real vectors and $\lambda^{(\alpha)}$ are scalar factors (possibly complex). Indeed, the components of $\boldsymbol{a}^{(\alpha)}$ may be taken to be proportional to the cofactors of the elements of any row of the real matrix $\{Q_{ik}(\mathbf{n}) - \rho v_\alpha^2\delta_{ik}\}$. Also, because $\boldsymbol{Q}(\mathbf{n})$ is symmetric, it follows that the eigenvectors $\boldsymbol{a}^{(1)}, \boldsymbol{a}^{(2)}, \boldsymbol{a}^{(3)}$ are orthogonal:

$$\boldsymbol{a}^{(\alpha)}\cdot\boldsymbol{a}^{(\beta)} = 0, \quad (\beta \neq \alpha). \quad (10.2.5)$$

Hence, corresponding to the direction of propagation $\mathbf{n}$, there are, in general, three linearly polarized homogeneous plane wave solutions, the amplitude vectors being mutually orthogonal. For a given wavelength $2\pi k^{-1}$, these are

$$\boldsymbol{u}^{(\alpha)} = \{\boldsymbol{a}^{(\alpha)} \exp ik(\mathbf{n}\cdot\boldsymbol{x} \pm v_\alpha t)\}^+, \quad (\alpha = 1, 2, 3;\ \text{no sum}). \qquad (10.2.6)$$

Alternatively, for a given angular frequency ω,

$$\boldsymbol{u}^{(\alpha)} = \{\boldsymbol{a}^{(\alpha)} \exp i\omega(\pm v_\alpha^{-1}\mathbf{n}\cdot\boldsymbol{x} - t)\}^+. \qquad (10.2.7)$$

10.2.1 Double roots of the secular equation – circularly polarized waves

A necessary and sufficient condition that a complex symmetric matrix have an isotropic eigenbivector is that the corresponding eigenvalue be at least double (section 3.2). Applying this to the special case of the real symmetric matrix $\boldsymbol{Q}(\mathbf{n})$, we immediately conclude that circularly polarized homogeneous waves may propagate only along those directions $\mathbf{n}$ such that the secular equation has a double or a triple root.

Because we deal here with the special case of a **real** symmetric matrix $\boldsymbol{q} = \boldsymbol{Q}(\mathbf{n})$, we provide a simpler, more transparent proof.

Suppose that for some $\mathbf{n}$ the secular equation has a double root, say $v_1^2 = v_2^2 \neq v_3^2$. The spectral decomposition of $\boldsymbol{q} = \boldsymbol{Q}(\mathbf{n})$ is

$$q_{ik} = \rho v_1^2 \delta_{ik} + \rho(v_3^2 - v_1^2) d_i d_k, \qquad (10.2.8)$$

where $\mathbf{d}$ is the unit eigenvector of $\boldsymbol{q}$ corresponding to the eigenvalue ρv_3^2. Thus if $\mathbf{a}$ and $\mathbf{b}$ are any two unit vectors forming an orthonormal triad with $\mathbf{d}$, we have

$$q_{ik} a_k = \rho v_1^2 a_i, \quad q_{ik} b_k = \rho v_1^2 b_i, \qquad (10.2.9)$$

and hence for arbitrary complex scalars α and β, we have

$$q_{ik}(\alpha a_k + \beta b_k) = \rho v_1^2(\alpha a_i + \beta b_i). \qquad (10.2.10)$$

Thus, the bivectors $\boldsymbol{A} = \alpha\mathbf{a} + \beta\mathbf{b}$ are eigenbivectors of the acoustical tensor $\boldsymbol{q} = \boldsymbol{Q}(\mathbf{n})$, and the corresponding waves are

$$\boldsymbol{u} = \{(\alpha\mathbf{a} + \beta\mathbf{b}) \exp ik(\mathbf{n}\cdot\boldsymbol{x} \pm v_1 t)\}^+. \qquad (10.2.11)$$

In general, because α and β are complex, such waves are elliptically polarized, with $(\alpha^+\mathbf{a} + \beta^+\mathbf{b})$ and $(\alpha^-\mathbf{a} + \beta^-\mathbf{b})$ as a pair of conjugate radii of the polarization ellipse. By taking $\beta\alpha^{-1}$ real, we obtain linearly polarized waves. By taking $\beta\alpha^{-1} = \pm i$, then the wave trains (10.2.11) are circularly polarized waves of opposite handedness.

Thus we conclude that, if for given **n**, the secular equation has a double root, then circularly polarized waves of opposite handedness may propagate along **n**. Moreover, if for given **n**, the secular equation has a triple root, (10.2.8) holds with $v_3^2 = v_1^2$, and hence any bivector is an eigenbivector of the acoustical tensor. In particular, circularly polarized waves may propagate along **n**.

Now, suppose that a circularly polarized homogeneous plane wave may propagate along a certain direction **n** with speed v. Then, if **a** and **b** are two unit orthogonal vectors in the plane of the polarization circle of the wave, it follows that

$$\boldsymbol{u} = \{(\mathbf{a} + i\mathbf{b}) \exp ik(\mathbf{n}\cdot\boldsymbol{x} - vt)\}^{+} \tag{10.2.12}$$

is a solution of the equations of motion. Hence, for the acoustical tensor $\boldsymbol{q} = \boldsymbol{Q}(\mathbf{n})$, we have

$$q_{ik}(a_k + ib_k) = \rho v^2(a_i + ib_i). \tag{10.2.13}$$

Because v^2 is a root of the secular equation (10.2.4), and therefore real, it follows from (10.2.13) that

$$q_{ik}a_k = \rho v^2 a_i, \quad q_{ik}b_k = \rho v^2 b_i. \tag{10.2.14}$$

Now **a** is orthogonal to **b**. Hence, the root v^2 is at least double, the eigenvalue ρv^2 corresponding to both **a** and **b**.

Thus, if a circularly polarized homogeneous plane wave propagates along **n**, then the corresponding root of the secular equation is at least double; if, for some **n**, the secular equation has a double or triple root, then a circularly polarized homogeneous plane wave may propagate along **n**.

Exercise 10.1

The linearly polarized wave $\boldsymbol{u} = \mathbf{a}\cos k(\mathbf{n}\cdot\boldsymbol{x} - vt)$, with **a** real, is said to be 'purely longitudinal' if $\mathbf{a} \times \mathbf{n} = \mathbf{0}$, and said to be 'purely transverse' if $\mathbf{a}\cdot\mathbf{n} = 0$. For purely longitudinal waves prove that the corresponding traction vector $\boldsymbol{t}_{(\mathbf{n})}$ satisfies $\boldsymbol{t}_{(\mathbf{n})} \times \mathbf{n} = \mathbf{0}$, whilst for purely transverse waves $\boldsymbol{t}_{(\mathbf{n})}\cdot\mathbf{n} = 0$.

10.3 Energy flux and energy density for homogeneous waves

Now, we consider the energy flux and energy density associated with the homogeneous wave (10.2.1).

Introducing (10.2.1) into the definitions (10.1.8) of the kinetic and stored energy densities, and taking means (denoted by a tilde) over a cycle $2\pi\omega^{-1}$, we obtain

$$\tilde{\kappa} = \tfrac{1}{4}\rho\omega^2 \boldsymbol{A}\cdot\bar{\boldsymbol{A}}, \quad \tilde{\sigma} = \tfrac{1}{4}k^2 c_{ijkl}A_i\bar{A}_k n_j n_l, \tag{10.3.1}$$

on using (10.1.2). Upon using the propagation condition (10.2.2), with $Q_{ik}(\mathbf{n})$ defined by (10.2.3), it follows that $\tilde{\kappa} = \tilde{\sigma}$, so that for the total mean energy density $\tilde{e}$, we have $\tilde{e} = 2\tilde{\kappa} = 2\tilde{\sigma}$.

Introducing (10.2.1) into the definition (10.1.9) of the energy flux, and taking the mean over a cycle, we obtain

$$\tilde{r}_i = \tfrac{1}{4}k^2 v c_{ijkl}(A_j\bar{A}_k + \bar{A}_j A_k)n_l. \tag{10.3.2}$$

Using (10.2.2) we note that

$$\tilde{\boldsymbol{r}}\cdot\mathbf{n} = v\tilde{e}, \tag{10.3.3}$$

in agreement with the general results obtained in section 8.3.

For the case of a simple root, v^2, of the secular equation, $\boldsymbol{A} = \lambda\boldsymbol{a}$, where $\boldsymbol{a}$ is a real vector and the wave is linearly polarized. Taking $\lambda = 1$ for simplicity, we note that $\tilde{e}$ and $\tilde{\boldsymbol{r}}$ reduce to

$$\tilde{e} = \tfrac{1}{2}\rho\omega^2 \boldsymbol{a}\cdot\boldsymbol{a} = \tfrac{1}{2}k^2 c_{ijkl}a_i a_k n_l n_j, \tag{10.3.4}$$

$$\tilde{r}_i = \tfrac{1}{2}k^2 v c_{ijkl}a_j a_k n_l. \tag{10.3.5}$$

For the case of a double root of the secular equation, $\boldsymbol{A} = \alpha\mathbf{a} + \beta\mathbf{b}$, where $\mathbf{a}$ and $\mathbf{b}$ are two orthogonal unit vectors, and α, β are arbitrary complex scalars. Denoting by e_1, $\boldsymbol{r}_1$ and by e_2, $\boldsymbol{r}_2$ the mean energy densities of the waves corresponding respectively to $\alpha = 1, \beta = 0$ and $\alpha = 0, \beta = 1$, we obtain from (10.3.1) and (10.3.2),

$$\tilde{e} = \alpha\bar{\alpha}e_1 + \beta\bar{\beta}e_2, \tag{10.3.6}$$

and

$$\tilde{\boldsymbol{r}} = \alpha\bar{\alpha}\boldsymbol{r}_1 + \beta\bar{\beta}\boldsymbol{r}_2 + (\bar{\alpha}\beta)^+\boldsymbol{r}_{12}, \tag{10.3.7}$$

where $\boldsymbol{r}_{12}$ is given by

$$(\boldsymbol{r}_{12})_i = \tfrac{1}{2}k^2 v c_{ijkl}(a_j b_k + a_k b_j)n_l. \tag{10.3.8}$$

From the propagation condition for the wave with amplitude $\mathbf{a}$, $c_{ijkl}a_k n_l n_j = \rho v^2 a_i$, and because $\mathbf{b}$ is orthogonal to $\mathbf{a}$, we have $c_{ijkl}b_i a_k n_l n_j = 0$, or equivalently, $c_{ijkl}a_i b_k n_l n_j = 0$. Hence $\boldsymbol{n}\cdot\boldsymbol{r}_{12} = 0$. Thus, the mean energy density $\tilde{e}$ for the elliptically polarized wave with amplitude $\boldsymbol{A} = \alpha\mathbf{a} + \beta\mathbf{b}$ is the sum of the mean energy densities for the two linearly polarized waves corresponding, respectively, to $\beta = 0$ and $\alpha = 0$. However, the mean energy flux $\tilde{\boldsymbol{r}}$ is **not**, in general, the sum of the mean energy fluxes for these two waves, because it includes an interaction term $(\bar{\alpha}\beta)^+ \boldsymbol{r}_{12}$, which is orthogonal to the propagation direction.

In order to analyse how the direction of the mean energy flux $\tilde{\boldsymbol{r}}$ varies when α and β are varied, we write $\alpha = a e^{i\lambda}$, $\beta = b e^{i\nu}$, with a, b, λ, ν, real. We consider the most general situation when $\boldsymbol{r}_1$, $\boldsymbol{r}_2$, $\boldsymbol{r}_{12}$ are linearly independent. Then, using oblique axes x, y, z along these vectors, (10.3.7) may be written

$$\tilde{r}_x = a^2 |\boldsymbol{r}_1|, \tag{10.3.9a}$$

$$\tilde{r}_y = b^2 |\boldsymbol{r}_2|, \tag{10.3.9b}$$

$$\tilde{r}_z = ab \cos(\nu - \lambda) |\boldsymbol{r}_{12}|. \tag{10.3.9c}$$

When $\nu - \lambda$ is varied, $\tilde{r}_z$ varies from $-ab|\boldsymbol{r}_{12}|$ to $+ab|\boldsymbol{r}_{12}|$, so that $\tilde{r}_z^2 \leqslant a^2 b^2 |\boldsymbol{r}_{12}|^2$, or, on eliminating a^2, b^2 from (10.3.9a) and (10.3.9b),

$$|\boldsymbol{r}_1|^2 |\boldsymbol{r}_2|^2 \tilde{r}_z^2 \leqslant |\boldsymbol{r}_{12}|^2 \tilde{r}_x \tilde{r}_y. \tag{10.3.10}$$

With the equality sign, this is the equation of an elliptical cone. Thus, when α and β are varied, that is, when a, b, λ, ν are varied, the mean energy flux $\tilde{\boldsymbol{r}}$ may take any direction on or inside this cone, or, more exactly, on or inside the semi-cone (because $\tilde{r}_x$ and $\tilde{r}_y$ do not change sign).

For circularly polarized waves, $\beta\alpha^{-1} = \pm i$, $(\bar{\alpha}\beta)^+ = 0$, and hence

$$\tilde{e} = \alpha\bar{\alpha}(e_1 + e_2), \quad \tilde{\boldsymbol{r}} = \alpha\bar{\alpha}(\boldsymbol{r}_1 + \boldsymbol{r}_2). \tag{10.3.11}$$

This mean energy flux is inside the semi-cone, because $\tilde{r}_z = 0$ and thus the inequality sign holds in (10.3.10). Thus circularly polarized waves are exceptional in that the mean energy flux is the vector sum of the mean energy flux vectors of the component linearly polarized waves.

Exercises 10.2

1. Determine the energy flux vector and energy density for purely longitudinal and for purely transverse waves, each of unit amplitude (see Exercise 10.1). For purely longitudinal waves, prove that $\boldsymbol{r} \times \boldsymbol{n} = \boldsymbol{0}$. In each case show that $\boldsymbol{r} \cdot \mathbf{n} = ve$.

2. For propagation along an arbitrary direction $\mathbf{n}$, let v_α be the phase velocities and $\boldsymbol{a}^{(\alpha)}$ the corresponding mutually orthogonal real eigenvectors of the acoustical tensor, $(\alpha = 1, 2, 3)$. Let $\boldsymbol{g}^{(\alpha)} = \tilde{\boldsymbol{r}}^{(\alpha)}(\tilde{e}^{(\alpha)})^{-1}$ be the energy flux velocities, or group velocities, corresponding to the three waves (section 8.2). Prove that

$$v_1 g_i^{(1)} + v_2 g_i^{(2)} + v_3 g_i^{(3)} = \frac{1}{\rho} c_{ijjl} n_l. \tag{10.3.12}$$

3. In the special case when, in (10.3.7), $\boldsymbol{r}_1$ and $\boldsymbol{r}_2$ are in the direction of $\mathbf{n}$, analyse how the direction of the mean energy flux $\tilde{\boldsymbol{r}}$ is varied when α and β are varied.

10.4 Incompressible and inextensible elastic materials

Some materials are subject to internal constraints. For example, rubber is generally taken to be incompressible, so that the volume of any piece of material may not change under deformation. Also, sometimes materials are reinforced with inextensible fibres, so that the lengths of material elements along the fibre direction may not change under deformation. Here we briefly consider these two types of internally constrained materials within the context of linearized elasticity theory.

10.4.1 Incompressible elastic materials

The constitutive equations for a homogeneous anisotropic incompressible elastic material are, in the context of linearized elasticity theory,

$$t_{ij} = -p\delta_{ij} + d_{ijkl} e_{kl}, \tag{10.4.1a}$$

$$e_{kk} = 0, \tag{10.4.1b}$$

where $p = p(\boldsymbol{x}, t)$ is a hydrostatic pressure to be determined from the equations of motion and boundary conditions, and where d_{ijkl} are elastic constants with the same symmetries as c_{ijkl} has in (10.1.2). The internal constraint is that the dilatation, e_{kk}, be zero because there are no changes in material volumes in any deformation. Inserting (10.4.1a) in (10.1.4), yields the equations of motion

$$\rho \partial_t^2 u_i = -p_{,i} + d_{ijkl} u_{k,lj}. \tag{10.4.2}$$

Time-harmonic homogeneous plane waves are solutions in the form

$$p = \{P \exp ik(\mathbf{n}\cdot\boldsymbol{x} - vt)\}^+, \quad \boldsymbol{u} = \{\boldsymbol{A} \exp ik(\mathbf{n}\cdot\boldsymbol{x} - vt)\}^+. \qquad (10.4.3)$$

From the equations of motion (10.4.2) and the incompressibility condition (10.4.1b), we obtain

$$ik^{-1}Pn_i + Q_{ik}(\mathbf{n})A_k = \rho v^2 A_i, \qquad (10.4.4a)$$

$$\mathbf{n}\cdot\boldsymbol{A} = 0, \qquad (10.4.4b)$$

where the symmetric tensor $\boldsymbol{Q}(\mathbf{n})$ has components

$$Q_{ik}(\mathbf{n}) = d_{ijkl}n_j n_l. \qquad (10.4.5)$$

Multiplying (10.4.4a) by n_i, and using (10.4.4b), it follows that

$$P = ikn_i Q_{ik} A_k. \qquad (10.4.6)$$

On eliminating the pressure amplitude P, (10.4.4a) gives the propagation conditions

$$(\delta_{ip} - n_i n_p)Q_{pk}A_k = \rho v^2 A_i, \quad \mathbf{n}\cdot\boldsymbol{A} = 0, \qquad (10.4.7)$$

or, on using (10.4.4b),

$$Q'_{ik}A_k = \rho v^2 A_i, \quad \mathbf{n}\cdot\boldsymbol{A} = 0, \qquad (10.4.8)$$

where

$$Q'_{ik} = (\delta_{ip} - n_i n_p)Q_{pr}(\delta_{rk} - n_r n_k). \qquad (10.4.9)$$

The reason for writing the propagation condition in terms of $\boldsymbol{Q}'$ is that it is symmetric, whereas in (10.4.7) the tensor acting upon $\boldsymbol{A}$ is not.

Clearly, $\mathbf{n}$ is an eigenvector of $\boldsymbol{Q}'$ corresponding to a zero eigenvalue. Because $\boldsymbol{Q}'$ is real and symmetric, the two other eigenvalues ρv_1^2, ρv_2^2 are both real and assuming $v_1^2 \neq v_2^2$, the corresponding amplitudes are $\boldsymbol{A}^{(1)} = \lambda^{(1)}\boldsymbol{a}^{(1)}$, $\boldsymbol{A}^{(2)} = \lambda^{(2)}\boldsymbol{a}^{(2)}$, where $\boldsymbol{a}^{(1)}$ and $\boldsymbol{a}^{(2)}$ are real vectors orthogonal to each other and to $\mathbf{n}$. Thus, in general, two purely transverse linearly polarized waves may propagate in every direction $\mathbf{n}$ in a homogeneous anisotropic incompressible elastic material.

10.4.2 Inextensible elastic materials

The constitutive equations for a homogeneous anisotropic elastic material, inextensible in the direction of the unit vector $\mathbf{l}$ are, in the context of linearized elasticity theory,

$$t_{ij} = -\tau l_i l_j + d_{ijkl} e_{kl}, \qquad (10.4.10a)$$

$$e_{ij} l_i l_j = 0, \qquad (10.4.10b)$$

where $\tau = \tau(\boldsymbol{x}, t)$ is a tension, to be determined from the equations of motion and the boundary conditions, and where the elastic constants d_{ijkl} have the same symmetries as c_{ijkl} has in (10.1.2). The internal constraint is that the extension, $e_{ij} l_i l_j$, in the direction $\mathbf{l}$ be zero. Inserting (10.4.10a) into (10.1.4), yields the equations of motion

$$\rho \partial_t^2 u_i = -\tau_{,j} l_i l_j + d_{ijkl} u_{k,lj}. \qquad (10.4.11)$$

Time-harmonic homogeneous planes waves are solutions in the form

$$\tau = \{T \exp ik(\mathbf{n}\cdot\boldsymbol{x} - vt)\}^+, \quad \boldsymbol{u} = \{\boldsymbol{A} \exp ik(\mathbf{n}\cdot\boldsymbol{x} - vt)\}^+. \qquad (10.4.12)$$

From the equations of motion (10.4.11) and the inextensibility condition (10.4.10b), we now obtain

$$ik^{-1} T(\mathbf{l}\cdot\boldsymbol{n}) l_i + Q_{ik}(\boldsymbol{n}) A_k = \rho v^2 A_i, \qquad (10.4.13a)$$

$$(\mathbf{l}\cdot\mathbf{n})(\mathbf{l}\cdot\boldsymbol{A}) = 0, \qquad (10.4.13b)$$

where the symmetric tensor $\boldsymbol{Q}(\mathbf{n})$ is defined by (10.4.5).

We note that if the propagation direction is orthogonal to the direction $\mathbf{l}$ of inextensibility, $\mathbf{l}\cdot\mathbf{n} = 0$, and hence (10.4.13b) is satisfied and (10.4.13) reduces to the propagation condition (10.2.2) of an unconstrained material. Thus, corresponding to a propagation direction $\mathbf{n}$, orthogonal to $\mathbf{l}$, there are, in general, three linearly polarized homogeneous plane waves, the amplitude vectors being mutually orthogonal.

However, for any propagation direction $\mathbf{n}$ which is not orthogonal to $\mathbf{l}$, (10.4.13b) reduces to $\mathbf{l}\cdot\boldsymbol{A} = 0$. Then, using the same procedure as for the incompressible case, we obtain the propagation conditions

$$Q''_{ik} A_k = \rho v^2 A_i, \quad \mathbf{l}\cdot\boldsymbol{A} = 0, \qquad (10.4.14)$$

where $\boldsymbol{Q}''$ is the symmetric tensor with components

$$Q''_{ik} = (\delta_{ip} - l_i l_p) Q_{pr} (\delta_{rk} - l_r l_k). \qquad (10.4.15)$$

From this, we conclude that, in general, in every directon $\mathbf{n}$ which is not orthogonal to $\mathbf{l}$, two linearly polarized waves may propagate, with amplitude vectors orthogonal to each other and to $\mathbf{l}$.

Exercise 10.3

1. Determine the mean energy flux $\tilde{\boldsymbol{r}}$ and the mean energy density $\tilde{e}$:
   (a) for the wave train (10.4.3) with $\boldsymbol{A} = \mathbf{a}$ real (incompressible materials);
   (b) for the wave train (10.4.12) with $\boldsymbol{A} = \mathbf{a}$ real, $\mathbf{n}\cdot\mathbf{l} \neq 0$ (inextensible materials). In both cases check that $\tilde{\boldsymbol{r}}\cdot\mathbf{n} = v\tilde{e}$.

10.5 The slowness surface and group velocity

Assume now that the elastic constants c_{ijkl} satisfy

$$c_{ijkl}b_ib_kd_jd_l > 0, \quad \forall \boldsymbol{b}, \boldsymbol{d} \neq \mathbf{0}. \tag{10.5.1}$$

Then the equilibrium equations corresponding to the equations of motion (10.1.5) are said to be 'strongly elliptic'. This is a weaker restriction on the elastic constants than the requirement that the stored energy density $\sigma = \frac{1}{2}c_{ijkl}e_{ij}e_{kl}$ be positive definite.

Now, for a linearly polarized wave with $\boldsymbol{A}$ along the real vector $\mathbf{a}$, we have, from (10.2.2) and (10.2.3),

$$\rho v^2 \mathbf{a}\cdot\mathbf{a} = Q_{ik}(\mathbf{n})a_ia_k = c_{ijkl}a_ia_kn_jn_l, \tag{10.5.2}$$

and it follows from (10.5.1) that $\rho v^2 > 0$. Thus the phase speed, v, of the wave is real.

On a line along $\mathbf{n}$, through a given origin 0, the values v_1^{-1}, v_2^{-1}, v_3^{-1} of the three slownesses corresponding to the direction $\mathbf{n}$ may be marked off. Then, as $\mathbf{n}$ varies, the corresponding points fill out three closed surfaces, forming a surface of three sheets called the 'slowness surface'. Every vector $\boldsymbol{s} = v^{-1}\mathbf{n}$ having 0 as origin and a point of this surface as extremity is a possible slowness vector. Because the wave speeds are obtained from the secular equation (10.2.4) and because $\boldsymbol{Q}(\mathbf{n})$ is quadratic in $\mathbf{n}$, the slowness surface has for equation

$$\det\{\boldsymbol{Q}(\boldsymbol{s}) - \rho\mathbf{1}\} = 0, \tag{10.5.3}$$

where $\boldsymbol{s} = v^{-1}\mathbf{n}$ denotes the slowness vector.

In order to write separate equations for the three sheets of the slowness surface (10.5.3), the secular equation (10.2.4) has to be solved for v^2, yielding the three roots $v = \mathscr{V}_1(\mathbf{n})$, $v = \mathscr{V}_2(\mathbf{n})$, $v = \mathscr{V}_3(\mathbf{n})$, (say), where v denotes the positive square root of v^2. Because $\boldsymbol{Q}(\mathbf{n})$ is quadratic in $\mathbf{n}$, it follows that $\mathscr{V}_1, \mathscr{V}_2, \mathscr{V}_3$ are homogeneous functions of degree one in **n**. (To give sense to this statement, we allow **n**, in (10.2.4), to be any vector, not necessarily of unit length.) Thus, because $\boldsymbol{s} = v^{-1}\mathbf{n}$, the three sheets of the slowness have for equations

$$\mathscr{V}_\alpha(\boldsymbol{s}) = 1, \quad (\alpha = 1, 2, 3). \tag{10.5.4}$$

Also, $v = \omega k^{-1}$, and thus $\boldsymbol{s} = \omega^{-1}\boldsymbol{k}$, and the dispersion relations corresponding to the three waves are

$$\omega = \mathscr{V}_\alpha(\boldsymbol{k}), \quad (\alpha = 1, 2, 3). \tag{10.5.5}$$

The corresponding energy flux velocities denoted by $\boldsymbol{g}^{(\alpha)}$, which are also the group velocities (section 8.2), are given by

$$g_i^{(\alpha)} = \frac{\partial \mathscr{V}_\alpha(\boldsymbol{k})}{\partial k_i} = \frac{\partial \mathscr{V}_\alpha(\boldsymbol{s})}{\partial s_i}, \quad (\alpha = 1, 2\ 3), \tag{10.5.6}$$

on using the fact that $\mathscr{V}_\alpha$ is homogeneous of degree one and that its derivatives are thus homogeneous of degree zero. Now $\partial \mathscr{V}_\alpha/\partial \boldsymbol{s}$ is normal to the slowness surface $\mathscr{V}_\alpha(\boldsymbol{s}) = 1$, because $\partial \mathscr{V}_\alpha/\partial s_i$ are the components of the gradient of (10.5.4) in a space of coordinates s_i. Thus, it follows from (10.5.6) that the group velocity of a wave train propagating along **n** with phase speed $v = \mathscr{V}_\alpha(\mathbf{n})$ is along the normal to the slowness surface $\mathscr{V}_\alpha(\boldsymbol{s}) = 1$ at the 'point' $\boldsymbol{s} = v^{-1}\mathbf{n}$.

For some directions **n** the sheets of the slowness surface may touch or intersect. Circularly polarized waves of either handedness may propagate along such directions because touching or intersecting sheets correspond to double roots of the secular equation. Figures with slowness surfaces for various anisotropic elastic media may be found for instance in Musgrave (1970).

Exercises 10.4

1. Consider a linearly polarized wave with amplitude A along the real unit vector $\hat{\mathbf{a}}$. Using the propagation condition (10.2.2) with $A = \hat{\mathbf{a}}$, prove that $g_p = \partial\omega/\partial k_p = (\rho v)^{-1} c_{ijkp}\hat{a}_i\hat{a}_k n_j$. Using the results of section 10.3, check that $\boldsymbol{g} = \tilde{\boldsymbol{r}}\tilde{e}^{-1}$.
2. Using the propagation condition (10.4.8), with $A = \hat{\mathbf{a}}$, find $g_p = \partial\omega/\partial k_p$ in the case of incompressible materials.
3. For fixed **n**, construct the ellipsoid $\mathscr{E}(\mathbf{n})$: $q_{ik}x_i x_k = 1$, where $q_{ik} = Q_{ik}(\mathbf{n})$. In terms of this ellipsoid, give a geometrical interpretation to the possible directions of polarization of the homogeneous waves in the cases of (a) unconstrained elastic materials; (b)

incompressible materials; (c) inextensible materials (assuming $\mathbf{n}\cdot\mathbf{l} \neq 0$).

4. For the three waves propagating with speeds $v = \mathscr{V}_\alpha(\mathbf{n})$, $(\alpha = 1, 2, 3)$, in the direction $\mathbf{n}$ in an anisotropic elastic medium, prove that

$$\sum_{\alpha=1}^{3} \mathscr{V}_\alpha^2(\mathbf{n}) = \frac{1}{\rho} c_{ijil} n_j n_l.$$

Using this, show that the sum

$$\sum_{\alpha=1}^{3} \{\mathscr{V}_\alpha^2(\mathbf{n}) + \mathscr{V}_\alpha^2(\mathbf{m}) + \mathscr{V}_\alpha^2(\mathbf{p})\}$$

has the same value for every triad of mutually orthogonal unit vectors $\mathbf{n}, \mathbf{m}, \mathbf{p}$.

5. Let $V_{jl} = c_{ijil}$. With the strong ellipticity condition, $\boldsymbol{V}$ is positive definite. Assume that it has three distinct eigenvalues $\lambda > \mu > \nu$. Using the result of Exercise 10.4.4, find all the directions $\mathbf{n}$ such that

$$\sum_{\alpha=1}^{3} \mathscr{V}_\alpha^2(\mathbf{n}) = \frac{\gamma}{\rho},$$

where γ is a prescribed value, with $\nu \leqslant \gamma \leqslant \lambda$.

10.6 Isotropic elastic bodies. Homogeneous waves

For a homogeneous isotropic elastic material,

$$c_{ijkl} = \lambda\delta_{ij}\delta_{kl} + \mu(\delta_{ik}\delta_{jl} + \delta_{il}\delta_{jk}), \tag{10.6.1}$$

so that the constitutive equation (10.1.1) reads

$$t_{ij} = \lambda e_{kk}\delta_{ij} + 2\mu e_{ij}. \tag{10.6.2}$$

Here μ is called the 'shear modulus' and λ the 'Lamé constant'. Thus, the equations of motion (10.1.5) now read

$$\rho\partial_t^2 u_i = (\lambda + \mu)u_{k,ik} + \mu u_{i,jj}. \tag{10.6.3}$$

The acoustical tensor $\boldsymbol{Q}(\mathbf{n})$ defined by (10.2.3) is now given by

$$\begin{aligned} Q_{ik}(\mathbf{n}) &= (\lambda + \mu)n_i n_k + \mu n_j n_j \delta_{ik} \\ &= (\lambda + 2\mu)n_i n_k + \mu(m_i m_k + p_i p_k), \end{aligned} \tag{10.6.4}$$

where $\mathbf{m}, \mathbf{p}$ are unit vectors forming an orthonormal triad with $\mathbf{n}$: $\mathbf{m}\cdot\mathbf{n} = \mathbf{n}\cdot\mathbf{p} = \mathbf{p}\cdot\mathbf{m} = 0$. Thus,

$$Q_{ik}(\mathbf{n}) - \rho v^2\delta_{ik} = \{(\lambda + 2\mu) - \rho v^2\}n_i n_k + (\mu - \rho v^2)(m_i m_k + p_i p_k), \tag{10.6.5}$$

and hence, the secular equation (10.2.4) is

$$\det\{\boldsymbol{Q}(\mathbf{n}) - \rho v^2\mathbf{1}\} = (\lambda + 2\mu - \rho v^2)(\mu - \rho v^2)^2 = 0. \tag{10.6.6}$$

It has the simple root

$$\rho v^2 = \lambda + 2\mu, \tag{10.6.7}$$

and the double root

$$\rho v^2 = \mu. \tag{10.6.8}$$

The slowness surface consists of the two concentric spheres $(\lambda + 2\mu)\boldsymbol{s}\cdot\boldsymbol{s} = \rho$, and $\mu\boldsymbol{s}\cdot\boldsymbol{s} = \rho$.

Corresponding to the simple root (10.6.7), $\boldsymbol{A} = \alpha\mathbf{n}$, so that the wave is

$$\boldsymbol{u} = \mathbf{n}\left\{\alpha \exp ik\left(\mathbf{n}\cdot\boldsymbol{x} - \left(\frac{\lambda + 2\mu}{\rho}\right)^{1/2} t\right)\right\}^+, \tag{10.6.9}$$

where α is an arbitrary (complex) constant. The wave is longitudinal – sometimes called a 'P-wave' where 'P' refers to 'Push–Pull' because it is longitudinal, or to 'Primary' because in the case of seismic disturbances it is the first to arrive from a distant earthquake source.

Corresponding to the double root (10.6.8), $\boldsymbol{A}$ is an arbitrary bivector orthogonal to $\mathbf{n}$. Thus $\boldsymbol{A} = \beta\mathbf{m} + \gamma\mathbf{p}$, so that the wave is

$$\boldsymbol{u} = \left\{(\beta\mathbf{m} + \gamma\mathbf{p}) \exp ik\left(\mathbf{n}\cdot\boldsymbol{x} - \left(\frac{\mu}{\rho}\right)^{1/2} t\right)\right\}^+, \tag{10.6.10}$$

where β and γ are arbitrary (complex) constants. The wave is transverse – sometimes called an 'S-wave' where 'S' refers to 'Shear–Shake' because it is transverse, or to 'Secondary' because it is the second to arrive from a distant earthquake source. In general ($\beta\gamma^{-1}$ complex), the wave is elliptically polarized. When $\beta\gamma^{-1}$ is real it is linearly polarized, and, when $\beta\gamma^{-1} = \pm i$, it is circularly polarized. Because the slowness surface consists of a pair of concentric spheres, it follows that the mean energy flux vectors for these waves is along $\mathbf{n}$, their direction of propagation.

10.7 Transversely isotropic materials. Homogeneous waves

Here we consider materials possessing an axis of rotational symmetry. These are called 'transversely isotropic'. For wave propagation, this is the next simplest case after isotropic media, because the secular equation may be factored. Taking the z-axis of a rectangular cartesian coordinate system $0xyz$ along the symmetry axis, the constitutive equations may be written (Love, 1927)

$$\begin{aligned} t_{11} &= d_{11}u_{1,1} + d_{12}u_{2,2} + d_{13}u_{3,3}, & t_{12} &= \tfrac{1}{2}(d_{11} - d_{12})(u_{1,2} + u_{2,1}), \\ t_{22} &= d_{12}u_{1,1} + d_{11}u_{2,2} + d_{13}u_{3,3}, & t_{13} &= d_{44}(u_{1,3} + u_{3,1}), \\ t_{33} &= d_{13}(u_{1,1} + u_{2,2}) + d_{33}u_{3,3}, & t_{23} &= d_{44}(u_{2,3} + u_{3,2}), \end{aligned} \quad (10.7.1)$$

where d_{11}, d_{12}, d_{13}, d_{33}, d_{44} are five material constants. These equations also describe linear elastic crystals with hexagonal symmetry (Fedorov, 1968). The equations of motion (10.1.5) now read

$$\begin{aligned} &d_{11}u_{1,11} + \tfrac{1}{2}(d_{11} - d_{12})u_{1,22} + d_{44}u_{1,33} + \tfrac{1}{2}(d_{11} + d_{12})u_{2,12} \\ &\quad + (d_{13} + d_{44})u_{3,13} = \rho\partial_t^2 u_1, \\ &\tfrac{1}{2}(d_{11} + d_{12})u_{1,12} + \tfrac{1}{2}(d_{11} - d_{12})u_{2,11} + d_{11}u_{2,22} + d_{44}u_{2,33} \\ &\quad + (d_{13} + d_{44})u_{3,23} = \rho\partial_t^2 u_2, \\ &(d_{13} + d_{44})(u_{1,13} + u_{2,23}) + d_{44}(u_{3,11} + u_{3,22}) + d_{33}u_{3,33} = \rho\partial_t^2 u_3. \end{aligned} \quad (10.7.2)$$

Insertion of (10.2.1) into (10.7.2) leads to the propagation condition (10.2.2) and the secular equation (10.2.4), where now the acoustical tensor $\boldsymbol{Q}(\mathbf{n})$ is given by

$$\begin{aligned} Q_{11}(\mathbf{n}) &= d_{11}n_1^2 + \tfrac{1}{2}(d_{11} - d_{12})n_2^2 + d_{44}n_3^2, \\ Q_{22}(\mathbf{n}) &= \tfrac{1}{2}(d_{11} - d_{12})n_1^2 + d_{11}n_2^2 + d_{44}n_3^2, \\ 2Q_{12}(\mathbf{n}) &= (d_{11} + d_{12})n_1n_2, \\ Q_{13}(\mathbf{n}) &= (d_{13} + d_{44})n_1n_3, \quad Q_{23}(\mathbf{n}) = (d_{13} + d_{44})n_2n_3, \\ Q_{33}(\mathbf{n}) &= d_{44}(n_1^2 + n_2^2) + d_{33}n_3^2. \end{aligned} \quad (10.7.3)$$

One root of the secular equation is

$$\rho v^2 = \tfrac{1}{2}(d_{11} - d_{12})(n_1^2 + n_2^2) + d_{44}n_3^2, \quad (10.7.4)$$

and the corresponding eigenvector of $Q_{ik}(\mathbf{n})$ is, up to a scalar factor,

$$\mathbf{a} = \mathbf{n} \times \mathbf{k} = (n_2, -n_1, 0), \quad (10.7.5)$$

where $\mathbf{k}$ denotes the unit vector along the symmery axis $0z$.

The two other roots ρv_2^2, ρv_3^2, of the secular equation are the solutions of the biquadratic equation

$$(\rho v^2)^2 - \{(d_{11} + d_{44})(n_1^2 + n_2^2) + (d_{33} + d_{44})n_3^2\}\rho v^2 \\ + d_{11}d_{44}(n_1^2 + n_2^2)^2 + \{d_{11}d_{33} + d_{44}^2 - (d_{13} + d_{44})^2\}(n_1^2 + n_2^2)n_3^2 \\ + d_{33}d_{44}n_3^4 = 0. \qquad (10.7.6)$$

Thus, the slowness surface consists of an ellipsoid (corresponding to (10.7.4)), and of a double surface symmetrical about the axis $0z$.

In order to study the propagation in a direction orthogonal to the symmetry axis, we take $\mathbf{n}$ along $0x$ without loss in generality. Then, the solutions are the longitudinal wave

$$\boldsymbol{u} = \mathbf{i}\left\{\alpha \exp ik\left(x - \left(\frac{d_{11}}{\rho}\right)^{1/2} t\right)\right\}^+, \qquad (10.7.7)$$

and the two transverse waves

$$\boldsymbol{u} = \mathbf{j}\left\{\beta \exp ik\left(x - \left(\frac{(d_{11} - d_{12})}{2\rho}\right)^{1/2} t\right)\right\}^+,$$

$$\boldsymbol{u} = \mathbf{k}\left\{\gamma \exp ik\left(x - \left(\frac{d_{44}}{\rho}\right)^{1/2} t\right)\right\}^+, \qquad (10.7.8)$$

where $\mathbf{i}, \mathbf{j}, \mathbf{k}$ are the unit vectors along the axes x, y, z, respectively, and α, β, γ are arbitrary constants.

Similarly, for propagation along the symmetry axis $0z$, the solutions are the longitudinal wave

$$\boldsymbol{u} = \mathbf{k}\left\{\gamma \exp ik\left(z - \left(\frac{d_{33}}{\rho}\right)^{1/2} t\right)\right\}^+, \qquad (10.7.9)$$

and the transverse wave

$$\boldsymbol{u} = \left\{(\alpha\mathbf{i} + \beta\mathbf{j}) \exp ik\left(z - \left(\frac{d_{44}}{\rho}\right)^{1/2} t\right)\right\}^+, \qquad (10.7.10)$$

corresponding to the double root, $\rho v^2 = d_{44}$, of the secular equation. Here also α, β, γ are arbitrary constants. In general the wave (10.7.10) is elliptically polarized. It is circularly polarized for $\beta\alpha^{-1} = \pm\mathrm{i}$.

For propagation in a general direction $\mathbf{n}$, the waves are neither purely longitudinal, nor purely transverse (Fedorov, 1968; Musgrave, 1970; Chadwick, 1989).

10.7.1 Circularly polarized homogeneous waves

In order that circularly polarized waves may propagate in a direction $\mathbf{n}$, the corresponding secular equation must possess a double root. In the present case, this may occur if either the biquadratic (10.7.6) possesses a real double root, or if the root (10.7.4) is also a root of the biquadratic.

Now, the discriminant, Δ, of the biquadratic (10.7.6) is

$$\Delta = \{(d_{11} - d_{44})(n_1^2 + n_2^2) - (d_{33} - d_{44})n_3^2\}^2 + 4(d_{13} + d_{44})^2(n_1^2 + n_2^2)n_3^2, \quad (10.7.11)$$

and is thus always strictly positive, and so (10.7.6) does not possess a real double root. The only possibility for circularly polarized waves is that the root (10.7.4) is also a root of the biquadratic (10.7.6). On inserting (10.7.4) into (10.7.6), we find that either

$$n_1^2 + n_2^2 = 0, \quad (10.7.12a)$$

or

$$n_1^2 + n_2^2 = \frac{R}{(1 + R)} = n_3^2 R, \quad (10.7.12b)$$

where R is given by

$$(d_{11} + d_{12})(d_{11} - d_{12} - 2d_{44})R = 2\{(d_{11} + d_{12})(d_{33} - d_{44}) - 2(d_{13} + d_{44})^2\}. \quad (10.7.13)$$

In case (10.7.12a), $\mathbf{n} = \pm \mathbf{k}$, so that we retrieve the fact that circularly polarized waves may propagate along the symmetry axis ((10.7.10) with $\beta\alpha^{-1} = \pm \mathrm{i}$).

Because $0 \leqslant n_3^2 \leqslant 1$, case (10.7.12b) is feasible only if $R > 0$. Thus, if $R < 0$, the symmetry axis is the only direction along which circularly polarized waves may propagate. Whilst if $R > 0$, there is then also a circular cone of directions, with axis $0z$ and semi-angle ψ, given by $\tan \psi = R^{1/2}$, such that circularly polarized waves may propagate along these directions. Their speed of propagation is given by

$$\rho v^2 = \{\tfrac{1}{2}R(d_{11} - d_{12}) + d_{44}\}n_3^2. \quad (10.7.14)$$

The possibilities $R < 0$ and $R > 0$ both occur in practice. For instance, for zinc $R < 0$, whilst for beryl $R > 0$.

When $R > 0$, corresponding to the double root (10.7.14) of the secular equation, we obtain the wave solution

$$\boldsymbol{u} = \{(\alpha \boldsymbol{a} + \beta \boldsymbol{b}) \exp \mathrm{i}k(\boldsymbol{n}\cdot\boldsymbol{x} - vt)\}^{+}, \tag{10.7.15}$$

where α, β are arbitrary (complex) numbers, and $\boldsymbol{a}, \boldsymbol{b}$ are given by

$$\boldsymbol{a} = (n_2, -n_1, 0), \quad \mathbf{b} = \frac{1}{(S^2 + R)^{1/2}}(Sn_1, Sn_2, -Rn_3), \tag{10.7.16}$$

with

$$S = \frac{2(d_{13} + d_{44})}{(d_{11} + d_{12})}. \tag{10.7.17}$$

We note that $\boldsymbol{a}\cdot\boldsymbol{b} = 0$ and $\boldsymbol{a}\cdot\boldsymbol{a} = \boldsymbol{b}\cdot\boldsymbol{b} = R(1 + R)^{-1}$. In general the wave (10.7.15) is elliptically polarized. For $\beta\alpha^{-1} = \pm \mathrm{i}$, it is circularly polarized. The unit normal $\mathbf{s}$ (say) to the plane of the amplitude bivector is along $\boldsymbol{a} \times \boldsymbol{b}$. We find

$$\mathbf{s} = \frac{1 + R}{R}\boldsymbol{a} \times \boldsymbol{b} = \frac{1}{n_3(S^2 + R)^{1/2}}(n_1, n_2, Sn_3). \tag{10.7.18}$$

The angle η (say) between the direction of propagation $\mathbf{n}$ and $\mathbf{s}$ is given by

$$\cos \eta = (S + R)\{(S^2 + R)(1 + R)\}^{-1/2} \tag{10.7.19}$$

Exercise 10.5

Particularize the results obtained here to the case of isotropic materials, and retrieve the conclusion that in an isotropic elastic material, circularly polarized waves may propagate in any direction.

10.8 The acoustical tensor for inhomogeneous waves

Now, with the notation of section 6.3, we assume that the displacement is given by

$$\boldsymbol{u} = \{\boldsymbol{A} \exp \mathrm{i}\omega(\boldsymbol{S}\cdot\boldsymbol{x} - t)\}^{+}, \tag{10.8.1}$$

where we recall that $\boldsymbol{S}$ denotes the slowness bivector.

Inserting $\boldsymbol{A} \exp \mathrm{i}\omega(\boldsymbol{S}\cdot\boldsymbol{x} - t)$ into the equations of motion (10.1.5) leads to the propagation condition

$$Q_{ik}(\boldsymbol{S})A_k = \rho A_i, \tag{10.8.2}$$

where, corresponding to (10.2.13), we now have

$$Q_{ik}(\boldsymbol{S}) = c_{ijkl}S_jS_l. \tag{10.8.3}$$

Applying the DE method (section 6.3), we write $\boldsymbol{S} = N\boldsymbol{C}$, $\boldsymbol{C}$ being of the form (6.3.10), and the propagation condition (10.8.2) becomes

$$Q_{ik}(\boldsymbol{C})A_k = \rho N^{-2}A_i. \tag{10.8.4}$$

Hence, for each prescribed bivector $\boldsymbol{C} = m\hat{\mathbf{m}} + \mathrm{i}\hat{\mathbf{n}}$, the corresponding values of ρN^{-2} and $\boldsymbol{A}$ are the eigenvalues and the eigenbivectors of the complex matrix $\boldsymbol{Q}(\boldsymbol{C})$ which is called the 'acoustical tensor' corresponding to the prescribed bivector $\boldsymbol{C}$. From (10.1.2), it follows that it is symmetric. The equation

$$\det\{\boldsymbol{Q}(\boldsymbol{C}) - \rho N^{-2}\mathbf{1}\} = 0, \tag{10.8.5}$$

for the determination of the eigenvalues ρN^{-2} is again called the 'secular equation', and is a cubic in N^{-2}.

Let the roots of this equation be denoted by N_α^{-2}, $(\alpha = 1, 2, 3)$. Because $\boldsymbol{Q}(\boldsymbol{C})$ is complex and symmetric, it follows (sections 3.1–3.2) that, if there are no repeated roots, the corresponding eigenbivectors $\boldsymbol{A}^{(1)}$, $\boldsymbol{A}^{(2)}$, $\boldsymbol{A}^{(3)}$ (say) are orthogonal and not isotropic:

$$\boldsymbol{A}^{(\alpha)}\cdot\boldsymbol{A}^{(\beta)} = 0, \quad (\beta \neq \alpha); \quad \boldsymbol{A}^{(\alpha)}\cdot\boldsymbol{A}^{(\alpha)} \neq 0, \quad (\text{no sum}). \tag{10.8.6}$$

Thus (see section 2.4), in this case, the ellipse of $\boldsymbol{A}^{(\alpha)}$ (polarization ellipse of the wave corresponding to the root N_α^{-2}), when projected upon the plane of the ellipse of $\boldsymbol{A}^{(\beta)}$ (polarization ellipse of the wave corresponding to the root N_β^{-2}, $\beta \neq \alpha$), is an ellipse similar and similarly situated to the ellipse of $\boldsymbol{A}^{(\beta)}$ when rotated through a quadrant. It follows that the ellipses of $\boldsymbol{A}^{(2)}$ and $\boldsymbol{A}^{(3)}$, when projected upon the plane of the ellipse of $\boldsymbol{A}^{(1)}$, are similar and similarly situated ellipses, both similar and similarly situated to the ellipse of $\boldsymbol{A}^{(1)}$ rotated through a quadrant.

10.8.1 Double roots of the secular equation. Circularly polarized waves

If, for given $\boldsymbol{C}$, there are no repeated roots of the secular equation, none of the eigenbivectors $\boldsymbol{A}^{(\alpha)}$ may be isotropic, so that circularly polarized inhomogeneous plane waves are not possible for the given $\boldsymbol{C}$.

On the other hand, if, for the given $\boldsymbol{C}$, at least two of the N_α^{-2} are equal, then the complex symmetric matrix $\boldsymbol{Q}(\boldsymbol{C})$ has an isotropic eigenbivector (section 3.2). Then, circularly polarized inhomogeneous plane waves are possible for this given $\boldsymbol{C}$. In sections 10.9 and 10.10 we will present examples of such waves in isotropic and transversely isotropic elastic materials.

Exercise 10.6

In dealing with inhomogeneous plane waves, we have seen that the mean energy density $\tilde{e}$ and the mean energy flux $\tilde{\boldsymbol{r}}$ have the forms $\tilde{e} = \hat{e}\exp(-2\omega\boldsymbol{S}^-\cdot\boldsymbol{x})$ and $\tilde{\boldsymbol{r}} = \hat{\boldsymbol{r}}\exp(-2\omega\boldsymbol{S}^-\cdot\boldsymbol{x})$, where $\hat{e}$ and $\hat{\boldsymbol{r}}$ are called the weighted mean energy flux and weighted mean energy density (section 8.3). Find $\hat{\boldsymbol{r}}$ and $\hat{e}$ for an inhomogeneous wave of slowness $\boldsymbol{S}$ and amplitude $\boldsymbol{A}$ propagating in an elastic material. Check that $\hat{\boldsymbol{r}}\cdot\boldsymbol{S}^+ = \hat{e}$ and $\hat{\boldsymbol{r}}\cdot\boldsymbol{S}^- = 0$ (8.3.19).

10.9 Isotropic elastic bodies. Inhomogeneous waves

Inserting (10.6.1) into (10.8.3) gives

$$Q_{ik}(\boldsymbol{S}) = (\lambda + \mu)S_iS_k + \mu(\boldsymbol{S}\cdot\boldsymbol{S})\delta_{ik}, \tag{10.9.1}$$

and the propagation condition is

$$(\lambda + \mu)(\boldsymbol{C}\cdot\boldsymbol{A})\boldsymbol{C} + \mu(\boldsymbol{C}\cdot\boldsymbol{C})\boldsymbol{A} = \rho N^{-2}\boldsymbol{A}. \tag{10.9.2}$$

The secular equation

$$\det[(\lambda + \mu)C_iC_k + \{\mu(\boldsymbol{C}\cdot\boldsymbol{C}) - \rho N^{-2}\}\delta_{ik}] = 0, \tag{10.9.3}$$

has the simple root

$$\rho N^{-2} = (\lambda + 2\mu)\boldsymbol{C}\cdot\boldsymbol{C} = (\lambda + 2\mu)(m^2 - 1), \tag{10.9.4}$$

and the double root

$$\rho N^{-2} = \mu\boldsymbol{C}\cdot\boldsymbol{C} = \mu(m^2 - 1). \tag{10.9.5}$$

Thus, for propagating waves, the prescribed bivector $\boldsymbol{C}$ may not be isotropic. Because N^{-2} given by (10.9.4) and (10.9.5) are real, it follows that in both cases, the planes of constant phase are orthogonal to the planes of constant amplitude.

Corresponding to the simple root (10.9.4), then, $\boldsymbol{A} = \alpha\boldsymbol{C}$, where α is an arbitrary scalar. Corresponding to the double root (10.9.5), $\boldsymbol{A}$

may be any bivector orthogonal to $\boldsymbol{C}$: $\boldsymbol{A} = \beta \boldsymbol{C}_{\perp} + \gamma \hat{\mathbf{m}} \times \hat{\mathbf{n}}$, where $\boldsymbol{C}_{\perp} = m^{-1}\hat{\mathbf{m}} + i\hat{\mathbf{n}}$ is the reciprocal of the bivector $\boldsymbol{C}$ (see section 2.5).

Thus, for each prescribed $\boldsymbol{C} = m\hat{\mathbf{m}} + i\hat{\mathbf{n}}$, we have an inhomogeneous wave, corresponding to the longitudinal or P wave, with slowness $\boldsymbol{S} = N\boldsymbol{C}$ and amplitude $\boldsymbol{A}$ given by

$$\boldsymbol{S} = \pm \frac{\rho^{1/2}(m\hat{\mathbf{m}} + i\hat{\mathbf{n}})}{\{(\lambda + 2\mu)(m^2 - 1)\}^{1/2}}, \quad \boldsymbol{A} = \alpha(m\hat{\mathbf{m}} + i\hat{\mathbf{n}}), \qquad (10.9.6)$$

and an inhomogeneous wave, corresponding to the transverse or S wave, with slowness $\boldsymbol{S} = N\boldsymbol{C}$ and amplitude $\boldsymbol{A}$ given by

$$\boldsymbol{S} = \pm \frac{\rho^{1/2}(m\hat{\mathbf{m}} + i\hat{\mathbf{n}})}{\{\mu(m^2 - 1)\}^{1/2}}, \quad \boldsymbol{A} = \beta(m^{-1}\hat{\mathbf{m}} + i\hat{\mathbf{n}}) + \gamma\hat{\mathbf{m}} \times \hat{\mathbf{n}}. \quad (10.9.7)$$

10.9.1 *Circularly polarized waves*

Because the root (10.9.5) is double, it follows that circularly polarized waves are possible for all choices of $\boldsymbol{C}$ for which $\boldsymbol{C}\cdot\boldsymbol{C} \neq 0$. Indeed, the wave (10.9.7) is circularly polarized when the arbitrary scalars γ and β are such that $\gamma\beta^{-1} = \pm(m^2 - 1)^{1/2}m^{-1}$. Then $\boldsymbol{A}$ is given by

$$\boldsymbol{A} = \beta\{m^{-1}\hat{\mathbf{m}} + i\hat{\mathbf{n}} \pm ((m^2 - 1)^{1/2}m^{-1})\hat{\mathbf{m}} \times \hat{\mathbf{n}}\}. \qquad (10.9.8)$$

The plane of the polarization circle passes through $\boldsymbol{S}^-$ (which is along $\hat{\mathbf{n}}$) and makes an angle η, given by $\tan\eta = \pm(m^2 - 1)^{1/2}$, with the plane of the bivector $\boldsymbol{S}$ (plane of $\hat{\mathbf{m}}$ and $\hat{\mathbf{n}}$). As m increases from 1 to ∞, this plane swings about $\boldsymbol{S}^-$ from the plane of $\boldsymbol{S}$ to a plane othogonal to the plane of $\boldsymbol{S}$.

10.9.2 *Rayleigh waves*

Linear combinations of the two solutions (10.9.6) and (10.9.7) are also solutions of the equations of motion. Now, we consider just such a combination which gives the solution of Rayleigh's problem of the propagation of waves over the free surface of a semi-infinite isotropic elastic material.

Let the material occupy the half-space $z \geqslant 0$, with free surface $z = 0$, and, without loss in generality, assume that the waves propagate along the x-axis. Also, we require that the waves be attenuated in the direction z, so that they are decaying with distance from the surface. Thus, we seek solutions of the form

$$\boldsymbol{u} = \{\boldsymbol{A}\exp(-\alpha z)\exp \mathrm{i}\kappa(x - ct)\}^+, \quad (\kappa > 0, \alpha > 0). \tag{10.9.9}$$

For these, $\omega = \kappa c$, and the slowness bivector is given by

$$\omega \boldsymbol{S} = \kappa \mathbf{i} + \mathrm{i}\alpha \mathbf{k}. \tag{10.9.10}$$

Then, from (10.9.5), $\mu \boldsymbol{S}\cdot\boldsymbol{S} = \rho$, $\boldsymbol{A}$ is orthogonal to $\boldsymbol{S}$, and, assuming that κ is given, we obtain the value of α, α_S (say) corresponding to the S-wave solution:

$$\mu(\kappa^2 - \alpha_S^2)\omega^{-2} = \rho, \quad \mathbf{S}\cdot\mathbf{A} = 0. \tag{10.9.11}$$

Then, with $v_S^2 = \mu\rho^{-1}$ (assumed positive) we have directly, from (10.9.11), that

$$\alpha_S^2 = \kappa^2\left(1 - \frac{c^2}{v_S^2}\right), \tag{10.9.12a}$$

$$\frac{A_1}{A_3} = -\frac{\mathrm{i}\alpha_S}{\kappa}. \tag{10.9.12b}$$

Similarly, from (10.9.4), $(\lambda + 2\mu)\boldsymbol{S}\cdot\boldsymbol{S} = \rho$, $\boldsymbol{A}$ is parallel to $\boldsymbol{S}$, and we obtain α_P corresponding to the P-wave solution. Then, with $v_P^2 = (\lambda + 2\mu)\rho^{-1}$ (assumed positive) we have directly that

$$\alpha_P^2 = \kappa^2\left(1 - \frac{c^2}{v_P^2}\right), \tag{10.9.13a}$$

$$\frac{A_1}{A_3} = -\frac{\mathrm{i}\kappa}{\alpha_P}. \tag{10.9.13b}$$

Hence, taking a linear combination of both waves, we note that the displacement field $\boldsymbol{u} = (u, v, w)$, given by

$$\begin{aligned} u &= \{(-\mathrm{i}\alpha_S D e^{-\alpha_S z} - \mathrm{i}\kappa B e^{-\alpha_P z})\exp \mathrm{i}\kappa(x - ct)\}^+, \\ v &= 0, \\ w &= \{(\kappa D e^{-\alpha_S z} + \alpha_P B e^{-\alpha_P z})\exp \mathrm{i}\kappa(x - ct)\}^+, \end{aligned} \tag{10.9.14}$$

is a solution of the equations of motion. Here κ is assumed to be prescribed (real), and the arbitrary constants B, D and the wave speed c are as yet undetermined. Of course, in order that $\alpha_P > 0$ and $\alpha_S > 0$ so that $\boldsymbol{u}$ decays with increasing z, we must have $c^2 < v_S^2 < v_P^2$, and we must take the positive square roots in (10.9.12a) and (10.9.13a) for α_P and α_S.

Now we wish to find B and D so that the surface $z=0$ is free of traction, which means that the traction vector $\boldsymbol{t}_{(z)}=(t_{zx}, t_{zy}, t_{zz})$ must be zero for $z=0$, and any time t. This leads to the system

$$(\kappa^2+\alpha_S^2)D+2\kappa\alpha_P B=0,$$
$$2\mu\kappa\alpha_S D-\{(\lambda+2\mu)\alpha_P^2-\lambda\kappa^2\}B=0. \tag{10.9.15}$$

Using (10.9.12) and (10.9.13) this system becomes

$$\left(2-\frac{c^2}{v_S^2}\right)D+2\left(1-\frac{c^2}{v_P^2}\right)^{1/2}B=0,$$
$$2\left(1-\frac{c^2}{v_S^2}\right)^{1/2}D+\left(2-\frac{c^2}{v_S^2}\right)B=0. \tag{10.9.16}$$

Hence, to obtain a nontrivial solution for B and D, the wave speed c must satisfy

$$\left(2-\frac{c^2}{v_S^2}\right)^2=4\left(1-\frac{c^2}{v_S^2}\right)^{1/2}\left(1-\frac{c^2}{v_P^2}\right)^{1/2}. \tag{10.9.17}$$

Note that $c=0$ is a root. Squaring both sides, and factoring out c^2, gives Rayleigh's cubic in c^2:

$$\left(\frac{c^2}{v_S^2}\right)^3-8\left(\frac{c^2}{v_S^2}\right)^2+\left(24-\frac{16v_S^2}{v_P^2}\right)\frac{c^2}{v_S^2}-16\left(1-\frac{v_S^2}{v_P^2}\right)=0. \tag{10.9.18}$$

It has been shown (Hayes and Rivlin, 1962) if $\mu>0$, $3\lambda+2\mu>0$, that there is only one real positive root $c^2=c_R^2$ (say) such that $c_R^2<v_S^2$ (to ensure decay of $\boldsymbol{u}$ with the distance z).

With $c^2=c_R^2$, we obtain $DB^{-1}=-(\alpha_P\alpha_S^{-1})^{1/2}$ from the system (10.9.16), on using (10.9.12) and (10.9.13). Inserting this into (10.9.14), and taking B real for simplicity, we obtain, for the displacement field of the Rayleigh wave,

$$\boldsymbol{u}(\boldsymbol{x},t)=\{\boldsymbol{A}(z)\exp i\kappa(x-c_R t)\}^+, \tag{10.9.19}$$

where the bivector $\boldsymbol{A}(z)=\boldsymbol{A}^+(z)+i\boldsymbol{A}^-(z)$ has its real and imaginary parts given by

$$\boldsymbol{A}^+(z)=-B\left(\frac{\alpha_P}{\alpha_S}\right)^{1/2}e^{-\alpha_S z}\{\kappa-(\alpha_P\alpha_S)^{1/2}e^{-(\alpha_P-\alpha_S)z}\}\mathbf{k},$$

$$\boldsymbol{A}^-(z)=-Be^{-\alpha_P z}\{\kappa-(\alpha_P\alpha_S)^{1/2}e^{(\alpha_P-\alpha_S)z}\}\mathbf{i}. \tag{10.9.20}$$

From (10.9.12) and (10.9.13), we note that $\alpha_P\alpha_S<\kappa^2$, and because

$v_P^2 > v_S^2$, that $\alpha_P > \alpha_S$. Hence, $A^+(z) \neq 0$ for any finite value of $z \geqslant 0$, whereas $A^-(z) = 0$ only once, when $z = z^*$, given by $(\alpha_P - \alpha_S)z^* = \ln(\kappa(\alpha_P\alpha_S)^{-1/2})$. At the free surface $z = 0$, the extremity of the vector $\boldsymbol{u}$ describes an ellipse, the ellipse of the bivector $\boldsymbol{A}(0)$. The major axis of this ellipse is orthogonal to the free surface, because $|\boldsymbol{A}^+(0)| > |\boldsymbol{A}^-(0)|$, on using $\alpha_P > \alpha_S$. Also, this ellipse is described in the retrograde sense. Indeed, considering, without loss in generality, the point $\boldsymbol{x} = 0$ on the free surface, we have $\boldsymbol{u}(0, 0) = \boldsymbol{A}^+(0) = -B(\alpha_P\alpha_S^{-1})^{1/2}\kappa\mathbf{k}$ and $\boldsymbol{u}(0, \frac{1}{2}\pi\omega^{-1}) = \boldsymbol{A}^-(0) = -B\kappa\mathbf{i}$, so that if $B\kappa > 0$ (< 0), then $\boldsymbol{u}(0, 0)$ is in the direction of decreasing (increasing) z, and $\boldsymbol{u}(0, \frac{1}{2}\pi\omega^{-1})$ is in the direction of decreasing (increasing) x. In either case the wave is retrograde. As $\boldsymbol{A}^-(z)\cdot\mathbf{i}$ changes sign at $z = z^*$, the ellipse changes from retrograde at $z = 0$ to prograde for $z > z^*$. At $z = z^*$, $\boldsymbol{A}^-(z^*) = \mathbf{0}$, and thus at $z = z^*$ the displacement field oscillates in the direction orthogonal to the free surface.

Using (8.4.14), it follows that for $\boldsymbol{g}$, the energy flux velocity for the combined motion of the two component inhomogeneous waves making up the displacement field (10.9.19), we have $\boldsymbol{g}\cdot\mathbf{i} = c_R$ and $\boldsymbol{g}\cdot\mathbf{k} = 0$, so that the mean energy propagates along the surface with the Rayleigh wave speed c_R.

Exercises 10.7

1. For $\boldsymbol{C} = 3\mathbf{j} + i\mathbf{i}$, determine the displacement field corresponding to the transverse wave (10.9.7). Describe the field. Obtain the corresponding circularly polarized waves.
2. Compute the weighted mean energy flux $\hat{\mathbf{r}}$ and weighted mean energy density $\hat{e}$ (see Exercise 10.6)

   (a) for the longitudinal wave (10.9.6);

   (b) for the transverse wave (10.9.7); analyse how the direction of $\hat{\mathbf{r}}$ is varied when β and γ are varied.

3. Examine the propagation of inhomogeneous plane waves in isotropic incompressible elastic materials.

10.10 Transversely isotropic materials. Inhomogeneous waves

For inhomogeneous waves in transversely isotropic materials, the acoustical tensor $\boldsymbol{Q}(\mathbf{n})$ given by (10.7.3) must be replaced by $\boldsymbol{Q}(\boldsymbol{C})$, which may be read off from (10.7.3) by replacing n_1, n_2, n_3 by C_1, C_2, C_3, respectively.

Hence, the secular equation (10.8.5) has the simple root (see (10.7.4))

$$\rho N^{-2} = \tfrac{1}{2}(d_{11} - d_{12})(C_1^2 + C_2^2) + d_{44}C_3^2, \qquad (10.10.1)$$

with corresponding eigenbivector, given, up to a scalar factor by

$$\boldsymbol{A} = \boldsymbol{C} \times \mathbf{k} = (C_2, -C_1, 0). \qquad (10.10.2)$$

The two other roots are solutions of the biquadratic equation (see (10.7.6))

$$\begin{aligned}(\rho N^{-2})^2 - \{(d_{11} + d_{44})(C_1^2 + C_2^2) + (d_{33} + d_{44})C_3^2\}\rho N^{-2} \\ + d_{11}d_{44}(C_1^2 + C_2^2)^2 + \{d_{11}d_{33} + d_{44}^2 - (d_{13} + d_{44})^2\} \\ \times (C_1^2 + C_2^2)C_3^2 + d_{33}d_{44}C_3^4 = 0. \qquad (10.10.3)\end{aligned}$$

We now consider some special cases.

10.10.1 The bivector $\boldsymbol{C}$ *is orthogonal to the symmetry axis* $\mathbf{k}$

If $C_3 = 0$, then the solutions are

$$\rho N_1^{-2} = \tfrac{1}{2}(d_{11} - d_{12})(C_1^2 + C_2^2), \quad \boldsymbol{A}^{(1)} = (C_2, -C_1, 0) = \boldsymbol{C} \times \mathbf{k}, \qquad (10.10.4a)$$

$$\rho N_2^{-2} = d_{11}(C_1^2 + C_2^2), \quad \boldsymbol{A}^{(2)} = (C_1, C_2, 0) = \boldsymbol{C}, \qquad (10.10.4b)$$

$$\rho N_3^{-2} = d_{44}(C_1^2 + C_2^2), \quad \boldsymbol{A}^{(3)} = (0, 0, 1) = \mathbf{k}. \qquad (10.10.4c)$$

Because $C_3 = 0$, we have $C_1^2 + C_2^2 = \boldsymbol{C}\cdot\boldsymbol{C} = m^2 - 1$, and it follows that N_1^{-2}, N_2^{-2}, N_3^{-2} are real and positive ($\boldsymbol{C}$ is assumed not to be isotropic). Hence, for these three waves, the planes of constant phase with unit normal $\hat{\mathbf{m}}$, are orthogonal to the planes of constant amplitude with unit normal $\hat{\mathbf{n}}$. Both (10.10.4a) and (10.10.4b) correspond to elliptically polarized waves, whilst (10.10.4c) corresponds to a linearly polarized wave, the direction of polarization being orthogonal to both the direction of propagation and the direction of attenuation.

Taking $\hat{\mathbf{m}} = \mathbf{i}$ and $\hat{\mathbf{n}} = \mathbf{j}$, without loss in generality, the displacement fields of the three wave trains are, respectively,

$$\boldsymbol{u}^{(1)} = -\mathrm{e}^{-\eta_1}(\sin\tau_1, m\cos\tau_1, 0), \qquad (10.10.5a)$$

$$\boldsymbol{u}^{(2)} = \mathrm{e}^{-\eta_2}(m\cos\tau_2, -\sin\tau_2, 0), \qquad (10.10.5b)$$

$$\boldsymbol{u}^{(3)} = \mathrm{e}^{-\eta_3}(0, 0, \cos\tau_3), \qquad (10.10.5c)$$

where

$$\eta_\alpha = \omega N_\alpha y, \quad \tau_\alpha = \omega(N_\alpha mx - t), \quad (\alpha = 1,2,3). \tag{10.10.6}$$

Note that $m(>1)$ may be arbitrarily prescribed.

10.10.2 The projection of **C** *onto the plane orthogonal to* **k** *is isotropic*

If $C_1^2 + C_2^2 = 0$, so that $C_2 = \pm iC_1$, the secular equation has the double root ($\rho N_1^{-2} = \rho N_2^{-2}$)

$$\rho N^{-2} = d_{44} C_3^2, \tag{10.10.7}$$

and the simple root

$$\rho N_3^{-2} = d_{33} C_3^2. \tag{10.10.8}$$

Assuming $(d_{11} + d_{12})(d_{33} - d_{44}) - 2(d_{13} + d_{44})^2 \neq 0$, the amplitude of the wave corresponding to the double root (10.10.7) is, up to a scalar factor, given by

$$\boldsymbol{A} = (C_1, C_2, 0) = \boldsymbol{C} - (\boldsymbol{C}\cdot\mathbf{k})\mathbf{k}. \tag{10.10.9}$$

Thus, this wave is circularly polarized in the plane orthogonal to the symmetry axis.

The amplitude of the wave corresponding to the simple root (10.10.8) is, up to a scalar factor, given by

$$\boldsymbol{A}^{(3)} = \left(C_1, C_2, \frac{d_{33} - d_{44}}{d_{13} + d_{44}} C_3 \right). \tag{10.10.10}$$

Thus, this wave is elliptically polarized, the projection of the polarization ellipse onto the xy-plane being a circle.

Because $C_1^2 + C_2^2 = 0$, we have $C_3^2 = \boldsymbol{C}\cdot\boldsymbol{C} = m^2 - 1$, so that C_3 is real ($m > 1$) and it follows that the roots (10.10.7) and (10.10.8) are real and positive. Hence, for both waves, the planes of constant phase are orthogonal to the planes of constant amplitude.

Now, recalling $\boldsymbol{C} = m\hat{\mathbf{m}} + i\hat{\mathbf{n}}$, we note that here $n_3 = \hat{\mathbf{n}}\cdot\mathbf{k} = 0$, and, without loss in generality, we take $\hat{\mathbf{n}} = \mathbf{j}$. Then $m\hat{\mathbf{m}}$ is in the xz-plane, and taking $C_3 = (m^2 - 1)^{1/2}$, we have

$$\boldsymbol{C} = (\pm 1, i, \sqrt{m^2 - 1}). \tag{10.10.11}$$

Hence, the displacement field $\boldsymbol{u}$ of the wave (10.10.7), (10.10.9), and the displacement field $\boldsymbol{u}^{(3)}$ of the wave (10.10.8), (10.10.10) are, respectively,

$$\boldsymbol{u} = \mathrm{e}^{-\eta}(\pm\cos\tau, -\sin\tau, 0), \tag{10.10.12}$$

$$\boldsymbol{u}^{(3)} = \mathrm{e}^{-\eta_3}\left(\pm\cos\tau_3, -\sin\tau_3, (m^2-1)^{1/2}\frac{(d_{33}-d_{44})\cos\tau_3}{(d_{13}+d_{44})}\right), \tag{10.10.13}$$

where

$$\eta = \left(\frac{\rho}{d_{44}(m^2-1)}\right)^{1/2}\omega y,$$

$$\tau = \omega\left\{\pm\left(\frac{\rho}{d_{44}(m^2-1)}\right)^{1/2}x + \left(\frac{\rho}{d_{44}}\right)^{1/2}z - t\right\}, \tag{10.10.14}$$

and

$$\eta_3 = \left(\frac{\rho}{d_{33}(m^2-1)}\right)^{1/2}\omega y,$$

$$\tau_3 = \omega\left\{\pm\left(\frac{\rho}{d_{33}(m^2-1)}\right)^{1/2}x + \left(\frac{\rho}{d_{33}}\right)^{1/2}z - t\right\}. \tag{10.10.15}$$

Again, note that $m > 1$ is arbitrary.

Exercise 10.8

Explore all the possibilities of having circularly polarized waves in transversely isotropic materials.

11

Plane waves in viscous fluids

In this chapter, the propagation of damped inhomogeneous plane waves in incompressible linearly viscous fluids is considered. The amplitude of the waves is assumed to be small so that the nonlinear contribution to the inertial term may be neglected in the equation of motion. Thus, the methods described in Chapter 6 may be applied.

11.1 Equations of motion

For a continuous medium, the balance of momentum, in the absence of body forces, reads

$$\rho(\partial_t v_i + v_{i,j}v_j) = t_{ij,j}, \tag{11.1.1}$$

where ρ is the density, v_i are the components of particle velocity and t_{ij} are the components of the symmetric Cauchy stress tensor.

We will assume that the nonlinear term $v_{i,j}v_j$ may be neglected in the expression for the acceleration, so that (11.1.1) becomes

$$\rho\partial_t v_i = t_{ij,j}. \tag{11.1.2}$$

For a homogeneous incompressible linearly viscous fluid, the constitutive equation is

$$t_{ij} = -p\delta_{ij} + \mu(v_{i,j} + v_{j,i}), \tag{11.1.3}$$

with the incompressibility constraint

$$\nabla \cdot \boldsymbol{v} = v_{i,i} = 0. \tag{11.1.4}$$

Here μ is the viscosity coefficient which is positive, and p is the pressure field.

The linearized equations of motion are obtained by introducing the constitutive equation (11.1.3) into the linearized balance of momentum (11.1.2). They read

$$\rho \partial_t v_i = -p_{,i} + \mu v_{i,jj}, \tag{11.1.5}$$

where the density ρ is a constant because the fluid is incompressible.

Multiplying (11.1.2) by v_i, we obtain the equation

$$\partial_t(\tfrac{1}{2}\rho \boldsymbol{v}\cdot\boldsymbol{v}) - (v_i t_{ij})_{,j} + t_{ij}v_{i,j} = 0. \tag{11.1.6}$$

Hence, an energy balance equation

$$\partial_t e + \nabla\cdot\boldsymbol{r} + d = 0 \tag{11.1.7}$$

holds with an energy density e, an energy flux vector $\boldsymbol{r}$ and an energy dissipation d given by

$$e = \tfrac{1}{2}\rho \boldsymbol{v}\cdot\boldsymbol{v}, \quad r_j = -v_i t_{ij}, \tag{11.1.8}$$

$$d = t_{ij}v_{i,j} = 2\mu D_{ij}D_{ij}, \tag{11.1.9}$$

where $D_{ij} = \tfrac{1}{2}(v_{i,j} + v_{j,i})$ is the strain rate tensor. Thus the energy density is simply the kinetic energy density, the energy flux is the mechanical energy flux due to stress, and the dissipation is due to the viscosity (it is always positive because $\mu > 0$).

11.2 The propagation condition

Here we consider damped inhomogeneous plane waves. Thus (section 6.4), we assume that the velocity and pressure fields are given by

$$\boldsymbol{v} = \{\boldsymbol{A}\exp \mathrm{i}\omega(\boldsymbol{S}\cdot\boldsymbol{x} - t)\}^+, \tag{11.2.1}$$

$$p = \{Q\exp \mathrm{i}\omega(\boldsymbol{S}\cdot\boldsymbol{x} - t)\}^+ + p_0, \tag{11.2.2}$$

where $\boldsymbol{S}$ is the slowness bivector, ω is the complex angular frequency, $\boldsymbol{A}$ is the amplitude bivector of the velocity field and Q is the complex amplitude of the pressure field. The pressure p_0 is the constant pressure in the absence of waves.

Inserting (11.2.1) into the equations of motion (11.1.5), and the incompressibility constraint (11.1.4) gives the propagation condition

$$\mathrm{i}Q\boldsymbol{S} = (\mathrm{i}\rho - \mu\omega\boldsymbol{S}\cdot\boldsymbol{S})\boldsymbol{A}, \tag{11.2.3}$$

with the constraint

$$\boldsymbol{S}\cdot\boldsymbol{A} = 0. \tag{11.2.4}$$

Taking the dot product of (11.2.3) by $\boldsymbol{S}$ yields

$$Q(\boldsymbol{S}\cdot\boldsymbol{S}) = 0, \tag{11.2.5}$$

so that there are two cases.

Either (a)

$$Q = 0, \tag{11.2.6a}$$

$$\mu\omega\boldsymbol{S}\cdot\boldsymbol{S} = \mathrm{i}\rho, \tag{11.2.6b}$$

$$\boldsymbol{S}\cdot\boldsymbol{A} = 0, \tag{11.2.6c}$$

or (b)

$$\boldsymbol{S}\cdot\boldsymbol{S} = 0, \tag{11.2.7a}$$

$$\boldsymbol{A}\cdot\boldsymbol{A} = 0, \tag{11.2.7b}$$

$$\boldsymbol{S}\cdot\boldsymbol{A} = 0, \tag{11.2.7c}$$

$$Q\boldsymbol{S} = \rho\boldsymbol{A}. \tag{11.2.7d}$$

In case (a) the increment in pressure is zero, and hence we call the corresponding solution the 'zero pressure wave'. In case (b) the slowness and amplitude bivectors are independent of the viscosity μ of the fluid, so that this same solution is valid for all values of μ. We call this the 'universal wave'.

We now consider in turn the two types of waves.

11.3 Zero pressure wave

For these solutions, the pressure is constant ($p = p_0$) and hence each component of the velocity field satisfies the equation

$$\nabla^2 v_i = \frac{\rho}{\mu}\partial_t v_i. \tag{11.3.1}$$

Also, we note that (11.2.6b) is precisely the complex dispersion relation corresponding to this equation. Thus, applying the DE method to this relation (see section 6.4), we obtain the wave bivector $\boldsymbol{K} = \omega\boldsymbol{S}$ in the form

$$\boldsymbol{K} = \pm\left(\frac{\rho\Omega}{\mu(m^2-1)}\right)^{1/2} \exp \mathrm{i}\left(\frac{\delta}{2}+\frac{\pi}{4}\right)\boldsymbol{C}, \tag{11.3.2}$$

where $\omega = \Omega e^{\mathrm{i}\delta}$ may be any prescribed complex number (of modulus Ω and argument δ), and where $\boldsymbol{C}$ may be any prescribed bivector $\boldsymbol{C} = m\hat{\mathbf{m}} + \mathrm{i}\hat{\mathbf{n}}$, with $\boldsymbol{C}\cdot\boldsymbol{C} \neq 0$, $m^2 \neq 1$.

If $\delta = 0$, so that there is no damping, it follows from (11.3.2) that the normals to the planes of constant phase and constant amplitude are along the equiconjugate radii of the ellipse of the bivector $\boldsymbol{C}$. For a given $\boldsymbol{C}$, if $\delta = 0$, then the acute angle between the two families of planes is the smallest possible, namely 2β, where $\beta = \tan^{-1}(m^{-1})$ (section 6.4, remark (a)). The angle between the two planes becomes $\frac{1}{2}\pi$ for $\delta = \frac{1}{2}\pi$, but then (11.2.1) and (11.2.2) no longer represent a propagating wave because the angular frequency $\omega = \mathrm{i}\Omega$ is purely imaginary. Any angle between 2β and $\frac{1}{2}\pi$ may be obtained by choosing the appropriate value of δ.

For zero pressure waves, the only constraint on the amplitude bivector $\boldsymbol{A}$ is that it satisfies $\boldsymbol{S}\cdot\boldsymbol{A} = 0$, or, equivalently, $\boldsymbol{C}\cdot\boldsymbol{A} = 0$. Hence, $\boldsymbol{A}$ is given by

$$\boldsymbol{A} = \alpha\boldsymbol{C}_\perp + \beta\hat{\mathbf{s}}, \tag{11.3.3}$$

where

$$\boldsymbol{C}_\perp = \frac{1}{m}\hat{\mathbf{m}} + \mathrm{i}\hat{\mathbf{n}}, \quad \hat{\mathbf{s}} = \hat{\mathbf{m}} \times \hat{\mathbf{n}}, \tag{11.3.4}$$

and where α and β are arbitrary complex numbers. For $\alpha = 0$, the wave is linearly polarized in the direction orthogonal to the plane of the bivector $\boldsymbol{C}$. For $\beta = 0$, the wave is elliptically polarized in the plane of $\boldsymbol{C}$. Because $\boldsymbol{C}\cdot\boldsymbol{A} = 0$, the ellipse of $\boldsymbol{A}$ is then similar and similarly situated to the ellipse of $\boldsymbol{C}$ when rotated through a quadrant.

Let $w^+(\boldsymbol{x}, t)$ and $w^-(\boldsymbol{x}, t)$ be defined by (6.4.7), so that $\boldsymbol{K}\cdot\boldsymbol{x} - \omega t = w^+ + \mathrm{i}w^-$. Then, using rectangular cartesian coordinate axes x, y, z along $\hat{\mathbf{m}}$, $\hat{\mathbf{n}}$, $\hat{\mathbf{s}}$ respectively, we have

$$w^+ = \pm\left(\frac{\rho\Omega}{\mu(m^2-1)}\right)^{1/2}\left\{mx\cos\left(\frac{\pi}{4}+\frac{\delta}{2}\right) - y\sin\left(\frac{\pi}{4}+\frac{\delta}{2}\right)\right\} - \Omega(\cos\delta)t,$$

$$w^- = \pm\left(\frac{\rho\Omega}{\mu(m^2-1)}\right)^{1/2}\left\{mx\cos\left(\frac{\pi}{4}+\frac{\delta}{2}\right) + y\cos\left(\frac{\pi}{4}+\frac{\delta}{2}\right)\right\} - \Omega(\sin\delta)t.$$

(11.3.5)

Because

$$\boldsymbol{v} = \{\boldsymbol{A}\mathrm{e}^{\mathrm{i}w^+}\mathrm{e}^{-w^-}\}^+ = \mathrm{e}^{-w^-}(\boldsymbol{A}^+\cos w^+ - \boldsymbol{A}^-\sin w^-), \tag{11.3.6}$$

we obtain the following general form for the zero pressure wave:

$$v_x = \frac{\mathrm{e}^{-w^-}}{m}(\alpha^+\cos w^+ - \alpha^-\sin w^+) = \frac{\mathrm{e}^{-w^-}}{m}(\alpha\mathrm{e}^{\mathrm{i}w^+})^+,$$

$$v_y = -\mathrm{e}^{-w^-}(\alpha^- \cos w^+ + \alpha^+ \sin w^+) = \mathrm{e}^{-w^-}(\mathrm{i}\alpha \mathrm{e}^{\mathrm{i}w^+})^+,$$

$$v_z = \mathrm{e}^{-w^-}(\beta^+ \cos w^+ - \beta^- \sin w^+) = \mathrm{e}^{-w^-}(\beta \mathrm{e}^{\mathrm{i}w^+})^+,$$

$$p = p_0, \tag{11.3.7}$$

where w^+ and w^- are given by (11.3.5).

Exercises 11.1

1. Let $\boldsymbol{A} = \alpha \boldsymbol{C}_\perp + \beta \hat{\mathbf{s}}$ and $\boldsymbol{A}' = \alpha' \boldsymbol{C}_\perp + \beta' \hat{\mathbf{s}}$ be two possible amplitude bivectors for zero pressure waves corresponding to a given $\boldsymbol{C}$. Find the condition that $\beta\alpha^{-1}$ and $\beta'\alpha'^{-1}$ must satisfy in order to have

$$\text{(a)}\ \boldsymbol{A}\cdot\boldsymbol{A}' = 0, \quad \text{(b)}\ \bar{\boldsymbol{A}}\cdot\boldsymbol{A}' = 0.$$

2. Let $\boldsymbol{A} = \alpha \boldsymbol{C}_\perp + \beta \hat{\mathbf{s}}$ be the amplitude bivector of a zero pressure wave. Find the values of $\beta\alpha^{-1}$ such that the wave be circularly polarized. For each value of $\beta\alpha^{-1}$, determine the plane of the polarization circle and the angle that this plane makes with the plane of $\boldsymbol{C}$.

11.4 Universal wave

The universal wave (11.2.7) is now considered. It is a circularly polarized wave which may propagate with an arbitrary isotropic slowness bivector.

Because $\boldsymbol{S}\cdot\boldsymbol{S} = \boldsymbol{A}\cdot\boldsymbol{A} = \boldsymbol{S}\cdot\boldsymbol{A} = 0$, it follows that the bivectors $\boldsymbol{A}$ and $\boldsymbol{S}$ are parallel (Theorem 2.2, section 2.6). Indeed, from (11.2.7d), we have $\boldsymbol{A} = (Q\rho^{-1})\boldsymbol{S}$, so that the equations (11.2.7) are all consistent, the pressure amplitude Q being an arbitrary complex number.

The bivector $\boldsymbol{C}$ is no longer arbitrary: it must be isotropic. Thus $m = 1$, and $\boldsymbol{C} = \hat{\mathbf{m}} + \mathrm{i}\hat{\mathbf{n}}$. Hence, recalling (6.4.4)–(6.4.6),

$$\boldsymbol{K} = k(\mathbf{h}^+ + \mathrm{i}\mathbf{h}^-), \tag{11.4.1}$$

where $k = \Omega T$ is an arbitrary real number, and where $\mathbf{h}^+$ and $\mathbf{h}^-$ are two arbitrary orthogonal unit vectors in the plane of $\boldsymbol{C}$.

Taking rectangular cartesian axes x, y, z along $\mathbf{h}^+$, $\mathbf{h}^-$, $\mathbf{h}^+ \times \mathbf{h}^-$, respectively, we have

$$\rho \boldsymbol{v} = k\exp(-ky+\omega^- t)\{(Q\omega^{-1})(\mathbf{h}^+ + \mathrm{i}\mathbf{h}^-)\exp \mathrm{i}(kx-\omega^+ t)\}^+,$$
$$p = \exp(-ky+\omega^- t)\{Q\exp \mathrm{i}(kx-\omega^+ t)\}^+ + p_0. \qquad (11.4.2)$$

The complex number $Q\omega^{-1}$ is arbitrary. Writing it in terms of its modulus and its argument, $Q\omega^{-1} = a e^{\mathrm{i}\gamma}$, we obtain the following explicit form of (11.4.2):

$$\rho v_x = ka\cos(kx-\omega^+ t+\gamma)\exp(-ky+\omega^- t), \qquad (11.4.3a)$$

$$\rho v_y = -ka\sin(kx-\omega^+ t+\gamma)\exp(-ky+\omega^- t), \qquad (11.4.3b)$$

$$v_z = 0, \qquad (11.4.3c)$$

$$p = a\{\omega^+\cos(kx-\omega^+ t+\gamma) - \omega^-\sin(kx-\omega^+ t+\gamma)\}\exp(-ky+\omega^- t) + p_0. \qquad (11.4.3d)$$

Here, α, k, γ, ω^+ and ω^- are arbitrary real numbers.

We note that the flow (11.4.3) is an irrotational plane flow.

Exercise 11.2

For the flow (11.4.3) compute the nonlinear term $v_{i,j}v_j$. Show that it is the gradient of a scalar function. Modify the expression (11.4.3d) for p in order to obtain an exact solution of the Navier–Stokes equations

$$\rho(\partial_t v_i + v_{i,j}v_j) = -p_{,i} + \mu v_{i,jj}, \quad \nabla\cdot\boldsymbol{v} = 0.$$

11.5 Energy flux, energy density and dissipation

11.5.1 General result

Here we consider energy propagation. Because the waves are not time-harmonic, taking the average over a period of time as in Chapter 8 is no longer meaningful. Also, damped inhomogeneous plane waves are not periodic in space. But clearly, they are periodic in w^+ (of period 2π) at fixed w^-, where w^+ and w^- are defined by (6.4.7). It makes sense to take a mean over a cycle of w^+. In this way we define a 'mean' energy flux, a 'mean' energy density and a 'mean' energy dissipation. These are functions of w^- and thus still depend upon $\boldsymbol{x}$ and t. Then, we define what we call the 'weighted mean energy flux', the 'weighted mean energy density' and 'weighted mean energy dissipation', which are independent of $\boldsymbol{x}$ and t. When ω is real, the mean over a cycle of w^+ reverts to the usual mean over a period introduced in Chapter 8.

First, we obtain a general result from the energy balance (11.1.7). Then, in sections 11.5.2 and 11.5.3 we discuss the cases of the zero pressure wave and of the universal wave.

Here, the velocity and stress field have the forms

$$
\begin{aligned}
v_i &= \tfrac{1}{2}\{A_i \exp i(\boldsymbol{K}\cdot\boldsymbol{x} - \omega t) + \text{c.c.}\} = \mathrm{e}^{-w^-}\{A_i \mathrm{e}^{\mathrm{i}w^+}\}^+, \\
t_{ij} &= -p_0\delta_{ij} + \tfrac{1}{2}\{B_{ij}\exp i(\boldsymbol{K}\cdot\boldsymbol{x} - \omega t) + \text{c.c.}\} \\
&= -p_0\delta_{ij} + \mathrm{e}^{-w^-}\{B_{ij}\mathrm{e}^{\mathrm{i}w^+}\}^+,
\end{aligned}
\tag{11.5.1}
$$

where c.c. stands for complex conjugate. Hence, the energy flux $\boldsymbol{r}$, energy density e, and energy dissipation d have the forms (compare with (8.3.8) and (8.3.9))

$$
\begin{aligned}
\boldsymbol{r} &= p_0\mathrm{e}^{-w^-}\{\boldsymbol{A}\mathrm{e}^{\mathrm{i}w^+}\}^+ + \mathrm{e}^{-2w^-}\{\boldsymbol{H}\mathrm{e}^{2\mathrm{i}w^+}\}^+ + \boldsymbol{\gamma}\mathrm{e}^{-2w^-} = \boldsymbol{r}(w^+, w^-), \\
e &= \mathrm{e}^{-2w^-}\{F\mathrm{e}^{2\mathrm{i}w^+}\}^+ + \alpha\mathrm{e}^{-2w^-} = e(w^+, w^-), \\
d &= \mathrm{e}^{-2w^-}\{G\mathrm{e}^{2\mathrm{i}w^+}\}^+ + \beta\mathrm{e}^{-2w^-} = d(w^+, w^-),
\end{aligned}
\tag{11.5.2}
$$

where $\boldsymbol{H}$ is a bivector, and F, G are complex numbers, but $\boldsymbol{\gamma}$ is a real vector and α, β are real numbers.

The expressions (11.5.2) are functions of w^+ and w^-. From (6.4.7), we note that $w^+ = w^+(\boldsymbol{x}, t)$ and $w^- = w^-(\boldsymbol{x}, t)$ are in general two independent variables. This is obvious when $\boldsymbol{K}^+$ and $\boldsymbol{K}^-$ have different directions (inhomogeneous waves), because then $\boldsymbol{K}^+\cdot\boldsymbol{x}$ and $\boldsymbol{K}^-\cdot\boldsymbol{x}$ are two independent space variables. When $\boldsymbol{K}^-$ is along $\boldsymbol{K}^+$ (homogeneous waves), w^+ and w^- are still independent, unless

$$\omega^+\boldsymbol{K}^- = \omega^-\boldsymbol{K}^+, \tag{11.5.3}$$

or, equivalently, using $\boldsymbol{K} = \omega\boldsymbol{S}$,

$$\{(\omega^+)^2 + (\omega^-)^2\}\boldsymbol{S}^- = \boldsymbol{0}. \tag{11.5.4}$$

Assuming $\omega \neq 0$ (we are interested in nonsteady solutions), we conclude that w^+ and w^- are independent variables, unless $\boldsymbol{S}^- = \boldsymbol{0}$. Obviously, for the universal wave, $\boldsymbol{S}^- \neq \boldsymbol{0}$, because $\boldsymbol{S}$ is isotropic. For the zero pressure wave, $\boldsymbol{S}^- = \boldsymbol{0}$ together with (11.2.6b), implies that ω be purely imaginary. Thus $\omega^+ = 0$, and then $\boldsymbol{K}^+ = \boldsymbol{0}$. We disregard this case because then w^+ is identically zero, and (11.2.1) and (11.2.2) no longer represent a propagating wave.

We may now define the 'mean' energy flux $\tilde{\boldsymbol{r}}$, the 'mean' energy density $\tilde{e}$, and the 'mean' energy dissipation $\tilde{d}$, by

$$\tilde{\boldsymbol{r}} = \frac{1}{2\pi}\int_0^{2\pi} \boldsymbol{r}(w^+, w^-)\,\mathrm{d}w^+,$$

$$\tilde{e} = \frac{1}{2\pi}\int_0^{2\pi} e(w^+, w^-)\,\mathrm{d}w^+,$$

$$\tilde{d} = \frac{1}{2\pi}\int_0^{2\pi} d(w^+, w^-)\,\mathrm{d}w^+, \tag{11.5.5}$$

where the integrations have to be carried out at fixed w^-. In fact, in a given plane of constant amplitude (which propagates), the damped inhomogeneous wave phenomenon is periodic in w^+, and we integrate over a cycle.

From the forms (11.5.2) of $\boldsymbol{r}$, e, d, we obtain

$$\tilde{\boldsymbol{r}} = \mathrm{e}^{-2w^-}\boldsymbol{\gamma}, \quad \tilde{e} = \mathrm{e}^{-2w^-}\alpha, \quad \tilde{d} = \mathrm{e}^{-2w^-}\beta. \tag{11.5.6}$$

The 'weighted mean' energy flux vector $\hat{\mathbf{r}}$, the 'weighted mean' energy $\hat{e}$, and the 'weighted mean' energy dissipation $\hat{d}$ are then defined and given by

$$\hat{\mathbf{r}} = \mathrm{e}^{2w^-}\tilde{\boldsymbol{r}} = \boldsymbol{\gamma}, \quad \hat{e} = \mathrm{e}^{2w^-}\tilde{e} = \alpha, \quad \hat{d} = \mathrm{e}^{2w^-}\tilde{d} = \beta. \tag{11.5.7}$$

The vector $\hat{\mathbf{r}}$ and the real scalars $\hat{e}$, $\hat{d}$ are constants (they do not depend on $\boldsymbol{x}$ and t).

Now, introducing (11.5.2) into the energy balance (11.1.7) and taking the mean over a cycle of w^+ at fixed w^-, we obtain, using (11.5.7),

$$\hat{\mathbf{r}}\cdot\boldsymbol{K}^- = \tfrac{1}{2}\hat{d} + \omega^-\hat{e}, \tag{11.5.8}$$

or, equivalently, because $\boldsymbol{K} = \Omega \mathrm{e}^{\mathrm{i}\delta}\boldsymbol{S}$,

$$\hat{\mathbf{r}}\cdot(\boldsymbol{S}^-\cos\delta + \boldsymbol{S}^+\sin\delta) = \frac{\hat{d}}{2\Omega} + \hat{e}\sin\delta. \tag{11.5.9}$$

For time-harmonic waves, ω is real, $\delta = 0$ and (11.5.9) reduces to

$$\hat{\mathbf{r}}\cdot\boldsymbol{S}^- = \frac{\hat{d}}{2\omega}. \tag{11.5.10}$$

In this case, we retrieve the result of Chapter 8 (Exercise 8.2.2).

11.5.2 Zero pressure wave

We now present the details for the two types of waves: the zero pressure wave and the universal wave (section 11.5.3). It will be shown that, in addition to the relation (11.5.8), $\hat{\mathbf{r}}$ and $\hat{e}$ always satisfy another relation, independent of the viscosity coefficient.

Consider the zero pressure wave for which

$$v_i = \{A_i \exp \mathrm{i}(\boldsymbol{K}\cdot\boldsymbol{x} - \omega t)\}^+,$$
$$t_{ij} = -p_0\delta_{ij} + \mu\{\mathrm{i}(A_iK_j + A_jK_i)\exp \mathrm{i}(\boldsymbol{K}\cdot\boldsymbol{x} - \omega t)\}^+. \qquad (11.5.11)$$

Thus, for the weighted mean energy flux, energy density and energy dissipation, we obtain

$$4\hat{\mathbf{r}} = -\mathrm{i}\mu\{(\boldsymbol{K}\cdot\bar{\boldsymbol{A}})\boldsymbol{A} - (\bar{\boldsymbol{K}}\cdot\boldsymbol{A})\bar{\boldsymbol{A}} + (\boldsymbol{A}\cdot\bar{\boldsymbol{A}})(\boldsymbol{K} - \bar{\boldsymbol{K}})\},$$
$$4\hat{e} = \rho\boldsymbol{A}\cdot\bar{\boldsymbol{A}},$$
$$2\hat{d} = \mu\{(\boldsymbol{K}\cdot\bar{\boldsymbol{A}})(\bar{\boldsymbol{K}}\cdot\boldsymbol{A}) + (\boldsymbol{A}\cdot\bar{\boldsymbol{A}})(\boldsymbol{K}\cdot\bar{\boldsymbol{K}})\}. \qquad (11.5.12)$$

Now, from (11.2.6c), we have $\boldsymbol{K}\cdot\boldsymbol{A} = 0$, so that

$$\hat{\mathbf{r}}\cdot\boldsymbol{K} = -\mathrm{i}\frac{\mu}{\rho}(\boldsymbol{K}\cdot\boldsymbol{K})\hat{e} + \frac{\mathrm{i}}{2}\hat{d}. \qquad (11.5.13)$$

Thus, on using (11.2.6b) which may also be written $\mu\boldsymbol{K}\cdot\boldsymbol{K} = \mathrm{i}\rho\omega^{-1}$, we obtain

$$\hat{\mathbf{r}}\cdot\boldsymbol{K} = \omega\hat{e} + \frac{\mathrm{i}}{2}\hat{d}. \qquad (11.5.14)$$

Taking the real and imaginary part of this complex relation gives

$$\hat{\mathbf{r}}\cdot\boldsymbol{K}^+ = \omega^+\hat{e}, \qquad (11.5.15a)$$

$$\hat{\mathbf{r}}\cdot\boldsymbol{K}^- = \omega^-\hat{e} + \tfrac{1}{2}\hat{d}. \qquad (11.5.15b)$$

The first of these is a new viscosity independent relation, whilst the second is the relation (11.5.8) obtained previously from the balance of energy.

When ω is real, (11.5.15a) reduces to $\hat{\mathbf{r}}\cdot\boldsymbol{S}^+ = \hat{e}$, which is the same as (8.3.19a) obtained in Chapter 8 for conservative systems.

Now, using (11.3.2)–(11.3.4), we have

$$\boldsymbol{A}\cdot\bar{\boldsymbol{A}} = \alpha\bar{\alpha}\left(\frac{1}{m^2} + 1\right) + \beta\bar{\beta},$$

$$\boldsymbol{K}\cdot\bar{\boldsymbol{K}} = \pm 2\bar{\alpha}\left(\frac{\rho\Omega}{\mu(m^2-1)}\right)^{1/2} \exp \mathrm{i}\left(\frac{\pi}{4}+\frac{\delta}{2}\right),$$

$$\boldsymbol{K}\cdot\bar{\boldsymbol{K}} = \frac{\rho\Omega}{\mu}\frac{m^2+1}{m^2-1}. \tag{11.5.16}$$

Introducing this in (11.5.12), we obtain

$$\hat{\boldsymbol{r}} = \alpha\bar{\alpha}\boldsymbol{r}_1 + \beta\bar{\beta}\boldsymbol{r}_2 + \boldsymbol{r}_{12}, \tag{11.5.17a}$$

$$\hat{e} = \alpha\bar{\alpha}e_1 + \beta\bar{\beta}e_2, \tag{11.5.17b}$$

$$\hat{d} = \alpha\bar{\alpha}d_1 + \beta\bar{\beta}d_2, \tag{11.5.17c}$$

where $\boldsymbol{r}_1$, $\boldsymbol{r}_2$, $\boldsymbol{r}_{12}$ are given by

$$\boldsymbol{r}_1 = \pm\frac{1}{2}\left(\frac{\mu\rho\Omega}{m^2-1}\right)^{1/2}\left\{m\left(\frac{3}{m^2}+1\right)\sin\left(\frac{\pi}{4}+\frac{\delta}{2}\right)\hat{\mathbf{m}} + \left(\frac{1}{m^2}+3\right)\cos\left(\frac{\pi}{4}+\frac{\delta}{2}\right)\hat{\mathbf{n}}\right\},$$

$$\boldsymbol{r}_2 = \pm\frac{1}{2}\left(\frac{\mu\rho\Omega}{m^2-1}\right)^{1/2}\left\{m\sin\left(\frac{\pi}{4}+\frac{\delta}{2}\right)\hat{\mathbf{m}} + \cos\left(\frac{\pi}{4}+\frac{\delta}{2}\right)\hat{\mathbf{n}}\right\} = \tfrac{1}{2}\mu\boldsymbol{K}^-,$$

$$\boldsymbol{r}_{12} = \pm\left(\frac{\mu\rho\Omega}{m^2-1}\right)^{1/2}\left\{(\alpha\bar{\beta})^-\sin\left(\frac{\pi}{4}+\frac{\delta}{2}\right) + (\bar{\alpha}\beta)^+\cos\left(\frac{\pi}{4}+\frac{\delta}{2}\right)\right\}\hat{\mathbf{m}}\times\hat{\mathbf{n}}. \tag{11.5.18}$$

Also, e_1, e_2 are given by

$$e_1 = \frac{\rho}{4}\frac{m^2+1}{m^2}, \quad e_2 = \frac{\rho}{4}, \tag{11.5.19}$$

and d_1, d_2 are given by

$$d_1 = \frac{\rho\Omega}{2}\frac{m^4+6m^2+1}{m^2(m^2-1)}, \quad d_2 = \frac{\rho\Omega}{2}\frac{m^2+1}{m^2-1}. \tag{11.5.20}$$

We note that $\boldsymbol{r}_1, e_1, d_1$ and $\boldsymbol{r}_2, e_2, d_2$ are the weighted mean energy flux, energy density and dissipation corresponding respectively to

$\alpha = 1$, $\beta = 0$ (elliptical polarization in the plane of $\mathbf{C}$) and to $\alpha = 0, \beta = 1$ (linear polarization in the direction orthogonal to the plane of $\mathbf{C}$). The weighted mean energy density $\hat{e}$ and weighted mean energy dissipation $\hat{d}$ for the general zero pressure wave (cf. (11.3.3) with $\alpha \neq 0$, $\beta \neq 0$) are the sum of the weighted mean energy densities and weighted mean energy dissipations for the elliptically polarized wave corresponding to $\beta = 0$, and the linearly polarized wave corresponding to $\alpha = 0$.

However, the weighted mean energy flux $\hat{\mathbf{r}}$ for the general zero pressure solution is not the sum of the weighted mean energy fluxes for these two waves because (11.5.17a) includes an interaction term $\mathbf{r}_{12}$. Owing to this interaction term, the weighted mean energy flux $\hat{\mathbf{r}}$ is not in general in the plane of $\mathbf{C}$ (plane of the wave bivector).

We also note that e_1, e_2, d_1, d_2 are independent of the viscosity coefficient μ and that $d_1(2\Omega)^{-1} > e_1$, and $d_2(2\Omega)^{-1} > e_2$, so that in general $\hat{d}(2\Omega)^{-1} > \hat{e}$.

In order to analyse how the direction of the energy flux $\hat{\mathbf{r}}$ varies when α and β are varied, we write $\alpha = a e^{i\lambda}$, $\beta = b e^{i\nu}$ with a, b, λ, ν real. Then, using rectangular cartesian coordinate axes x, y, z along $\hat{\mathbf{m}}, \hat{\mathbf{n}}, \hat{\mathbf{s}} = \hat{\mathbf{m}} \times \hat{\mathbf{n}}$, the components of $\hat{\mathbf{r}}$ given by 11.5.17a may be written

$$\hat{r}_x = \pm\frac{1}{2}\left(\frac{\mu\rho\Omega}{m^2-1}\right)^{1/2}\left\{a^2\left(\frac{3}{m^2}+1\right)+b^2\right\} m \sin\left(\frac{\pi}{4}+\frac{\delta}{2}\right), \qquad (11.5.21a)$$

$$\hat{r}_y = \pm\frac{1}{2}\left(\frac{\mu\rho\Omega}{m^2-1}\right)^{1/2}\left\{a^2\left(\frac{1}{m^2}+3\right)+b^2\right\} \cos\left(\frac{\pi}{4}+\frac{\delta}{2}\right), \qquad (11.5.21b)$$

$$\hat{r}_z = \pm\left(\frac{\mu\rho\Omega}{m^2-1}\right)^{1/2} ab \cos\left(\frac{\pi}{4}+\frac{\delta}{2}+\lambda-\nu\right). \qquad (11.5.21c)$$

When $\lambda - \nu$ is varied, $\hat{r}_z$ varies from

$$-\left(\frac{\mu\rho\Omega}{m^2-1}\right)^{1/2} ab \quad \text{to} \quad +\left(\frac{\mu\rho\Omega}{m^2-1}\right)^{1/2} ab,$$

so that $(m^2-1)\hat{r}_z^2 \leqslant \mu\rho\Omega a^2 b^2$, or, on eliminating a^2, b^2 from (11.5.21a) and (11.5.21b)

$$\left(\frac{m^2-1}{m}\right)^2 \hat{r}_z^2 \cos^2\delta \leqslant 8\left(\frac{m^2+1}{m}\right)\hat{r}_x \hat{r}_y \cos\delta - 2\hat{r}_x^2\left(\frac{1}{m^2}+3\right)(1-\sin\delta)$$
$$- 2\hat{r}_y^2(3+m^2)(1+\sin\delta). \qquad (11.5.22)$$

With the equality, this is the equation of an elliptical cone. This cone is symmetrical with respect to the plane of the wave bivector ($\hat{r}_z = 0$). The intersection with this plane consists of two straight lines along $\boldsymbol{r}_1$ and $\boldsymbol{r}_2$ (which is along $\boldsymbol{K}^-$). The principal axes of the cone are the bisectors of these straight lines, and the z-axis (which is along $\hat{\mathbf{s}}$).

Thus, when α and β are varied, that is when a, b, λ, ν are varied, the weighted mean energy flux vector may take any direction on or inside this cone, or, more exactly, on or inside the semi-cone (because $\hat{r}_x$ and $\hat{r}_y$ do not change sign). If $\alpha \neq 0$ and $\beta \neq 0$, then, when $\nu - \lambda = \frac{1}{4}\pi - \frac{1}{2}\delta$ or $\nu - \lambda = \frac{5}{4}\pi - \frac{1}{2}\delta$, the weighted mean energy flux $\hat{\mathbf{r}}$ lies on the semi-cone; otherwise it lies inside the semi-cone.

Exercises 11.3

1. Compute the weighted mean energy flux, energy density and dissipation for the circularly polarized waves obtained in Exercise 11.1.2.
2. Show that the direction of $\boldsymbol{K}^+$ may never lie on or inside the cone (11.5.22), so that the mean energy flux is never along the phase velocity.
3. On a drawing, sketch the ellipse of $\boldsymbol{C}$, and the directions of $\boldsymbol{r}_1, \boldsymbol{r}_2$ (which is parallel to $\boldsymbol{K}^-$) and $\boldsymbol{K}^+$ (use the + sign in (11.3.2) and (11.5.18)).

11.5.3 Universal wave

We now consider the universal wave. In this case

$$v_i = \frac{1}{\rho}\left\{\frac{Q}{\omega} K_i \exp i(\boldsymbol{K}\cdot\boldsymbol{x} - \omega t)\right\}^+,$$

$$t_{ij} = -p_0\delta_{ij} + \left\{\left(-Q\delta_{ij} + 2i\frac{\mu}{\rho\omega}QK_iK_j\right)\exp i(\boldsymbol{K}\cdot\boldsymbol{x} - \omega t)\right\}^+, \tag{11.5.23}$$

and hence for the weighted mean energy flux, energy density and energy dissipation, we obtain

$$4\hat{\mathbf{r}} = \frac{Q\bar{Q}}{\rho\Omega^2}\left\{\bar{\omega}\boldsymbol{K} + \omega\bar{\boldsymbol{K}} - 2i\frac{\mu}{\rho}(\boldsymbol{K}\cdot\bar{\boldsymbol{K}})(\boldsymbol{K} - \bar{\boldsymbol{K}})\right\},$$

$$4\hat{e} = \frac{Q\bar{Q}}{\rho\Omega^2} \boldsymbol{K}\cdot\bar{\boldsymbol{K}},$$

$$\hat{d} = \mu \frac{Q\bar{Q}}{\rho\Omega^2} (\boldsymbol{K}\cdot\bar{\boldsymbol{K}})^2. \tag{11.5.24}$$

Again, we see that (11.5.14) holds. Hence, for the universal wave, as well as for the zero pressure wave, the two viscosity independent relations (11.5.15) are valid.

Now, using (11.4.1), we obtain

$$\hat{\mathbf{f}} = \frac{Q\bar{Q}}{2\rho\Omega^2} k \left\{ \omega^+ \mathbf{h}^+ + \left(\omega^- + 4\frac{\mu}{\rho} k^2 \right) \mathbf{h}^- \right\},$$

$$\hat{e} = \frac{Q\bar{Q}}{2\rho\Omega^2} k^2, \quad \hat{d} = 4\mu \frac{Q\bar{Q}}{\rho\Omega^2} k^4. \tag{11.5.25}$$

Here the weighted mean energy flux $\hat{\mathbf{f}}$ lies in the plane of $\boldsymbol{C}$ (plane of the wave bivector). We note that at fixed amplitude the dissipation depends on the viscosity coefficient μ contrary to the case of the zero pressure wave.

Appendix

Spherical trigonometry

A spherical triangle ABC is a triangle whose sides are arcs of three great circles on a sphere (see Figure A.1). Thus, A, B, C are any three points on a sphere. Let O be the centre of this sphere. The sides are the arcs $a = \widehat{BC}$, $b = \widehat{CA}$, $c = \widehat{AB}$ of great circles measured in radians (arc lengths, if the sphere is assumed to be of unit radius). The angles $\hat{A}$, $\hat{B}$, $\hat{C}$ are the angles between the planes of the great circles forming the sides passing through the vertices A, B, C, respectively.

Here we first consider arbitrary triangles and then give special attention to the case when the triangle is right-angled, that is when one of the angles is $\frac{1}{2}\pi$.

A.1 Arbitrary triangles

First we derive the **fundamental formula**:

$$\cos a = \cos b \cos c + \sin b \sin c \cos \hat{A}. \tag{A.1}$$

Proof: Take the radius of the sphere to be of unit length. Let α be the plane passing through O, and orthogonal to $\boldsymbol{OA}$. Let $\boldsymbol{OB}'$ and $\boldsymbol{OC}'$ denote unit vectors along the projections of $\boldsymbol{OB}$ and $\boldsymbol{OC}$, respectively, onto the plane α (B' and C' are points on the sphere; see Figure A.1). Then,

$$\boldsymbol{OB}' = \frac{\boldsymbol{OB} - (\boldsymbol{OB}\cdot\boldsymbol{OA})\boldsymbol{OA}}{\sin c}, \quad \boldsymbol{OC}' = \frac{\boldsymbol{OC} - (\boldsymbol{OC}\cdot\boldsymbol{OA})\boldsymbol{OA}}{\sin b}. \tag{A.2}$$

Hence,

$$\cos \hat{A} = \boldsymbol{OB}'\cdot\boldsymbol{OC}' = \frac{\boldsymbol{OB}\cdot\boldsymbol{OC} - (\boldsymbol{OB}\cdot\boldsymbol{OA})(\boldsymbol{OC}\cdot\boldsymbol{OA})}{\sin b \sin c} = \frac{\cos a - \cos b \cos c}{\sin b \sin c}, \tag{A.3}$$

from which (A.1) follows immediately.

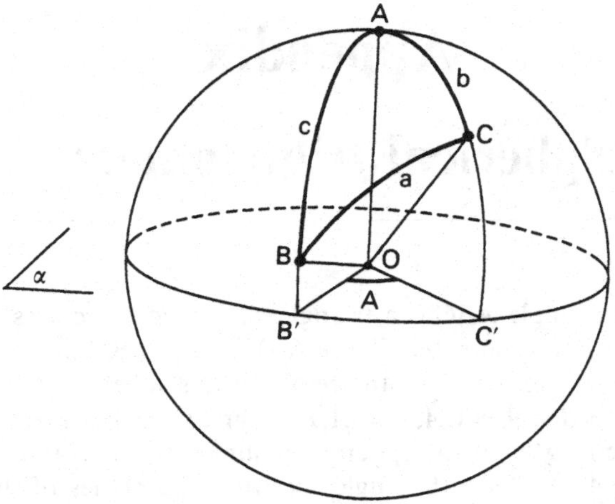

Figure A.1 *Arbitrary spherical triangle; the fundamental formula.*

From (A.1), we obtain, by cyclic permutations of a, b, c and $\hat{A}, \hat{B}, \hat{C}$,

$$\cos b = \cos c \cos a + \sin c \sin a \cos \hat{B}, \tag{A.4}$$

$$\cos c = \cos a \cos b + \sin a \sin b \cos \hat{C}. \tag{A.5}$$

We also note the **sine formula**:

$$\frac{\sin \hat{A}}{\sin a} = \frac{\sin \hat{B}}{\sin b} = \frac{\sin \hat{C}}{\sin c}. \tag{A.6}$$

Proof: From (A.3), we have

$$\frac{\sin^2 \hat{A}}{\sin^2 a} = \frac{1 - \cos^2 \hat{A}}{\sin^2 a} = \frac{\sin^2 b \sin^2 c - (\cos a - \cos b \cos c)^2}{\sin^2 a \sin^2 b \sin^2 c}, \tag{A.7}$$

and hence

$$\frac{\sin^2 \hat{A}}{\sin^2 a} = \frac{1 - \cos^2 a - \cos^2 b - \cos^2 c + 2 \cos a \cos b \cos c}{\sin^2 a \sin^2 b \sin^2 c}. \tag{A.8}$$

Then, (A.6) follows from the fact that the right-hand side of (A.8) remains unchanged by cyclic permutations of a, b, c.

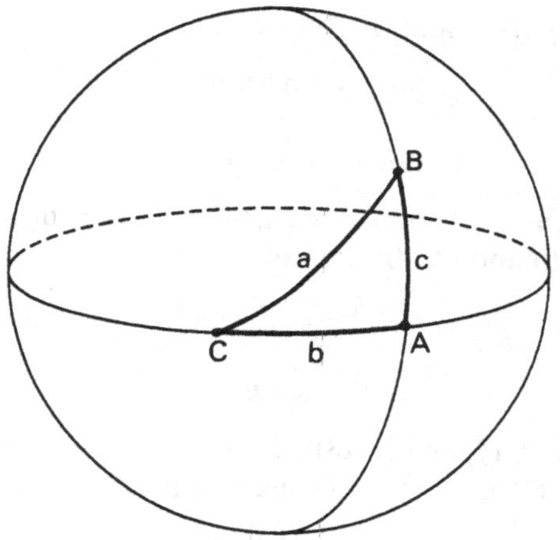

Figure A.2 *Right-angled spherical triangle* ($\hat{A} = \frac{1}{2}\pi$).

A.2 Right-angled triangles

We now consider the case when one of the angles, $\hat{A}$ (say), is $\frac{1}{2}\pi$ (Figure A.2). The fundamental formula (A.1), with $\hat{A} = \frac{1}{2}\pi$, yields

$$\cos a = \cos b \cos c. \tag{A.9}$$

The sine formula (A.6), with $\hat{A} = \frac{1}{2}\pi$, yields

$$\sin c = \sin a \sin \hat{C}, \tag{A.10}$$

$$\sin b = \sin a \sin \hat{B}. \tag{A.11}$$

Now, inserting $\cos c$, given by (A.5), into (A.9), we obtain

$$(\cos a)(1 - \cos^2 b) = \sin a \sin b \cos b \cos \hat{C}, \tag{A.12}$$

and hence, dividing by $\cos a \sin b \cos b$,

$$\tan b = \tan a \cos \hat{C}. \tag{A.13}$$

Then, similarly, using (A.4), we obtain

$$\tan c = \tan a \cos \hat{B}. \tag{A.14}$$

From the three basic formulae (A.9), (A.10) and (A.13), and using also (A.11) and (A.14), several other simple formulae may be derived. First, writing $\tan a = \sin a(\cos a)^{-1}$ in (A.13), and inserting $\cos a$ and

$\sin a$ as given by (A.9) and (A.10), we obtain

$$\tan c = \sin b \tan \hat{C}. \tag{A.15}$$

Similarly,

$$\tan b = \sin c \tan \hat{B}. \tag{A.16}$$

Writing $\tan c = \sin c(\cos c)^{-1}$ in (A.14) and inserting $\cos c$ and $\sin c$ as given by (A.9) and (A.10), we now obtain

$$\cos \hat{B} = \sin \hat{C} \cos b. \tag{A.17}$$

Similarly,

$$\cos \hat{C} = \sin \hat{B} \cos c. \tag{A.18}$$

Finally, from (A.17) and (A.18) we note that $\cos b = \cos \hat{B}(\sin \hat{C})^{-1}$, and $\cos c = \cos \hat{C}(\sin \hat{B})^{-1}$. Inserting this into (A.9) gives

$$\cos a = \cot \hat{B} \cos \hat{C}. \tag{A.19}$$

Answers to exercises

Chapter 1

1.1

Let the two radii be

$$\boldsymbol{r}_1 = a\cos\theta_1\mathbf{i} + b\sin\theta_1\mathbf{j},$$
$$\mathbf{r}_2 = a\cos\theta_2\mathbf{i} + b\sin\theta_2\mathbf{j}.$$

Their slopes are $ba^{-1}\tan\theta_1$ and $ba^{-1}\tan\theta_2$. Thus $\tan\theta_1\tan\theta_2 = -1$, and hence $\theta_1 = \theta_2 \pm \frac{1}{2}\pi$, so that $\boldsymbol{r}_1$ and $\boldsymbol{r}_2$ are conjugate.

1.2

1. (a) Let the equi-conjugate radii be of length R and let δ be the angle between them. Then from (1.1.5)

$$2R^2 = a^2 + b^2, \quad R = \{\tfrac{1}{2}(a^2 + b^2)\}^{1/2}, \quad R^2\sin\delta = ab.$$

Thus

$$\sin\delta = \frac{2ab}{a^2 + b^2}, \quad \tan\frac{\delta}{2} = \frac{b}{a}.$$

Remark The equi-conjugate radii are along the diagonals of the rectangle constructed upon the principal axes.

(b) With axes Ox' and Oy' along the equi-conjugate radii the equation of the ellipse is

$$x'^2 + y'^2 = R^2 = \tfrac{1}{2}(a^2 + b^2).$$

2. For some θ we have

$$\boldsymbol{c} = a\cos\theta\mathbf{i} + b\sin\theta\mathbf{j}, \quad \boldsymbol{d} = -a\sin\theta\mathbf{i} + b\cos\theta\mathbf{j}.$$

Hence, $\boldsymbol{r}$, given by (1.3.3), may be written as

$$\boldsymbol{r} = a\cos(\theta + \phi)\mathbf{i} + b\sin(\theta + \phi)\mathbf{j}.$$

Thus $(\theta + \phi)$ may be interpreted on the auxiliary circle as θ has been interpreted in secton 1.1. This yields the interpretation of ϕ shown in Figure 1.3.

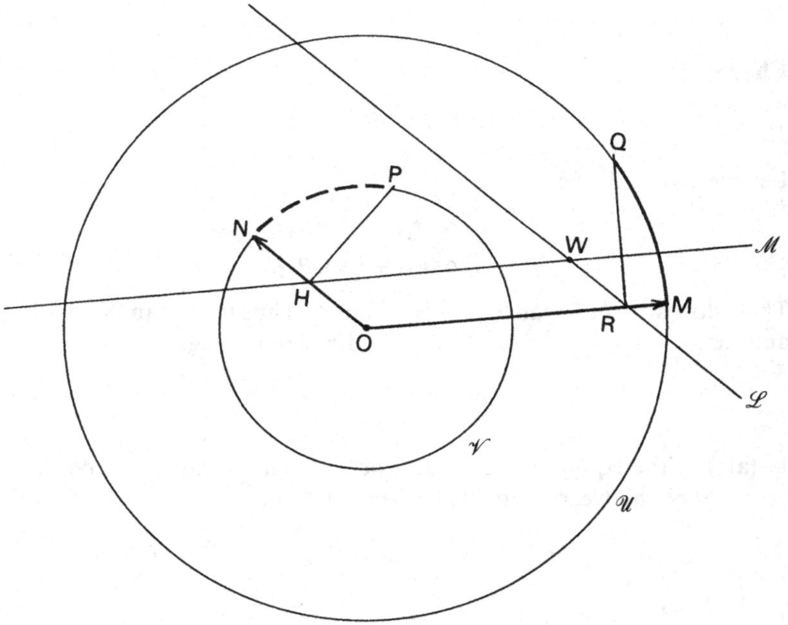

Construction of a point W of an ellipse with given conjugate radii.

3. Let $\boldsymbol{c} = \boldsymbol{OM}$, $\boldsymbol{d} = \boldsymbol{ON}$. With O as centre, draw two circles, $\mathscr{U}$ of radius c and $\mathscr{V}$ of radius d.

   Let Q be the point at angular distance ϕ from OM on $\mathscr{U}$. From Q drop the perpendicular QR on OM and through R draw the line $\mathscr{L}$ parallel to ON. The line $\mathscr{L}$ has equation $x' = c'\cos\phi$. Similarly, let P be the point at angular distance $\frac{1}{2}\pi - \phi$ from ON on $\mathscr{V}$. From P drop the perpendicular PH on ON and through H draw the line $\mathscr{M}$ parallel to OM. The line $\mathscr{M}$ has equation $y' = d'\sin\phi$. Since the coordinates of the point of intersection W of $\mathscr{L}$ and $\mathscr{M}$ are $(c'\cos\phi, d'\sin\phi)$, it follows that W is on the ellipse.

4. The area is

$$\iint \mathrm{d}x\,\mathrm{d}y = \int_0^1 \int_{\phi_1}^{\phi_2} ab\rho\,\mathrm{d}\rho\,\mathrm{d}\phi = \tfrac{1}{2}ab(\phi_2 - \phi_1),$$

where we have put $x = a\rho\cos\phi$, $y = b\rho\sin\phi$, $0 \leqslant \rho \leqslant 1$.

5. $\boldsymbol{r} = (2\boldsymbol{i} - 3\boldsymbol{j})\cos\phi + (4\boldsymbol{i} + \boldsymbol{j})\sin\phi$, or, alternatively,

$$x = 2\cos\phi + 4\sin\phi, \quad y = -3\cos\phi + \sin\phi \quad \text{(parametric)},$$

or

$$(x - 4y)^2 + (2y - 3x)^2 = 196 \quad \text{(cartesian)}.$$

6. $\boldsymbol{r}(\phi) = \boldsymbol{c}\cos\phi + \boldsymbol{d}\sin\phi$, $\boldsymbol{r}(\phi + \tfrac{1}{2}\pi) = -\boldsymbol{c}\sin\phi + \boldsymbol{d}\cos\phi$. Thus

$$|\boldsymbol{r}(\phi)|^2 + |\boldsymbol{r}(\phi + \tfrac{1}{2}\pi)|^2 = c^2 + d^2; \quad \boldsymbol{r}(\phi) \times \boldsymbol{r}(\phi + \tfrac{1}{2}\pi) = \boldsymbol{c} \times \boldsymbol{d}.$$

1.3

Because $\boldsymbol{\alpha}^{-1} = \boldsymbol{c} \otimes \boldsymbol{c} + \boldsymbol{d} \otimes \boldsymbol{d}$, and $\boldsymbol{\alpha}^{-2} = (\boldsymbol{c}\cdot\boldsymbol{c})\boldsymbol{c} \otimes \boldsymbol{c} + (\boldsymbol{d}\cdot\boldsymbol{d})\boldsymbol{d} \otimes \boldsymbol{d} + (\boldsymbol{c}\cdot\boldsymbol{d})(\boldsymbol{c} \otimes \boldsymbol{d} + \boldsymbol{d} \otimes \boldsymbol{c})$, we have $\operatorname{tr}\alpha^{-1} = \boldsymbol{c}\cdot\boldsymbol{c} + \boldsymbol{d}\cdot\boldsymbol{d}$, $[\operatorname{tr}(\boldsymbol{\alpha}^{-1})]^2 - \operatorname{tr}(\boldsymbol{\alpha}^{-2}) = 2\det(\boldsymbol{\alpha}^{-1}) = 2|\boldsymbol{c} \times \boldsymbol{d}|^2$.

1.4

The principal axes are the common conjugate directions of the ellipse and the circle $\boldsymbol{x}^{\mathrm{T}}\boldsymbol{x} = 1(\boldsymbol{\beta} = \mathbf{1})$.

Chapter 2

2.1

1. $\boldsymbol{r} = (2\mathbf{i} + 3\mathbf{j})\cos\phi + (-6\mathbf{i} + \mathbf{j})\sin\phi$, or, alternatively $x = 2\cos\phi - 6\sin\phi$, $y = 3\cos\phi + \sin\phi$, or $(x + 6y)^2 + (-3x + 2y)^2 = 400$.
2. The sum of the squares of the projections of the pairs of the conjugate radii of an ellipse onto an arbitrary direction is constant.
3. $\boldsymbol{a} = 2(10^{1/2})\mathbf{i}$, $\boldsymbol{b} = 10^{1/2}\mathbf{j}$.
4. $\boldsymbol{a} = 5(2^{-1/2})(\mathbf{i} - \mathbf{j})$, $\boldsymbol{b} = 5(2^{-1/2})(\mathbf{i} + \mathbf{j})$.

2.2

1. $\boldsymbol{A}_{\perp} = \frac{1}{25}(4\mathbf{i} + 3\mathbf{j} - 5\mathbf{k}) + \mathrm{i}\frac{1}{75}(31\mathbf{i} - 8\mathbf{j} + 5\mathbf{k})$.

2.

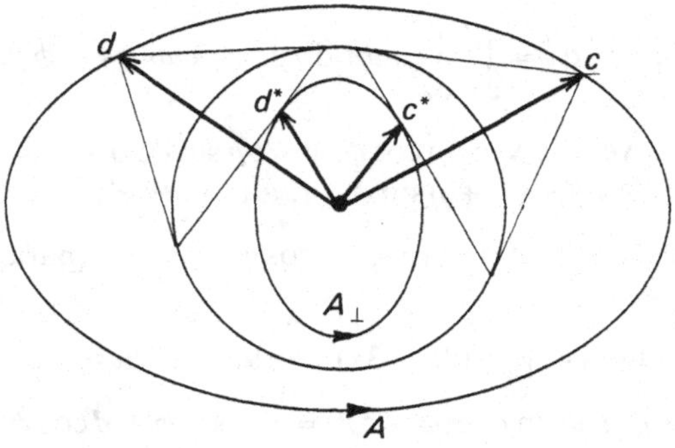

Construction of the conjugate radii $\boldsymbol{c}^*$, $\boldsymbol{d}^*$ *of the ellipse of* $\boldsymbol{A}_\perp$.

2.3

1. If $\boldsymbol{A}$ is isotropic, then from (2.6.8), $\boldsymbol{A}$ is parallel to $\boldsymbol{A}_\perp$. If $\boldsymbol{A}$ is parallel to $\boldsymbol{A}_\perp$, $\boldsymbol{A} = \alpha \boldsymbol{A}_\perp$ (say), and then using the fact (2.5.3) that $\boldsymbol{A}\cdot\boldsymbol{A}_\perp = 0$, it follows that $\boldsymbol{A}\cdot\boldsymbol{A} = 0$.
2. Without loss, $\boldsymbol{A} = \alpha(\mathbf{i} + \mathrm{i}\mathbf{j})$, $\boldsymbol{B} = \beta\mathbf{i} + \gamma\mathbf{j}$, for some scalars α, β, γ. $\boldsymbol{A}\cdot\boldsymbol{B} = 0$ leads to $\beta = -\mathrm{i}\gamma$ and $\boldsymbol{B} = -\mathrm{i}\gamma\boldsymbol{A}$.
3. $\cos\theta = \hat{\mathbf{m}}\cdot\hat{\mathbf{n}}$ (θ is the angle between $\hat{\mathbf{m}}$ and $\hat{\mathbf{n}}$).

2.4

1. $\boldsymbol{A}\times\boldsymbol{B} = 13^{1/2}\hat{\boldsymbol{p}} + \mathrm{i}\mathbf{i}$, $\hat{\mathbf{p}} = 13^{-1/2}(-2\mathbf{j} + 3\mathbf{k})$. The unit normal $\hat{\mathbf{n}}$ to the plane of the ellipse of $\boldsymbol{A}\times\boldsymbol{B}$ is $\hat{\mathbf{n}} = 13^{-1/2}(3\mathbf{j} + 2\mathbf{k})$. The projection of $\boldsymbol{A}$ upon plane of ellipse of $\boldsymbol{A}\times\boldsymbol{B}$ is

$$\boldsymbol{A}' = \boldsymbol{A} - (\boldsymbol{A}\cdot\hat{\mathbf{n}})\hat{\mathbf{n}} = 2\mathbf{i} - \frac{2\mathrm{i}\hat{\mathbf{p}}}{13^{1/2}}.$$

Also

$$\boldsymbol{B}' = \frac{3}{13^{1/2}}\hat{\mathbf{p}} + 3\mathrm{i}\mathbf{i} = \tfrac{3}{2}\mathrm{i}\boldsymbol{A}'.$$

Let $x, y, z, \hat{z}$ denote coordinates along $\mathbf{i}$, $\mathbf{j}$, $\mathbf{k}$, $\hat{\mathbf{p}}$. Then ellipses are

$$\boldsymbol{A}\times\boldsymbol{B}\colon x^2 + \frac{\hat{z}^2}{13} = 1, \quad 3y + 2z = 0; \quad \boldsymbol{A}\colon \frac{x^2}{4} + y^2 = 1, z = 0;$$

$$A'\colon x^2 + 13\hat{z}^2 = 4, 3y + 2z = 0; \quad B\colon \frac{x^2}{9} + z^2 = 1, y = 0;$$

$$B'\colon x^2 + 13\hat{z}^2 = 9,\ 3y + 2z = 0.$$

2. The projection, A', of A upon the plane orthogonal to $\hat{\mathbf{n}}$ is a circle.

2.5

1. For (2.8.3):

$$(A \times B)\cdot(C \times D) = \{(C \times D) \times A\}\cdot B = \{(A\cdot C)D - (A\cdot D)C\}\cdot B$$
$$= (A\cdot C)(B\cdot D) - (A\cdot D)(B\cdot C).$$

For (2.8.4):

$$(A \times B) \times (C \times D) = (A \times B\cdot D)C - D(A \times B\cdot C)$$
$$= (A\cdot C \times D)B - A(B\cdot C \times D).$$

For (2.8.5):

$$\{D \times (A \times B)\} \times C = (C\cdot D)A \times B - D(A \times B\cdot C)$$
$$= \{(B\cdot D)A - (A\cdot D)B\} \times C$$
$$= (B\cdot D)A \times C - (A\cdot D)B \times C.$$

2. $A, B, C = A \times B$ are linearly independent if and only if $A \times B\cdot C = (A \times B)\cdot(A \times B) \neq 0$.

3. $A_\perp \cdot A_\perp = \dfrac{4A\cdot A}{(A \times \bar{A})\cdot(A \times \bar{A})}.$

4. $A \times A_\perp = \dfrac{2A\cdot A}{(A \times \bar{A})\cdot(A \times \bar{A})} A \times \bar{A}.$

5. Using (2.8.3), we have $(B \times A)\cdot(B \times A) = 0$. But $(B \times A)\cdot A = 0$. Hence $B \times A = \lambda A$ with $\lambda(A\cdot\bar{A}) = A \times \bar{A}\cdot B$.

2.6

1. Write $X = \lambda A + \mu\bar{A} + \nu(A \times \bar{A})$. Then $A\cdot X = 0$ leads to $\mu = 0$ and

$X \times A = A$ leads to $\nu A \cdot \bar{A} = 1$. Thus,

$$X = \frac{\bar{A} \times A}{A \cdot \bar{A}} + \lambda A.$$

Clearly $X \cdot X = (A \times \bar{A}) \cdot (A \times \bar{A})(A \cdot \bar{A})^{-2} = -1$.

2. (a) We have $\alpha X \cdot A = A \cdot B$. Using this, and taking the cross product of the equation with A, leads to $(A \cdot A + \alpha^2)X = A \cdot B \alpha^{-1} A + \alpha B + A \times B$.
   (b) The compatibility condition is $(A \cdot B)A + \alpha^2 B + \alpha A \times B = \mathbf{0}$. Because $\alpha^2 = -A \cdot A$, this also reads $A \times (A \times B) + \alpha A \times B = \mathbf{0}$.
   (c) $X = (2\alpha)^{-1} B - (2\alpha^3)^{-1}(A \cdot B)A$ is a solution. All the solutions are obtained by adding to this solution the solutions of the homogeneous equation (2.9.7).
3. If $A \cdot C \neq 0$,

$$X = \frac{C \times B + \alpha A}{A \cdot C}.$$

If $A \cdot C = 0$, we have the compatibility condition $\alpha(A \cdot \bar{A}) = (\bar{A} \times B) \cdot C$. When this condition is satisfied, the solutions are

$$X = \frac{\bar{A} \times B}{A \cdot \bar{A}} + \lambda A.$$

2.7

1. $B \cdot C = C \cdot A = 0$ implies $C = \lambda(A \times B) = \lambda(-\mathbf{i} - \mathrm{i}\mathbf{j} + \mathbf{k})$. $A^* = -\mathrm{i}\mathbf{j} + \mathbf{k}$, $B^* = -\mathrm{i}\mathbf{i} + \mathbf{j}$, $C^* = \lambda^{-1}(-\mathbf{i} - \mathrm{i}\mathbf{j} + \mathbf{k})$.
2. $A^* = \frac{1}{2}(\mathbf{i} + \mathrm{i}\mathbf{j})$, $B^* = \frac{1}{2}(\mathbf{i} + \mathrm{i}\mathbf{j}) + \mathbf{k}$, $C^* = -\mathrm{i}\mathbf{j} - \mathbf{k}$, $C \cdot C = 0$, $A^* \cdot A^* = C^* \cdot C^* = 0$.

2.8

1. $C_1 = \frac{1}{2}(3 + 2\mathrm{i})(\mathbf{i} - \mathrm{i}\mathbf{j})$, $C_2 = \frac{1}{2}(7 + 4\mathrm{i})(\mathbf{i} + \mathrm{i}\mathbf{j})$.
2. $C_1 = \mathrm{e}^{-\mathrm{i}\theta}(a\mathbf{i} + \mathrm{i}b\mathbf{j})$, $C_2 = \lambda \mathrm{e}^{-\mathrm{i}\theta}(b\mathbf{i} - \mathrm{i}a\mathbf{j})$, $\lambda = (a+b)(a-b)^{-1}$, or $\lambda = -(a-b)(a+b)^{-1}$. This is illustrated below.
3. The real vector A may be decomposed into two coplanar bivectors whose ellipses have opposite handedness. Each ellipse has a

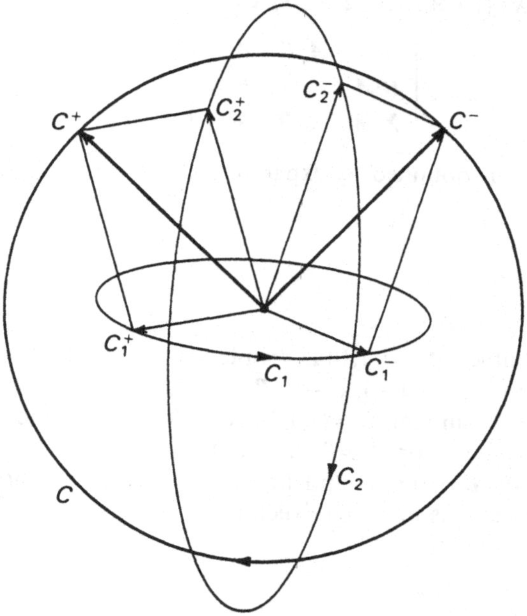

Decomposition (2.12.18) of an isotropic bivector.

principal axis along the direction of $\boldsymbol{A}$. If $k = b$ one ellipse becomes a circle. If $a - k = b$ the other ellipse becomes a circle.

4. The real vector $\boldsymbol{A}$ may be decomposed into two coplanar bivectors whose ellipses have opposite handedness with the same aspect ratio and the ellipses have a principal axis along the direction of $\boldsymbol{A}$. The major axis of one ellipse coincides with the minor axis of the other ellipse.

2.9

1. Use (2.8.6).
2. $\boldsymbol{A} \otimes \bar{\boldsymbol{A}}_\perp = \boldsymbol{c} \otimes \boldsymbol{c}^* + \boldsymbol{d} \otimes \boldsymbol{d}^* + \mathrm{i}(\boldsymbol{d} \otimes \boldsymbol{c}^* - \boldsymbol{c} \otimes \boldsymbol{d}^*)$. But $(\boldsymbol{c}^*, \boldsymbol{d}^*, \mathbf{k})$ is the triad reciprocal to $(\boldsymbol{c}, \boldsymbol{d}, \mathbf{k})$, and thus, $\boldsymbol{c} \otimes \boldsymbol{c}^* + \boldsymbol{d} \otimes \boldsymbol{d}^* + \mathbf{k} \otimes \mathbf{k} = \mathbf{1}$.
3. For any bivector $\boldsymbol{X}$, we have

$$X^{\mathrm{T}}\{(\boldsymbol{A}\times\boldsymbol{B})\otimes(\boldsymbol{A}\times\boldsymbol{B})\}X=(\boldsymbol{A}\times\boldsymbol{B}\cdot\boldsymbol{X})^2$$

$$=\begin{vmatrix} \boldsymbol{A}\cdot\boldsymbol{A} & \boldsymbol{A}\cdot\boldsymbol{B} & \boldsymbol{A}\cdot\boldsymbol{X} \\ \boldsymbol{B}\cdot\boldsymbol{A} & \boldsymbol{B}\cdot\boldsymbol{B} & \boldsymbol{B}\cdot\boldsymbol{X} \\ \boldsymbol{X}\cdot\boldsymbol{A} & \boldsymbol{X}\cdot\boldsymbol{B} & \boldsymbol{X}\cdot\boldsymbol{X} \end{vmatrix}.$$

The result is obtained by expanding this determinant.

Chapter 3

3.1

1. $\lambda=2$ (double), $\boldsymbol{A}=(\mathrm{i},1,0)$ isotropic. $\mu=\mathrm{i}$ (simple), $\boldsymbol{B}=(0,0,1)$.
2. $\lambda=\mathrm{i}$ (double), $\boldsymbol{A}=(-\mathrm{i},2,-\mathrm{i})$, $\boldsymbol{B}=(-1,0,1)$ and linear combinations. $\mu=1$ (simple), $\boldsymbol{C}=(1,\mathrm{i},1)$. $\boldsymbol{A}\pm\mathrm{i}\boldsymbol{B}$ are isotropic.
3. $\lambda=1$ (triple). For $b\neq 0$: $\boldsymbol{A}=(\mathrm{i},1,0)$. For $b=0$ and $a\neq 0$: $\boldsymbol{A}=(1,\mathrm{i},0)$, $\boldsymbol{B}=(0,0,1)$ and linear combinations. For $a=b=0$: every bivector is an eigenbivector.

3.2

1. $\boldsymbol{Q}=\boldsymbol{A}\otimes\boldsymbol{A}-\boldsymbol{B}\otimes\boldsymbol{B}+\mathrm{i}\boldsymbol{C}\otimes\boldsymbol{C}=\begin{bmatrix} -\mathrm{i} & -\mathrm{i} & \mathrm{i} \\ -\mathrm{i} & 3 & -\mathrm{i} \\ \mathrm{i} & -\mathrm{i} & -\mathrm{i} \end{bmatrix}.$

2. $\boldsymbol{Q}=\boldsymbol{1}-2\boldsymbol{C}\otimes\boldsymbol{C}-\mathrm{i}\gamma\boldsymbol{D}\otimes\boldsymbol{D}=\begin{bmatrix} -\mathrm{i}\frac{\gamma}{2} & \sqrt{2}\frac{\gamma}{2} & 1-\mathrm{i}\frac{\gamma}{2} \\ \sqrt{2}\frac{\gamma}{2} & 1+\mathrm{i}\gamma & \sqrt{2}\frac{\gamma}{2} \\ 1-\mathrm{i}\frac{\gamma}{2} & \sqrt{2}\frac{\gamma}{2} & -\mathrm{i}\frac{\gamma}{2} \end{bmatrix},$

   with $\gamma\neq 0$ arbitrary (complex).
3. λ: triple eigenvalue. Eigenvectors: $\boldsymbol{D}=\frac{1}{2}(2^{1/2})(\mathrm{i}-1,0,\mathrm{i}+1)$ and its scalar multiples. $\boldsymbol{D}=\boldsymbol{A}+\mathrm{i}\boldsymbol{B}$ with $\boldsymbol{A}=\frac{1}{2}(2^{1/2})(-1,0,1)$, $\boldsymbol{B}=\frac{1}{2}(2^{1/2})\,(1,0,1)$.
   Note: $\boldsymbol{A}\cdot\boldsymbol{A}=\boldsymbol{B}\cdot\boldsymbol{B}=1$, $\boldsymbol{A}\cdot\boldsymbol{B}=0$, $\boldsymbol{C}=\boldsymbol{A}\times\boldsymbol{B}=(0,1,0)$.
   $\gamma=\boldsymbol{A}^{\mathrm{T}}\boldsymbol{Q}\boldsymbol{B}=0$, $\delta=\boldsymbol{C}^{\mathrm{T}}\boldsymbol{Q}\boldsymbol{A}=-\mathrm{i}\frac{1}{2}(2^{1/2})$, $\boldsymbol{D}'=-\frac{1}{2}\mathrm{i}(\mathrm{i}-1,0,\mathrm{i}+1)$.
   $\boldsymbol{A}'=\frac{1}{4}(2^{1/2})(\mathrm{i}\boldsymbol{A}+3\boldsymbol{B})=\frac{1}{4}(3-\mathrm{i},0,3+\mathrm{i})$.
   $\boldsymbol{B}'=\frac{1}{4}(2^{1/2})(-3\boldsymbol{A}+\mathrm{i}\boldsymbol{B})=\frac{1}{4}(3+\mathrm{i},0,-3+\mathrm{i})$.
   $\boldsymbol{C}'=\boldsymbol{C}=(0,1,0)$.

$$T = \begin{bmatrix} \frac{1}{4}(3-\mathrm{i}) & \frac{1}{4}(3+\mathrm{i}) & 0 \\ 0 & 0 & 1 \\ \frac{1}{4}(3+\mathrm{i}) & \frac{1}{4}(-3+\mathrm{i}) & 0 \end{bmatrix}.$$

$$\text{Check: } T^{\mathrm{T}}QT = \begin{bmatrix} \lambda & 0 & 1 \\ 0 & \lambda & \mathrm{i} \\ 1 & \mathrm{i} & \lambda \end{bmatrix}.$$

3.3

1. $\lambda = 1 + \mathrm{i}, \quad A = (\sqrt{2}, 0, \mathrm{i}),$
   $\mu = \mathrm{i}, \quad B = (0, 1, 0),$
   $\nu = -1 + \mathrm{i}, \quad C = (-\mathrm{i}, 0, \sqrt{2}),$
   $M_\lambda = (\sqrt{2}, -1, 2\mathrm{i}), \quad P_\lambda = (\sqrt{2}, 1, 2\mathrm{i}),$
   $M_\mu = (\sqrt{2} + \mathrm{i}, 0, -\sqrt{2} + \mathrm{i}), \quad P_\mu = (\sqrt{2} - \mathrm{i}, 0, \sqrt{2} + \mathrm{i}),$
   $\mathrm{M}_\nu = (-2\mathrm{i}, 1, \sqrt{2}), \quad P_\nu = (2\mathrm{i}, 1, -\sqrt{2}).$
2. $\lambda = 1$ (double), $\mathrm{D} = (\frac{1}{2}\sqrt{2}, \mathrm{i}, \frac{1}{2}\sqrt{2})$, isotropic (Exercise 3.1.2);
   $\mu = -1$ (simple), $C = (\frac{1}{2}\sqrt{2}, 0, -\frac{1}{2}\sqrt{2});$
   $D = A + \mathrm{i}B$ with $A = (\frac{1}{2}\sqrt{2}, 0, \frac{1}{2}\sqrt{2}),\ B = (0, 1, 0);$
   $\gamma = A^{\mathrm{T}}QB = 2 \Rightarrow D' = \sqrt{2}D = \exp(\mathrm{i}\Phi)D$ with $\Phi = -\mathrm{i}\ln\sqrt{2};$
   $A' = \frac{3}{4}\sqrt{2}A + \mathrm{i}\frac{1}{4}\sqrt{2}B = (\frac{3}{4}, \mathrm{i}\frac{1}{4}\sqrt{2}, \frac{3}{4}),$
   $B' = -\mathrm{i}\frac{1}{4}\sqrt{2}A + \frac{3}{4}\sqrt{2}B = (-\frac{1}{4}\mathrm{i}, \frac{3}{4}\sqrt{2}, -\frac{1}{4}\mathrm{i}).$
   Thus from (3.4.17) and (3.4.23),
   $M_\lambda = (-\frac{1}{2}\sqrt{2} + \mathrm{i}\{1 - \frac{1}{2}\sqrt{2}\}, 1 - \mathrm{i}, -\frac{1}{2}\sqrt{2} - \mathrm{i}\{1 + \frac{1}{2}\sqrt{2}\}),$
   $P_\lambda = (\frac{1}{2}\sqrt{2} + \mathrm{i}\{1 + \frac{1}{2}\sqrt{2}\}, -1 + \mathrm{i}, \frac{1}{2}\sqrt{2} - \mathrm{i}\{1 - \frac{1}{2}\sqrt{2}\}),$
   $M_\mu = (1 - \mathrm{i}, \sqrt{2}\{1 - \mathrm{i}\}, 1 - \mathrm{i}), P_\mu = (1, \sqrt{2}\mathrm{i}, 1).$
3. λ: triple eigenvalue. $D = \frac{1}{2}\sqrt{2}(\mathrm{i} - 1, 0, \mathrm{i} + 1)$, isotropic;
   $D' = -\frac{1}{2}\mathrm{i}(\mathrm{i} - 1, 0, \mathrm{i} + 1),$
   $C' = (0, 1, 0)$ (Exercise 3.1.3).
   Thus, from (3.4.26), $M_\mu = (0, 1, 0)$, $P_\mu = (1 + \mathrm{i}, 0, 1 - \mathrm{i})$.
4. $M \cdot M \neq 0$, $P \cdot P = 0$, $M \cdot P = 2$. Case (b)(ii). The eigenbivectors are given by (3.4.24). Thus
   $P = (1, \mathrm{i}, 0)$, eigenvalue $\lambda = 1 + \frac{1}{2}M \cdot P = 2$ (double);
   $P \times M = (\mathrm{i}, -1, -2\mathrm{i})$, eigenvalue $\mu = 1$ (simple).
5. $M \cdot M = P \cdot P = P \cdot M = 0$. Case (c)(ii). The eigenbivectors are $M = P = (1, \mathrm{i}, 0)$, and any bivector orthogonal to $M = P = (1, \mathrm{i}, 0)$. $\lambda = 1$ is a triple eigenvalue.

3.4

1. $\boldsymbol{w} = (2 - a, 0, a - 2)$, $\boldsymbol{Q}\boldsymbol{w} = \boldsymbol{0}$.
   For $a = 2$, $\boldsymbol{Q}$ has three orthogonal eigenbivectors with real direction:
   $$(-1, 0, 1), \quad (1, -1, 1), \quad (1, 2, 1).$$
   For $a \neq 2$, $\boldsymbol{Q}$ has just one eigenbivector with a real direction (defined up to a scalar factor):
   $$(-1, 0, 1).$$
2. $\boldsymbol{w} = (0, b, 0)$, $\boldsymbol{Q}\boldsymbol{w} = b(a, 2 + \mathrm{i}, 0)$.
   For $b = 0$, $\boldsymbol{Q}$ has three orthogonal eigenbivectors with real direction.
   For $b \neq 0$, $a = 0$, $\boldsymbol{Q}$ has just one eigenbivector with a real direction (defined up to a scalar factor).
   For $b \neq 0$, $a \neq 0$, $\boldsymbol{Q}$ has no eigenbivectors with real direction.
3. For $a_{12} = a_{13} = a_{23} = 0$ ($\boldsymbol{Q}^+$ diagonal), $\boldsymbol{Q}$ has three orthogonal eigenbivectors with real direction.
   For $a_{12} = a_{13} = 0 \neq a_{23}$, or $a_{12} = a_{23} = 0 \neq a_{13}$, or $a_{13} = a_{23} = 0 \neq a_{12}$, $\boldsymbol{Q}$ has just one eigenbivector with a real direction (defined up to a scalar factor). In the other cases, $\boldsymbol{Q}$ has no eigenbivectors with real direction.

Chapter 4

4.1

1. Eigenvalues: $\lambda = e^a$, $\lambda^{-1} = e^{-a}$, $\varepsilon = -1$.
   For $a \neq 0$:
   $$\begin{aligned} &\lambda = e^a \text{ (simple)}, && \boldsymbol{D} = (1, -\mathrm{i}, 0) \text{ isotropic;} \\ &\lambda^{-1} = e^{-a} \text{ (simple)}, && \boldsymbol{E} = (1, \mathrm{i}, 0) \text{ isotropic;} \\ &\varepsilon = -1 \text{ (simple)}, && \boldsymbol{C} = (0, 0, 1). \end{aligned}$$
   For $a = 0$:
   $\lambda = \lambda^{-1} = 1$ (double), $\boldsymbol{A} = (1, 0, 0)$,
   $\boldsymbol{B} = (0, 1, 0)$ and any linear combination;
   $\varepsilon = -1$ (simple), $\boldsymbol{C} = (0, 0, 1)$.
2. Eigenvalue: $\varepsilon = +1$ (triple); eigenbivectors: $\boldsymbol{D} = (\frac{1}{2}\sqrt{2}, \mathrm{i}, \frac{1}{2}\sqrt{2})$ and scalar multiples (isotropic).

4.2

1.

$$R = \begin{bmatrix} 1 & -\mathrm{i} & -1 \\ \frac{1}{2}\mathrm{i} & \frac{3}{2} & -\mathrm{i} \\ \frac{1}{2} & \frac{1}{2}\mathrm{i} & 1 \end{bmatrix}.$$

2. $\varepsilon = +1$ triple eigenvalue. Eigenbivectors: $\boldsymbol{D} = \boldsymbol{A} + \mathrm{i}\boldsymbol{B}$ (and scalar multiples), with
$\boldsymbol{A} = \frac{1}{2}\sqrt{2}(1, 0, 1)$, $\boldsymbol{B} = (0, 1, 0)$. Note $\boldsymbol{A}\cdot\boldsymbol{A} = \boldsymbol{B}\cdot\boldsymbol{B} = 1$, $\boldsymbol{A}\cdot\boldsymbol{B} = 0$.
$\boldsymbol{C} = \boldsymbol{A} \times \boldsymbol{B} = \frac{1}{2}\sqrt{2}(-1, 0, 1)$, $\delta = \varepsilon \boldsymbol{A}^{\mathrm{T}}\boldsymbol{R}\boldsymbol{C} = 2\mathrm{i}$,
$\boldsymbol{A}' = \frac{3}{4}\mathrm{i}\boldsymbol{A} - \frac{5}{4}\boldsymbol{B} = \frac{1}{8}(3\sqrt{2}\mathrm{i}, -10, 3\sqrt{2}\mathrm{i})$,
$\boldsymbol{B}' = \frac{5}{4}\boldsymbol{A} + \frac{3}{4}\mathrm{i}\boldsymbol{B} = \frac{1}{8}(5\sqrt{2}, 6\mathrm{i}, 5\sqrt{2})$, $\boldsymbol{C}' = \boldsymbol{A}' \times \boldsymbol{B}' = \boldsymbol{C}$.

$$T = \begin{bmatrix} \frac{3}{8}\sqrt{2}\mathrm{i} & \frac{5}{8}\sqrt{2} & -\frac{1}{2}\sqrt{2} \\ -\frac{5}{4} & \frac{3}{4}\mathrm{i} & 0 \\ \frac{3}{8}\sqrt{2}\mathrm{i} & \frac{5}{8}\sqrt{2} & \frac{1}{2}\sqrt{2} \end{bmatrix}. \text{ Check } T^{\mathrm{T}}RT = \begin{bmatrix} \frac{1}{2} & -\frac{1}{2}\mathrm{i} & 1 \\ -\frac{1}{2}\mathrm{i} & \frac{3}{2} & \mathrm{i} \\ -1 & -\mathrm{i} & 1 \end{bmatrix}.$$

4.3

1. $\boldsymbol{\Omega} = (2, 2\mathrm{i}, 2)$.

$$\begin{aligned} &\lambda = -2\mathrm{i} \text{ (simple)}, && \boldsymbol{D} = \sqrt{2}(1, \mathrm{i}, 0), \text{ isotropic;} \\ &\lambda = +2\mathrm{i} \text{ (simple)}, && \boldsymbol{E} = \sqrt{2}(0, -\mathrm{i}, -1), \text{ isotropic;} \\ &\lambda = 0 \text{ (simple)}, && \boldsymbol{C} = (1, \mathrm{i}, 1), \boldsymbol{C}\cdot\boldsymbol{C} = 1. \end{aligned}$$

$$T = \begin{bmatrix} \frac{1}{2}\sqrt{2} & -\mathrm{i}\frac{1}{2}\sqrt{2} & 1 \\ 0 & \sqrt{2} & \mathrm{i} \\ -\frac{1}{2}\sqrt{2} & -\mathrm{i}\frac{1}{2}\sqrt{2} & 1 \end{bmatrix} \quad \text{(for instance).}$$

2. $\boldsymbol{\Omega} = (1, \mathrm{i}\sqrt{2}, 1)$, $\lambda = 0$ (triple), $\boldsymbol{D} \parallel \boldsymbol{\Omega}$ isotropic.
Write $\boldsymbol{D} = \boldsymbol{A} + \mathrm{i}\boldsymbol{B}$ with $\boldsymbol{A} = (1, 0, 0)$, $\boldsymbol{B} = (0, \sqrt{2}, -\mathrm{i})$.
Then $\boldsymbol{C} = (0, \mathrm{i}, \sqrt{2})$, $\delta = \mathrm{i} = \exp(\mathrm{i}\frac{1}{2}\pi)$, $\boldsymbol{A}' = -\boldsymbol{B}$, $\boldsymbol{B}' = \boldsymbol{A}$.

$$T = \begin{bmatrix} 0 & 1 & 0 \\ -\sqrt{2} & 0 & \mathrm{i} \\ \mathrm{i} & 0 & \sqrt{2} \end{bmatrix}.$$

4.4

1.

$$\tilde{\boldsymbol{W}} = \Theta\begin{bmatrix}0 & -1 & 0\\ 1 & 0 & 0\\ 0 & 0 & 0\end{bmatrix}, \quad \tilde{\boldsymbol{W}}^3 = -\Theta^3\begin{bmatrix}0 & -1 & 0\\ 1 & 0 & 0\\ 0 & 0 & 0\end{bmatrix}, \ldots,$$

$$\tilde{\boldsymbol{W}}^2 = -\Theta^2\begin{bmatrix}1 & 0 & 0\\ 0 & 1 & 0\\ 0 & 0 & 0\end{bmatrix}, \quad \tilde{\boldsymbol{W}}^4 = \Theta^4\begin{bmatrix}1 & 0 & 0\\ 0 & 1 & 0\\ 0 & 0 & 0\end{bmatrix}, \ldots.$$

Hence,

$$\exp\tilde{\boldsymbol{W}} = \mathbf{1} + \sin\Theta\begin{bmatrix}0 & -1 & 0\\ 1 & 0 & 0\\ 0 & 0 & 0\end{bmatrix} + (\cos\Theta - 1)\begin{bmatrix}1 & 0 & 0\\ 0 & 1 & 0\\ 0 & 0 & 0\end{bmatrix}$$

$$= \begin{bmatrix}\cos\Theta & -\sin\Theta & 0\\ \sin\Theta & \cos\Theta & 0\\ 0 & 0 & 1\end{bmatrix}.$$

2.

$$\tilde{\boldsymbol{W}} = \delta\begin{bmatrix}0 & 0 & 1\\ 0 & 0 & \mathrm{i}\\ -1 & -\mathrm{i} & 0\end{bmatrix}, \quad \tilde{\boldsymbol{W}}^2 = \delta^2\begin{bmatrix}-1 & -\mathrm{i} & 0\\ -\mathrm{i} & 1 & 0\\ 0 & 0 & 0\end{bmatrix},$$

$$\tilde{\boldsymbol{W}}^3 = \tilde{\boldsymbol{W}}^4 = \cdots = \mathbf{0}.$$

Hence,

$$\exp\tilde{\boldsymbol{W}} = 1 + \tilde{\boldsymbol{W}} + \tfrac{1}{2}\tilde{\boldsymbol{W}}^2 = \begin{bmatrix}1 - \frac{1}{2}\delta^2 & -\mathrm{i}\frac{1}{2}\delta^2 & \delta\\ -\mathrm{i}\frac{1}{2}\delta^2 & 1 + \frac{1}{2}\delta^2 & \mathrm{i}\delta\\ -\delta & -\mathrm{i}\delta & 1\end{bmatrix}.$$

3. Let $\delta = 1$ in answer 4.4.2.

4.5

1. $\boldsymbol{C} = (1, \mathrm{i}, 1)$, $\Theta = \mathrm{i}\ln 2$, $\boldsymbol{\Omega}' = \frac{2}{3}\mathrm{i}\boldsymbol{C} = (-\frac{2}{3}\mathrm{i}, \frac{2}{3}, -\frac{2}{3}\mathrm{i})$,

$$\boldsymbol{W}' = \begin{bmatrix}0 & \frac{2}{3}\mathrm{i} & \frac{2}{3}\\ -\frac{2}{3}\mathrm{i} & 0 & \frac{2}{3}\mathrm{i}\\ -\frac{2}{3} & -\frac{2}{3}\mathrm{i} & 0\end{bmatrix}.$$

2. $\Omega' = B' - iA' = (\sqrt{2}, 2i, \sqrt{2})$,

$$W' = \begin{bmatrix} 0 & -\sqrt{2} & 2i \\ \sqrt{2} & 0 & -\sqrt{2} \\ -2i & \sqrt{2} & 0 \end{bmatrix}.$$

Chapter 5

5.1

1. $(g^*g^T)_{ip} = \frac{1}{2}e_{ikm}e_{jln}g_{kl}g_{mn}g_{pj}$. But $e_{jln}g_{pj}g_{kl}g_{mn} = (\det g)e_{pkm}$. Hence $(g^*g^T)_{ip} = \frac{1}{2}(\det g)e_{ikm}e_{pkm} = (\det g)\delta_{ip}$.
2. By (5.1.3), for every P, Q, we have

$$\begin{aligned}(g_1g_2)^*(P \times Q) &= g_1g_2P \times g_1g_2Q \\ &= g_1^*(g_2P \times g_2Q) = g_1^*g_2^*(P \times Q).\end{aligned}$$

3. For every P, Q, we have

$$\begin{aligned}&\{(A\cdot A)\mathbf{1} - A \otimes A\}^*(P \times Q) \\ &= \{(A\cdot A)P - A(A\cdot P)\} \times \{(A\cdot A)Q - A(A\cdot Q)\} \\ &= (A\cdot A)^2 P \times Q + (A\cdot A)A \times \{(A\cdot Q)P - (A\cdot P)Q\} \\ &= (A\cdot A)^2 P \times Q + (A\cdot A)A \times \{A \times (P \times Q)\} = (A\cdot A)(A \otimes A)(P \times Q).\end{aligned}$$

4. For every P, Q, we have $W^*(P \times Q) = WP \times WQ = (\Omega \times P) \times (\Omega \times Q)$

$$= \Omega\{\Omega\cdot(P \times Q)\} = (\Omega \otimes \Omega)(P \times Q).$$

5.

$$\begin{aligned}\operatorname{tr} g^* &= \tfrac{1}{2}e_{ikm}e_{iln}g_{kl}g_{mn} = \tfrac{1}{2}(\delta_{kl}\delta_{mn} - \delta_{kn}\delta_{ml})g_{kl}g_{mn} \\ &= \tfrac{1}{2}(g_{kk}g_{mm} - g_{kl}g_{lk}) = \tfrac{1}{2}\{(\operatorname{tr} g)^2 - \operatorname{tr}(g^2)\}.\end{aligned}$$

$g^*g^T = (\det g)\mathbf{1}$, and hence $(\det g^*)(\det g) = (\det g)^3$.

5.2

1. The systems

$$\begin{cases} A\cdot A = 0, \\ A^T\alpha_1 A = 0, \end{cases} \quad \text{and} \quad \begin{cases} A\cdot A = 0, \\ A^T\alpha_2 A = 0, \end{cases}$$

are equivalent if and only if the equations of one are linear combinations of the equations of the other, that is if and only if there exists $c_1 \neq 0$, $c_2 \neq 0$ and c_0, such that

$$A^{\mathrm{T}}(c_1\boldsymbol{\alpha}_1 - c_2\boldsymbol{\alpha}_2)A = c_0 A\cdot A, \text{ for every bivector } A.$$

Hence $c_1\boldsymbol{\alpha}_1 - c_2\boldsymbol{\alpha}_2 = c_0\mathbf{1}$.

2. $$\boldsymbol{x}^{\mathrm{T}}\boldsymbol{\alpha}\boldsymbol{x} = (\lambda+\nu-\mu)(x^2+y^2+z^2) + (\lambda-\mu)\left\{\frac{\mu-\nu}{\lambda-\nu}x^2 - y^2 - \frac{\lambda-\mu}{\lambda-\nu}z^2\right\}$$
$$+(\mu-\nu)\left\{\frac{\mu-\nu}{\lambda-\nu}x^2 + y^2 - \frac{\lambda-\mu}{\lambda-\nu}z^2\right\}.$$

But

$$\mu(A_h\cdot\boldsymbol{x})(A_k\cdot\boldsymbol{x}) = \left(\left(\frac{\mu-\nu}{\lambda-\nu}\right)^{1/2} x + \mathrm{i}y\right)^2 - \frac{\lambda-\mu}{\lambda-\nu}z^2,$$

and

$$\mu(\bar{A}_h\cdot\boldsymbol{x})(A_k\cdot\boldsymbol{x}) = \frac{\mu-\nu}{\lambda-\nu}x^2 - \left(\mathrm{i}y + \left(\frac{\lambda-\mu}{\lambda-\nu}\right)^{1/2} z\right)^2.$$

Hence,

$$\boldsymbol{x}^{\mathrm{T}}\boldsymbol{\alpha}\boldsymbol{x} = (\lambda+\nu-\mu)\boldsymbol{x}\cdot\boldsymbol{x} + \frac{\mu}{2}(\lambda-\mu)\{(A_h\cdot\boldsymbol{x})(A_k\cdot\boldsymbol{x}) + (\bar{A}_h\cdot\boldsymbol{x})(\bar{A}_k\cdot\boldsymbol{x})\}$$
$$+\frac{\mu}{2}(\mu-\nu)\{(\bar{A}_h\cdot\boldsymbol{x})(A_k\cdot\boldsymbol{x}) + (A_h\cdot\boldsymbol{x})(\bar{A}_k\cdot\boldsymbol{x})\},$$

from which the result follows immediately.

For the spheroid

$$\frac{x^2}{a^2} + \frac{y^2+z^2}{b^2} = 1,$$

we have

$$\boldsymbol{\alpha} = a^{-2}\mathbf{1} - \frac{(a^{-2}-b^{-2})}{2}b^{-2}(A_x\otimes\bar{A}_x + \bar{A}_x\otimes A_x).$$

For the spheroid

$$\frac{x^2+y^2}{a^2} + \frac{z^2}{c^2} = 1,$$

we have

$$\boldsymbol{\alpha}=c^{-2}\mathbf{1}+\frac{(a^{-2}-c^{-2})}{2}a^{-2}(\boldsymbol{A}_z\otimes\bar{\boldsymbol{A}}_z+\bar{\boldsymbol{A}}_z\otimes\boldsymbol{A}_z).$$

3.

$$\boldsymbol{\alpha h}=\mu\boldsymbol{h}+\frac{\lambda-\nu}{2}(\boldsymbol{h}\cdot\boldsymbol{k})\boldsymbol{h}+\frac{\lambda-\nu}{2}\boldsymbol{k},$$

$$\boldsymbol{\alpha k}=\mu\boldsymbol{k}+\frac{\lambda-\nu}{2}(\boldsymbol{h}\cdot\boldsymbol{k})\boldsymbol{k}+\frac{\lambda-\nu}{2}\boldsymbol{h}.$$

But

$$\boldsymbol{h}\cdot\boldsymbol{k}=\frac{\lambda+\nu-2\mu}{\lambda-\nu}.$$

4. From (5.6.5), $\mu\boldsymbol{\alpha}^{-1}=\mathbf{1}-\frac{\lambda-\mu}{2}(\boldsymbol{\alpha}^{-1}\boldsymbol{h}\otimes\boldsymbol{k}+\boldsymbol{\alpha}^{-1}\boldsymbol{k}\otimes\boldsymbol{h})$. But, from Exercise 5.2.3,

$2\boldsymbol{h}=(\lambda+\nu)\boldsymbol{\alpha}^{-1}\boldsymbol{h}+(\lambda-\nu)\boldsymbol{\alpha}^{-1}\boldsymbol{k},\quad 2\boldsymbol{k}=(\lambda-\nu)\boldsymbol{\alpha}^{-1}\boldsymbol{h}+(\lambda+\nu)\boldsymbol{\alpha}^{-1}\boldsymbol{k}.$

Inserting this into the expression for $\mu\boldsymbol{\alpha}^{-1}$ gives the result.

5. Using the results of Exercises 5.2.3 and 5.2.4, we have for every $\boldsymbol{x}$,

$$\begin{aligned}(\lambda+\nu)\mu\boldsymbol{x}^{\mathrm{T}}\boldsymbol{\alpha}^{-1}\boldsymbol{x}+\lambda\nu(\lambda-\nu)(\boldsymbol{x}^{\mathrm{T}}\boldsymbol{\alpha}^{-1}\boldsymbol{h})(\boldsymbol{x}^{\mathrm{T}}\boldsymbol{\alpha}^{-1}\boldsymbol{k})&=(\lambda+\nu)\boldsymbol{x}\cdot\boldsymbol{x}\\ &-\frac{(\lambda-\nu)^2(\lambda+\nu)}{4}\{(\boldsymbol{x}^{\mathrm{T}}\boldsymbol{\alpha}^{-1}\boldsymbol{h})^2+(\boldsymbol{x}^{\mathrm{T}}\boldsymbol{\alpha}^{-1}\boldsymbol{k})^2\}\\ &-\frac{(\lambda-\nu)(\lambda^2+\nu^2)}{2}(\boldsymbol{x}^{\mathrm{T}}\boldsymbol{\alpha}^{-1}\boldsymbol{h})(\boldsymbol{x}^{\mathrm{T}}\boldsymbol{\alpha}^{-1}\boldsymbol{k}),\end{aligned}$$

$$\begin{aligned}4(\boldsymbol{x}\cdot\boldsymbol{h})(\boldsymbol{x}\cdot\boldsymbol{k})&=\{(\lambda+\nu)\boldsymbol{x}^{\mathrm{T}}\boldsymbol{\alpha}^{-1}\boldsymbol{h}+(\lambda-\nu)\boldsymbol{x}^{\mathrm{T}}\boldsymbol{\alpha}^{-1}\boldsymbol{k}\}\{(\lambda-\nu)\boldsymbol{x}^{\mathrm{T}}\boldsymbol{\alpha}^{-1}\boldsymbol{h}\\ &\quad+(\lambda+\nu)\boldsymbol{x}^{\mathrm{T}}\boldsymbol{\alpha}^{-1}\boldsymbol{k}\}\\ &=(\lambda^2-\nu^2)\{(\boldsymbol{x}^{\mathrm{T}}\boldsymbol{\alpha}^{-1}\boldsymbol{h})^2+(\boldsymbol{x}^{\mathrm{T}}\boldsymbol{\alpha}^{-1}\boldsymbol{k})^2\}\\ &\quad+2(\lambda^2+\nu^2)(\boldsymbol{x}^{\mathrm{T}}\boldsymbol{\alpha}^{-1}\boldsymbol{h})(\boldsymbol{x}^{\mathrm{T}}\boldsymbol{\alpha}^{-1}\boldsymbol{k}).\end{aligned}$$

Hence,

$$\begin{aligned}&(\lambda+\nu)\mu\boldsymbol{x}^{\mathrm{T}}\boldsymbol{\alpha}^{-1}\boldsymbol{x}+\lambda\nu(\lambda-\nu)(\boldsymbol{x}^{\mathrm{T}}\boldsymbol{\alpha}^{-1}\boldsymbol{h})(\boldsymbol{x}^{\mathrm{T}}\boldsymbol{\alpha}^{-1}\boldsymbol{k})\\ &\quad=(\lambda+\nu)\boldsymbol{x}\cdot\boldsymbol{x}-(\lambda-\nu)(\boldsymbol{x}\cdot\boldsymbol{h})(\boldsymbol{x}\cdot\boldsymbol{k}),\end{aligned}$$

from which the identity follows.

Chapter 6

6.1

1. (a) $\omega^2 = \beta^2 + \alpha^2 k^2$, (b) $\omega^2 = \dfrac{\alpha^2 k^2}{1+\beta^2 k^2}$, (c) $\omega = \dfrac{k^2}{\alpha}$.
2. (a) $v^2 = \dfrac{\alpha^2\omega^2}{\omega^2-\beta^2}$, $(\omega^2 > \beta^2)$, (b) $v^2 = \alpha^2 - \beta^2\omega^2$, $\left(\omega^2 < \dfrac{\alpha^2}{\beta^2}\right)$,

   (c) $v^2 = \dfrac{\omega}{\alpha}$, $(\omega > 0)$.

6.2

1. $(k^2\mathbf{1} - \boldsymbol{k}\otimes\boldsymbol{k} - \omega^2\boldsymbol{\beta})\boldsymbol{A} = \mathbf{0}$,
   or $(\mathbf{1} - \mathbf{n}\otimes\mathbf{n} - v^2\boldsymbol{\beta})\boldsymbol{A} = \mathbf{0}$.
   Clearly, $v^2 = 0$, $\boldsymbol{A} = \mathbf{n}$ is a solution. We have $v^2\bar{\boldsymbol{A}}^{\mathrm{T}}\boldsymbol{\beta}\boldsymbol{A} = (\mathbf{n}\times\bar{\boldsymbol{A}})\cdot(\mathbf{n}\times\boldsymbol{A})$, and hence the two other values of v^2 are positive.

6.3

1. $k^2 = \mathrm{i}\alpha\omega$.
   For ω real and positive, $k = \pm(\tfrac{1}{2}\alpha\omega)^{1/2}(1+\mathrm{i})$. With the + sign,

$$u(\boldsymbol{x},t) = \exp\left(-\left(\frac{\alpha\omega}{2}\right)^{1/2}\mathbf{n}\cdot\boldsymbol{x}\right)\left\{A\exp\mathrm{i}\left(\left(\frac{\alpha\omega}{2}\right)^{1/2}\mathbf{n}\cdot\boldsymbol{x} - \omega t\right)\right\}.$$

   For k real, $\omega = -\mathrm{i}\alpha k^{-2}$ and thus $\omega^+ = 0$ (no propagation of damped waves).
2. $k^2c^2 = (\omega + \mathrm{i}\alpha)^2$, $kc = \pm(\omega + \mathrm{i}\alpha)$.
   For ω real, $k^+c = \pm\omega$, $k^-c = \pm\alpha$. With the + sign,

$$u(\boldsymbol{x},t) = \exp\left(-\frac{\alpha}{c}\mathbf{n}\cdot\boldsymbol{x}\right)\left\{A\exp\mathrm{i}\omega\left(\frac{1}{c}\mathbf{n}\cdot\boldsymbol{x} - t\right)\right\}^+ \quad \text{(attenuated wave).}$$

   Note that $\omega k^-(k^+)^{-1} = \alpha > 0$.
   For k real, $\omega^+ = \pm kc$, $\omega^- = -\alpha$. With the + sign for ω^+,

$$u(\boldsymbol{x},t) = \exp(-\alpha t)\{A\exp ik(\mathbf{n}\cdot\boldsymbol{x} - ct)\}^+ \quad \text{(damped wave).}$$

   Note that $\omega^- = -\alpha < 0$.

6.4

1. For $m = 1$, the ellipse of $\boldsymbol{C}$ is a circle, and the planes of constant phase are always orthogonal to the planes of constant amplitude. Assuming $m \neq 1$, the planes of constant phase are orthogonal to the planes of constant amplitude if and only if N^{-2} is real (positive or negative). From (6.3.21), the condition for N^{-2} to be real is $\hat{\boldsymbol{m}}^{\mathrm{T}}\boldsymbol{\alpha}\hat{\mathbf{n}} = 0$.

   Hence, $\hat{\mathbf{m}}$ and $\hat{\mathbf{n}}$ are along the principal axes of a section of the ellipsoid $\boldsymbol{x}^{\mathrm{T}}\boldsymbol{\alpha}\boldsymbol{x} = 1$. The ellipse of $\boldsymbol{C}$ must be chosen to be coaxial with a section of this ellipsoid.

2. $\alpha^2\omega^2\boldsymbol{S}\cdot\boldsymbol{S} = \omega^2 - \beta^2$, $N^{-2} = \dfrac{\alpha^2\omega^2(m^2-1)}{\omega^2-\beta^2}$,

$$\boldsymbol{S} = \pm\frac{1}{\alpha\omega}\left(\frac{\omega^2-\beta^2}{m^2-1}\right)^{1/2}(m\hat{\mathbf{m}} + \mathrm{i}\hat{\mathbf{n}}) \text{ for } \omega^2 > \beta^2,$$

$$\boldsymbol{S} = \pm\frac{1}{\alpha\omega}\left(\frac{\beta^2-\omega^2}{m^2-1}\right)^{1/2}(\mathrm{i}m\hat{\mathbf{m}} - \hat{\mathbf{n}}) \text{ for } \omega^2 < \beta^2.$$

3. $\omega^2\boldsymbol{S}\cdot\boldsymbol{S} = \mathrm{i}\alpha\omega$, $N^{-2} = -\dfrac{\mathrm{i}\omega(m^2-1)}{\alpha}$,

$$\boldsymbol{S} = \pm\left(\frac{\alpha}{\omega(m^2-1)}\right)^{1/2}\exp(\mathrm{i}\tfrac{1}{4}\pi)(m\hat{\mathbf{m}} + \mathrm{i}\hat{\mathbf{n}}), \text{ (taking } \omega > 0),$$

$$\boldsymbol{S}^+ = \pm\left(\frac{\alpha}{2\omega(m^2-1)}\right)^{1/2}(m\hat{\mathbf{m}} - \hat{\mathbf{n}}),$$

$$\boldsymbol{S}^- = \pm\left(\frac{\alpha}{2\omega(m^2-1)}\right)^{1/2}(m\hat{\mathbf{m}} + \hat{\mathbf{n}}).$$

6.5

1. Propagation condition: $\boldsymbol{A} \times \boldsymbol{S} + \boldsymbol{A} = \mathbf{0}$. Let $\boldsymbol{S} = N\boldsymbol{C} = N(m\hat{\mathbf{m}} + \mathrm{i}\hat{\mathbf{n}})$. The propagation condition becomes: $\boldsymbol{A} \times \boldsymbol{C} + N^{-1}\boldsymbol{A} = \mathbf{0}$. Using the results of Example 2.2 (section 2.9), we have nontrivial solutions $\boldsymbol{A}$, when $N^{-2} = -\boldsymbol{C}\cdot\boldsymbol{C}$. Then $\boldsymbol{A} = \alpha\boldsymbol{C} \times (\bar{\boldsymbol{C}} + N\boldsymbol{C} \times \bar{\boldsymbol{C}})$. Thus $N = \pm\mathrm{i}(m^2-1)^{-1/2}$, $\boldsymbol{S} = \pm\mathrm{i}(m^2-1)^{-1/2}(m\hat{\mathbf{m}} + \mathrm{i}\hat{\mathbf{n}})$, $\boldsymbol{A}$ is parallel to the bivector $\pm m(m^2-1)^{-1/2}(m^{-1}\hat{\mathbf{m}} + \mathrm{i}\hat{\mathbf{n}}) - \hat{\mathbf{m}} \times \hat{\mathbf{n}}$.
2. Propagation condition: $\{(\boldsymbol{S}\cdot\boldsymbol{S})\mathbf{1} - \boldsymbol{S}\otimes\boldsymbol{S} - \boldsymbol{\beta}\}\boldsymbol{A} = \mathbf{0}$. Let $\boldsymbol{S} = N\boldsymbol{C}$,

$\boldsymbol{C}\cdot\boldsymbol{C}=0$. The propagation condition becomes $\boldsymbol{C}(\boldsymbol{C}\cdot\boldsymbol{A})+N^{-2}\boldsymbol{\beta}\boldsymbol{A}=\boldsymbol{0}$. Thus $\boldsymbol{A}$ is parallel to $\boldsymbol{\beta}^{-1}\boldsymbol{C}$ and $N^{-2}=-\boldsymbol{C}^{\mathrm{T}}\boldsymbol{\beta}^{-1}\boldsymbol{C}$.

6.6

1. The propagation condition yields $c^2\boldsymbol{S}\cdot\boldsymbol{S}=1+\mathrm{i}\alpha\Omega^{-1}=1+\mathrm{i}\alpha\Omega^{-1}\mathrm{e}^{-\mathrm{i}\delta}$. Hence

$$N^2=c^{-2}(m^2-1)^{-1}\{1+\alpha\Omega^{-1}\sin\delta+\mathrm{i}\alpha\Omega^{-1}\cos\delta\},$$

and thus

$$T^2=c^{-2}(m^2-1)^{-1}\sqrt{1+2\alpha\Omega^{-1}\sin\delta+(\alpha\Omega^{-1})^2},$$

$$\tan 2\phi=\frac{\alpha\Omega^{-1}\cos\delta}{1+\alpha\Omega^{-1}\sin\delta}.$$

For the most acute θ, $\phi=\frac{1}{4}\pi-\delta$ so that $\tan 2\phi=(\tan 2\delta)^{-1}$, and hence the most acute θ occurs when δ satisfies $2\sin^2\delta+\alpha\Omega^{-1}\sin\delta-1=0$.

2. The propagation condition yields $\boldsymbol{K}^{\mathrm{T}}\boldsymbol{\alpha}\boldsymbol{K}=\omega^2\boldsymbol{S}^{\mathrm{T}}\boldsymbol{\alpha}\boldsymbol{S}=\mathrm{i}\beta\omega$. Hence,

$$N^{-2}=\frac{\Omega}{\beta}(\boldsymbol{C}^{\mathrm{T}}\boldsymbol{\alpha}\boldsymbol{C})\exp\mathrm{i}(\delta-\tfrac{1}{2}\pi),$$

and thus

$$T^{-2}=\frac{\Omega}{\beta}\{(\boldsymbol{C}^{\mathrm{T}}\boldsymbol{\alpha}\boldsymbol{C})(\bar{\boldsymbol{C}}^{\mathrm{T}}\boldsymbol{\alpha}\bar{\boldsymbol{C}})\}^{1/2}$$

$$e^{-2\mathrm{i}\phi}=\boldsymbol{C}^{\mathrm{T}}\boldsymbol{\alpha}\boldsymbol{C}\exp[\mathrm{i}(\delta-\tfrac{1}{2}\pi)]\{(\boldsymbol{C}^{\mathrm{T}}\boldsymbol{\alpha}\boldsymbol{C})(\bar{\boldsymbol{C}}^{\mathrm{T}}\boldsymbol{\alpha}\bar{\boldsymbol{C}})\}^{-1/2}.$$

Note: $\boldsymbol{C}^{\mathrm{T}}\boldsymbol{\alpha}\boldsymbol{C}=m^2\hat{\mathbf{m}}^{\mathrm{T}}\boldsymbol{\alpha}\hat{\mathbf{m}}-\hat{\mathbf{n}}^{\mathrm{T}}\boldsymbol{\alpha}\hat{\mathbf{n}}+2\mathrm{i}m\hat{\mathbf{m}}^{\mathrm{T}}\boldsymbol{\alpha}\hat{\mathbf{n}}$. Let v be the argument of $\boldsymbol{C}^{\mathrm{T}}\boldsymbol{\alpha}\boldsymbol{C}$: $\tan v=2m\hat{\mathbf{m}}^{\mathrm{T}}\boldsymbol{\alpha}\hat{\mathbf{n}}\{m^2\hat{\mathbf{m}}^{\mathrm{T}}\boldsymbol{\alpha}\hat{\mathbf{m}}-\hat{\mathbf{n}}^{\mathrm{T}}\boldsymbol{\alpha}\hat{\mathbf{n}}\}^{-1}$. Then, $\phi=\frac{1}{4}\pi-\frac{1}{2}(\delta+v)$, or $\frac{5}{4}\pi-\frac{1}{2}(\delta+v)$.

Chapter 7

7.1

1. (a) $\delta_2-\delta_1=0$ or π;
   (b) $a_1=a_2$ and $\delta_2-\delta_1=\pm\frac{1}{2}\pi$.

7.2

1. (a) $\tan\alpha = \frac{1}{3}\sqrt{3}$, $\alpha = \frac{1}{3}\pi$, $\delta = \frac{1}{2}\pi$;
$\sin 2\beta = \sin 2\alpha \sin\delta = \frac{1}{2}\sqrt{3}$, $\beta = \frac{1}{6}\pi$, $I = 4a^2$;
$\cos 2\chi = (\cos 2\alpha)(\cos 2\beta)^{-1} = -1$, $\chi = \frac{1}{2}\pi$.
(b) $\tan\alpha = 1$, $\alpha = \frac{1}{4}\pi$, $\delta = \frac{1}{3}\pi$;
$\beta = \frac{1}{6}\pi$, $\chi = \frac{1}{4}\pi$, $I = 2a^2$.
2. (a) $\beta = 0$, $\chi = 0$;
(b) $\beta = 0$, $\chi = \pm\frac{1}{2}\pi$;
(c) $\beta = \frac{1}{4}\pi$ or $\beta = -\frac{1}{4}\pi$ (χ undefined).
3. $\beta_* = -\beta$, $\chi_* = \chi + \frac{1}{2}\pi$;
$\boldsymbol{A}'' = (\cos\beta\cos\chi - \mathrm{i}\sin\beta\sin\chi)\mathbf{i} + (\cos\beta\sin\chi + \mathrm{i}\sin\beta\cos\chi)\mathbf{j}$,
$\boldsymbol{A}''_* = (-\cos\beta\sin\chi + \mathrm{i}\sin\beta\cos\chi)\mathbf{i} + (\cos\beta\cos\chi + \mathrm{i}\sin\beta\sin\chi)\mathbf{j}$.

7.3

1. (a) $I = 4a^2$, $Q = -2a^2$, $U = 0$, $V = 2\sqrt{3}a^2$.
(b) $I = 2a^2$, $Q = 0$, $U = a^2$, $V = \sqrt{3}a^2$.
2. Stokes parameters corresponding to $\boldsymbol{P}'$:
$s_0^{(1)} = I'^{(1)} = 1$, $s_1^{(1)} = Q'^{(1)} = \cos 2\beta'\cos 2\chi'$, $s_2^{(1)} = U'^{(1)} = \cos 2\beta'\sin 2\chi'$,
$s_3^{(1)} = V'^{(1)} = \sin 2\beta'$;
Stokes parameters corresponding to $\boldsymbol{Q}'$:
$s_0^{(2)} = I'^{(2)} = 1$, $s_1^{(2)} = Q'^{(2)} = -\cos 2\beta'\cos 2\chi'$,
$s_2^{(2)} = U'^{(2)} = -\cos 2\beta'\sin 2\chi'$, $s_3^{(2)} = V'^{(2)} = -\sin 2\beta'$;
$$\left.\begin{aligned} I^{(1)} &= \tfrac{1}{2}(II'^{(1)} + QQ'^{(1)} + UU'^{(1)} + VV'^{(1)}) = \tfrac{1}{2}s_A s_A^{(1)},\\ I^{(2)} &= \tfrac{1}{2}(II'^{(2)} + QQ'^{(2)} + UU'^{(2)} + VV'^{(2)}) = \tfrac{1}{2}s_A s_A^{(2)}.\end{aligned}\right. \quad (A = 0, 1, 2, 3)$$
3. $H_{\alpha\beta} = A_\alpha \bar{A}_\beta$, $H'^{(1)}_{\alpha\beta} = P'_\alpha \bar{P}'_\beta$, $H'^{(2)}_{\alpha\beta} = Q'_\alpha \bar{Q}'_\beta$,
$I^{(1)} = \lambda'\bar{\lambda}' = A_\alpha \bar{P}'_\alpha \bar{A}_\beta P'_\beta = \mathrm{tr}(\boldsymbol{H}\boldsymbol{H}'^{(1)})$,
$I^{(2)} = \mu'\bar{\mu}' = A_\alpha \bar{Q}'_\alpha \bar{A}_\beta Q'_\beta = \mathrm{tr}(\boldsymbol{H}\boldsymbol{H}'^{(2)})$.
Check:
$$\boldsymbol{H}'^{(1)} = \frac{1}{2}\begin{pmatrix} I'^{(1)} + Q'^{(1)} & U'^{(1)} - \mathrm{i}V'^{(1)} \\ U'^{(1)} + \mathrm{i}V'^{(1)} & I'^{(1)} - Q'^{(1)} \end{pmatrix};$$
using (7.3.21), computation of $I^{(1)} = \mathrm{tr}(\boldsymbol{H}\boldsymbol{H}'^{(1)})$ yields the result of Exercise 7.3.2.

4. Compute the traces using (7.3.19) and (7.3.23).

5. $I_q = 1$, $\beta_q = 0$, $\chi_q = 0$ (linear polarization along the x-axis);
$I_u = 1$, $\beta_u = 0$, $\chi_u = \frac{1}{4}\pi$ (linear polarization tilted at $\frac{1}{4}\pi$ to the positive x-axis);
$I_v = 1$, $\beta_v = \frac{1}{4}\pi$, χ_v indeterminate (right-handed circular polarization).
6. From (7.3.22), we have $A_\alpha \bar{A}_1 = H_{\alpha 1}$ and $A_\alpha \bar{A}_2 = H_{\alpha 2}$, so that, up to a scalar factor, the components of the bivector A are given by the elements of the first (or the second) column of the hermitian matrix $\boldsymbol{H}$.
Thus $A = \alpha\{(I+Q)\mathbf{i} + (U + \mathrm{i}V)\mathbf{j}\} = \beta\{(U - \mathrm{i}V)\mathbf{i} + (I - Q)\mathbf{j}\}$. But $A \cdot \bar{A} = I$ and hence $2\alpha\bar{\alpha} = (1+Q)^{-1}$, $2\beta\bar{\beta} = (\mathrm{I} - \mathrm{Q})^{-1}$. Hence,

$$A = \frac{e^{\mathrm{i}\phi}}{(2(I+Q))^{1/2}}\{(I+Q)\mathbf{i} + (U+\mathrm{i}V)\mathbf{j}\},\ (\phi \text{ arbitrary}),$$

or

$$A = \frac{e^{\mathrm{i}\psi}}{(2(I-Q))^{1/2}}\{(U-\mathrm{i}V)\mathbf{i} + (I-Q)\mathbf{j}\},\ (\psi \text{ arbitrary}).$$

(Note that ϕ and ψ are related through $\tan(\psi - \phi) = VU^{-1}$).

7.4

1. $\tan\alpha = \dfrac{|a_R - a_L|}{a_R + a_L}$, $\delta = \pm\dfrac{\pi}{2}$,
$2\beta = \pm 2\alpha$, $2\chi = 0$ or π.
Locus: meridian passing through P_0 (intersection of the sphere with the QV-plane).
2. Equation of the straight line $P_0^* P$:

$$\frac{q+1}{Q+1} = \frac{u}{U} = \frac{v}{V}$$

(where q, u, v are the coordinates of a generic point of the straight line).
Intersection P' with the UV-plane has coordinates:

$$u = \frac{U}{Q+1},\ v = \frac{V}{Q+1}.$$

Using (7.3.21) with $a_1^2 + a_2^2 = 1$, we have $Q + 1 = 2a_1^2$, and

$U+\mathrm{i}V=2a_1a_2e^{\mathrm{i}\delta}$. Hence,

$$u+\mathrm{i}v=\frac{a_2}{a_1}e^{\mathrm{i}\delta}=\frac{a_2e^{\mathrm{i}\delta_2}}{a_1e^{\mathrm{i}\delta_1}}=\frac{A_2}{A_1}.$$

7.5

1. $\tilde{s}_A\gamma_{AB}\tilde{s}_B=s_C\mu_{AC}\gamma_{AB}\mu_{BD}s_D=0$ under the condition $s_A\gamma_{AB}s_B=0$. Thus, introducing a multiplier λ, $s_C(\mu_{AC}\gamma_{AB}\mu_{BD}-\lambda\gamma_{CD})s_D=0$, for every s_C. Hence $\mu_{AC}\gamma_{AB}\mu_{BD}=\lambda\gamma_{CD}$.
2. The Pauli matrices have the property: $\mathbf{\Pi}_i\mathbf{\Pi}_j=\delta_{ij}\mathbf{1}+i\varepsilon_{ijk}\mathbf{\Pi}_k$ $(i,j,k=1,2,3)$. Computing the left-hand side we obtain the following results:
all indices 0: $\frac{1}{2}\operatorname{tr}\mathbf{\Pi}_0=1$, three indices 0: $\frac{1}{2}\operatorname{tr}\mathbf{\Pi}_i=0$,
two indices 0: $\frac{1}{2}\operatorname{tr}(\mathbf{\Pi}_i\mathbf{\Pi}_j)=\delta_{ij}$, one index 0: $\frac{1}{2}\operatorname{tr}(\mathbf{\Pi}_i\mathbf{\Pi}_j\mathbf{\Pi}_k)=ie_{ijk}$,
no index 0:

$$\begin{aligned}\tfrac{1}{2}\operatorname{tr}(\mathbf{\Pi}_i\mathbf{\Pi}_j\mathbf{\Pi}_k\mathbf{\Pi}_l)&=\tfrac{1}{2}\operatorname{tr}(\delta_{ij}\delta_{kl}\mathbf{1}+ie_{ijp}\delta_{kl}\mathbf{\Pi}_p+ie_{klq}\delta_{ij}\mathbf{\Pi}_q-e_{ijp}e_{klq}\mathbf{\Pi}_p\mathbf{\Pi}_q)\\&=\delta_{ij}\delta_{kl}-e_{ijp}e_{klp}=\delta_{ij}\delta_{kl}-\delta_{ik}\delta_{jl}+\delta_{il}\delta_{kj}.\end{aligned}$$

Then, it is easily checked that the right-hand side gives the same results.
3. From (7.5.28), we have $\mu_{AB}s_B^{(1)}=\alpha\bar{\alpha}s_A^{(1)}$, $\mu_{AB}s_B^{(2)}=\beta\bar{\beta}s_B^{(2)}$.
4. $\sigma_1=\dfrac{Q}{I}=\dfrac{1-\zeta\bar{\zeta}}{1+\zeta\bar{\zeta}}$, $\sigma_2=\dfrac{U}{I}=\dfrac{\zeta+\bar{\zeta}}{1+\zeta\bar{\zeta}}$, $\sigma_3=\dfrac{V}{I}=\mathrm{i}\dfrac{\bar{\zeta}-\zeta}{1+\zeta\bar{\zeta}}$.
5. $$\boldsymbol{J}=\frac{1}{\zeta^{(1)}-\zeta^{(2)}}\begin{bmatrix}\beta\zeta^{(1)}-\alpha\zeta^{(2)} & \alpha-\beta\\ -(\alpha-\beta)\zeta^{(1)}\zeta^{(2)} & \alpha\zeta^{(1)}-\beta\zeta^{(2)}\end{bmatrix}.$$
6. $\bar{\tilde{\boldsymbol{A}}}\cdot\tilde{\boldsymbol{A}}=\bar{\boldsymbol{A}}^{\mathrm{T}}\bar{\boldsymbol{J}}^{\mathrm{T}}\boldsymbol{J}\boldsymbol{A}=\bar{\boldsymbol{A}}\cdot\boldsymbol{A}$ for every $\boldsymbol{A}$, and hence $\bar{\boldsymbol{J}}^{\mathrm{T}}\boldsymbol{J}=\mathbf{1}$.
$\mu_{A0}=\frac{1}{2}\operatorname{tr}(\mathbf{\Pi}_A\boldsymbol{J}\bar{\boldsymbol{J}}^{\mathrm{T}})=\frac{1}{2}\operatorname{tr}\mathbf{\Pi}_A=\delta_{A0}$, $\mu_{0B}=\frac{1}{2}\operatorname{tr}(\boldsymbol{J}\mathbf{\Pi}_B\bar{\boldsymbol{J}}^{\mathrm{T}})=\frac{1}{2}\operatorname{tr}\mathbf{\Pi}_B=\delta_{0B}$.
Then, from (7.5.8), it follows that μ_{ij} is orthogonal. Thus, (7.5.13) reduces to $\tilde{\sigma}_i=\mu_{ij}\sigma_j$ with μ_{ij} orthogonal: rotation on the Poincaré sphere.

7.6

1. Let 2θ and 2ϕ be the longitude and latitude of the point P_1 on the Poincaré sphere. Then $s_A^{(1)}=(1,\cos 2\phi\cos 2\theta,\cos 2\phi\sin 2\theta,\sin 2\phi)$. We have $\tilde{I}=\frac{1}{2}s_B^{(1)}s_B$, where s_B are the Stokes parameters of

the incident wave given by (7.6.6). Hence

$$\tilde{I} = \tfrac{1}{2} + \tfrac{1}{2}\cos 2\phi \cos 2\beta \cos 2(\chi - \theta) + \tfrac{1}{2}\sin 2\phi \sin 2\beta.$$

But, in the spherical triangle PNP_1 (N: north pole), $\widehat{NP}_1 = \tfrac{1}{2}\pi - 2\phi$, $\widehat{NP} = \tfrac{1}{2}\pi - 2\beta$, and the angle $\hat{N} = 2(\chi - \theta)$, so that (7.6.8) yields

$$\cos \widehat{P_1P} = \sin 2\phi \sin 2\beta + \cos 2\phi \cos 2\beta \cos 2(\chi - \theta).$$

Hence $\tilde{I} = \tfrac{1}{2}(1 + \cos(\widehat{P_1P})) = \cos^2 \tfrac{1}{2}(\widehat{P_1P})$.

2. (a) Using (7.6.11) and (7.6.13), we get

$$\boldsymbol{J} = \frac{1}{2}\begin{bmatrix} 1 & -\mathrm{i} \\ \mathrm{i} & 1 \end{bmatrix}, \quad \boldsymbol{\mu} = \frac{1}{2}\begin{bmatrix} 1 & 0 & 0 & 1 \\ 0 & 0 & 0 & 0 \\ 0 & 0 & 0 & 0 \\ 1 & 0 & 0 & 1 \end{bmatrix}.$$

(b) Using (7.6.11) and (7.6.13), we get

$$\boldsymbol{J} = \frac{2^{1/2}}{2}\begin{bmatrix} 1 + \mathrm{i}\cos 2\theta & \mathrm{i}\sin 2\theta \\ \mathrm{i}\sin 2\theta & 1 - \mathrm{i}\cos 2\theta \end{bmatrix},$$

$$\boldsymbol{\mu} = \begin{bmatrix} 1 & 0 & 0 & 0 \\ 0 & \cos^2 2\theta & \sin 2\theta \cos 2\theta & -\sin 2\theta \\ 0 & \sin 2\theta \cos 2\theta & \sin^2 2\theta & \cos 2\theta \\ 0 & \sin 2\theta & -\cos 2\theta & 0 \end{bmatrix}.$$

3. From (7.6.11) with $\boldsymbol{P}\cdot\bar{\boldsymbol{P}} = 1$, we have

$$\boldsymbol{J} = e^{\mathrm{i}\varphi}\boldsymbol{H} + e^{\mathrm{i}\varphi}\boldsymbol{H}',$$

where $\boldsymbol{H}$ and $\boldsymbol{H}'$ are the hemitian matrices $H_{\alpha\beta} = P_\alpha \bar{P}_\beta$, $H'_{\alpha\beta} = P'_\alpha \bar{P}'_\beta$. But $\boldsymbol{H} = \tfrac{1}{2}(s_0^{(1)}\boldsymbol{\Pi}_0 + s_i^{(1)}\boldsymbol{\Pi}_i)$, $\boldsymbol{H}' = \tfrac{1}{2}(s_0^{(1)}\boldsymbol{\Pi}_0 - s_i^{(1)}\boldsymbol{\Pi}_i)$, and $s_0^{(1)} = 1$.
Hence, $\boldsymbol{J} = \cos\varphi\boldsymbol{1} + \mathrm{i}\sin\varphi s_i^{(1)}\boldsymbol{\Pi}_i$.

4. $\boldsymbol{J} = [\boldsymbol{P}\cdot\bar{\boldsymbol{P}}]^{-1}(e^k\boldsymbol{P}\otimes\bar{\boldsymbol{P}} + \mathrm{e}^{-k}\bar{\boldsymbol{P}}'\otimes\boldsymbol{P}')$,
and using the same procedure as for Exercise 7.6.3, with $\boldsymbol{P}\cdot\bar{\boldsymbol{P}} = 1$, $\boldsymbol{J} = (\cosh k)\boldsymbol{\Pi}_0 + (\sinh k)s_i^{(1)}\boldsymbol{\Pi}_i$, $(i = 1, 2, 3)$. From (7.5.28), on using $s_A^{(2)} = \gamma_{AB}s_B^{(1)}$, we find $\mu_{AB} = \delta_{AB} + \tfrac{1}{2}(\mathrm{e}^{2k} - 1)s_A^{(1)}s_B^{(1)} + \tfrac{1}{2}(\mathrm{e}^{-2k} - 1)\cdot \gamma_{AC}s_C^{(1)}\gamma_{BD}s_D^{(1)}$, and thus $\mu_{00} = \cosh 2k$, $\mu_{i0} = \mu_{0i} = (\sinh 2k)s_i^{(1)}$, $\mu_{ij} = \delta_{ij} + (\cosh 2k - 1)\, s_i^{(1)}s_j^{(1)}$.

Chapter 8

8.1

These results are still valid: $\boldsymbol{\nu}$ and β are independent of ω. The coefficient of $\boldsymbol{\nu}$ in equation (8.3.8) is replaced by

$$f(\omega)\overline{f(\omega)}\exp[\mathrm{i}(\omega\boldsymbol{S}-\bar{\omega}\bar{\boldsymbol{S}})\cdot\boldsymbol{x}-\mathrm{i}(\omega-\bar{\omega})t],$$

with a similar replacement in the coefficient of β in equation (8.3.9). Equation (8.3.12) is replaced by

$$\boldsymbol{\nu}\cdot(\omega\boldsymbol{S}-\bar{\omega}\bar{\boldsymbol{S}})=(\omega-\bar{\omega})\beta.$$

Because this holds for arbitrary ω, it holds for $\omega+a$ where a is an arbitrary real constant. Hence,

$$\boldsymbol{\nu}\cdot\boldsymbol{S}=\boldsymbol{\nu}\cdot\bar{\boldsymbol{S}},$$

and thus we recover (8.3.17):

$$\boldsymbol{\nu}\cdot\boldsymbol{S}=\beta.$$

8.2

1. Let the rate of energy dissipation per unit volume per unit time be denoted by d. Then in place of (8.1.1) we have

$$\frac{\partial}{\partial t}\iiint_V \mathrm{e}\,\mathrm{d}v=-\iint_S \boldsymbol{r}\cdot\boldsymbol{n}\,\mathrm{d}s-\iiint_V d\,\mathrm{d}v,$$

and hence, in place of (8.1.2) we have

$$\nabla\cdot\boldsymbol{r}+\frac{\partial e}{\partial t}+d=0.$$

2. The form of d is similar to that of e in (8.3.9), with B and β replaced by D and δ (say), respectively. The mean value of d, say $\tilde{d}$, is then

$$\tilde{d}=f\bar{f}\delta\exp\mathrm{i}\omega(\boldsymbol{S}-\bar{\boldsymbol{S}})\cdot\boldsymbol{x}.$$

Then, in place of (8.3.13) we have

$$\mathrm{i}\omega\boldsymbol{\nu}\cdot(\boldsymbol{S}-\bar{\boldsymbol{S}})=-\delta,\quad\text{or}\quad \boldsymbol{\nu}\cdot\boldsymbol{S}^-=\frac{\delta}{2\omega},$$

and hence

$$\tilde{\boldsymbol{r}}\cdot\boldsymbol{S}^- = \frac{\tilde{d}}{2\omega}.$$

Note that, for dissipative systems, the assumption, introduced in section 8.3, that the propagation condition does not involve ω is in general not valid. However, the derivation of (8.3.13) remains valid even when the amplitude bivector $\boldsymbol{A}$ of the wave is frequency dependent, in which case $\boldsymbol{\Gamma}$, $\boldsymbol{v}$, B, and β, all depend on ω. It is thus justified to use this type of derivation in the case of dissipative systems. However, the derivation of (8.3.19) uses explicitly the fact that these quantities do not depend on ω, and thus it may not be generalized to the case of dissipative systems.

Chapter 9

9.1

1. $\mathrm{tr}(\boldsymbol{\Pi}\boldsymbol{\kappa}^{-1}) = (\boldsymbol{C}\cdot\boldsymbol{C})\mathrm{tr}(\boldsymbol{\kappa}^{-1}) - \boldsymbol{C}^{\mathrm{T}}\boldsymbol{\kappa}^{-1}\boldsymbol{C} = I_Q$, $\mathrm{tr}(\boldsymbol{\Pi}_*\boldsymbol{\kappa}_*^{-1}) = (\boldsymbol{C}\cdot\boldsymbol{C})$ $(\det\boldsymbol{\kappa})^{-1}\mathrm{tr}\{(\boldsymbol{C}\otimes\boldsymbol{C})\boldsymbol{\kappa}\} = (\det\boldsymbol{\kappa})^{-1}(\boldsymbol{C}\cdot\boldsymbol{C})(\boldsymbol{C}^{\mathrm{T}}\boldsymbol{\kappa}\boldsymbol{C}) = II_Q$, $\det(\boldsymbol{\Pi}\boldsymbol{\kappa}^{-1}) = 0 = III_Q$.

9.2

1. From the propagation condition (9.2.13), $\mu_0\boldsymbol{E}^{\mathrm{T}}\boldsymbol{\kappa}\bar{\boldsymbol{E}} = (\boldsymbol{S}\cdot\boldsymbol{S})(\boldsymbol{E}\cdot\bar{\boldsymbol{E}}) - (\boldsymbol{S}\cdot\bar{\boldsymbol{E}})(\boldsymbol{S}\cdot\boldsymbol{E})$. Hence, (9.5.9) becomes $2\mu_0\hat{e} = (\boldsymbol{S}\cdot\boldsymbol{S}^+)(\boldsymbol{E}\cdot\bar{\boldsymbol{E}}) - (\boldsymbol{S}\cdot\bar{\boldsymbol{E}})\ (\boldsymbol{S}^+\cdot\boldsymbol{E})$. Thus, from (9.5.8), $\mu_0\hat{\boldsymbol{r}}\cdot\boldsymbol{S} = 2\mu_0\hat{e}$. Hence, $\hat{\boldsymbol{r}}\cdot\boldsymbol{S}^+ = \hat{e}$ and $\hat{\boldsymbol{r}}\cdot\boldsymbol{S}^- = 0$, because $\hat{\boldsymbol{r}}$ and $\hat{e}$ are real.
2. From (9.5.7), $\boldsymbol{S}\cdot(\boldsymbol{E}\times\bar{\boldsymbol{H}})$ is real. Hence, $\boldsymbol{S}^+\cdot(\boldsymbol{E}\times\bar{\boldsymbol{H}})^- + \boldsymbol{S}^-\cdot(\boldsymbol{E}\times\bar{\boldsymbol{H}})^+ = 0$. But $\hat{\boldsymbol{r}}\cdot\boldsymbol{S}^- = \frac{1}{2}(\boldsymbol{E}\times\bar{\boldsymbol{H}})^+\cdot\boldsymbol{S}^- = 0$.

9.3

1. Taking the cross product of $\hat{\mathbf{n}}$ with (9.2.7), we obtain (since $\boldsymbol{C} = \hat{\mathbf{n}}$), $N\hat{\mathbf{n}}\times(\hat{\mathbf{n}}\times\boldsymbol{H}) = -N\boldsymbol{H} = -\hat{\mathbf{n}}\times\boldsymbol{D}$, and hence $\boldsymbol{H} = N^{-1}\hat{\mathbf{n}}\times\boldsymbol{D}$. Also, $\tilde{\boldsymbol{r}} = \frac{1}{2}\boldsymbol{E}\times\boldsymbol{H} = \frac{1}{2}\boldsymbol{\kappa}^{-1}\boldsymbol{D}\times(N^{-1}\hat{\mathbf{n}}\times\boldsymbol{D}) = \frac{1}{2}N^{-1}\{(\boldsymbol{D}^{\mathrm{T}}\boldsymbol{\kappa}^{-1}\boldsymbol{D})\hat{\mathbf{n}} - \boldsymbol{D}(\hat{\mathbf{n}}^{\mathrm{T}}\boldsymbol{\kappa}^{-1}\boldsymbol{D})\}$.
2. $\boldsymbol{H} = N^{-1}\hat{\mathbf{n}}\times\boldsymbol{D}$. Thus

$$\boldsymbol{H}_{1,2} = N_{1,2}^{-1}\{(\sin\phi_1)^{-1}\hat{\mathbf{n}}\times\mathbf{h} \pm (\sin\phi_2)^{-1}\hat{\mathbf{n}}\times\mathbf{k}\}.$$

$$\begin{aligned}\boldsymbol{H}_1\times\boldsymbol{D}_2 &= N_1^{-1}\{(\hat{\mathbf{n}}\times\hat{\mathbf{h}}^*) + (\hat{\mathbf{n}}\times\hat{\mathbf{k}}^*)\}\times(\hat{\mathbf{h}}^* - \hat{\mathbf{k}}^*)\\ &= N_1^{-1}\{-\hat{\mathbf{n}} - \hat{\mathbf{n}}(\hat{\mathbf{h}}^*\cdot\hat{\mathbf{k}}^*) + \hat{\mathbf{n}}(\hat{\mathbf{h}}^*\cdot\hat{\mathbf{k}}^*) + \hat{\mathbf{n}}\} = 0.\end{aligned}$$

Analogously $\boldsymbol{H}_2 \times \boldsymbol{D}_1 = \boldsymbol{0}$.

3. $\lambda > \mu > \nu$.

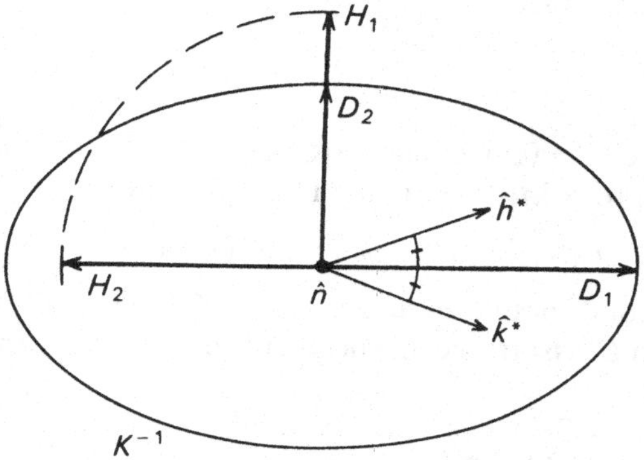

Biaxial crystals: homogeneous waves propagating in an arbitrary direction.

(Here $\boldsymbol{D}_1$. $\boldsymbol{D}_2$ are normalized by choosing $\boldsymbol{D}_1^{\mathrm{T}}\boldsymbol{\kappa}^{-1}\boldsymbol{D}_1 = \boldsymbol{D}_2^{\mathrm{T}}\boldsymbol{\kappa}^{-1}\boldsymbol{D}_2 = 1$.)

4. $\lambda = \mu > \nu$.

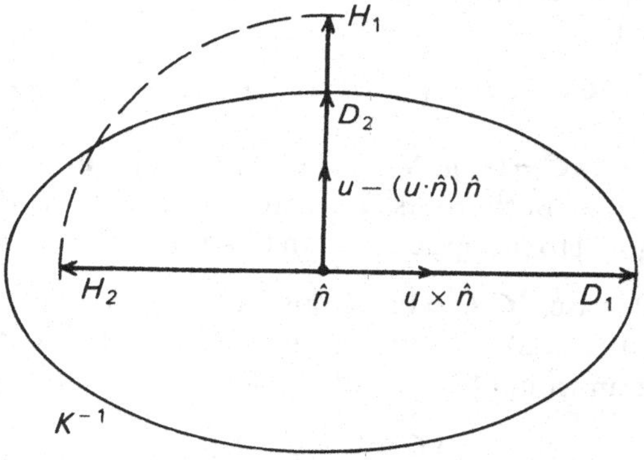

Uniaxial crystals: homogeneous waves propagating in an arbitrary direction.

(Here $\boldsymbol{D}_1$, $\boldsymbol{D}_2$ are normalized by choosing $\boldsymbol{D}_1^{\mathrm{T}}\boldsymbol{\kappa}^{-1}\boldsymbol{D}_1 = \boldsymbol{D}_2^{\mathrm{T}}\boldsymbol{\kappa}^{-1}\boldsymbol{D}_2 = 1$.)

9.4

(a) $4\hat{e} = \boldsymbol{C}^{\mathrm{T}}\boldsymbol{\kappa}^{-1}\bar{\boldsymbol{C}} + \mu_0^{-1}N\bar{N}\{(\boldsymbol{C}\cdot\bar{\boldsymbol{C}})(\boldsymbol{C}^{\mathrm{T}}\boldsymbol{\kappa}^{-2}\bar{\boldsymbol{C}}) - (\boldsymbol{C}^{\mathrm{T}}\boldsymbol{\kappa}^{-1}\bar{\boldsymbol{C}})^2\}$,

$2\hat{\boldsymbol{r}} = \mu_0^{-1}\{(\boldsymbol{C}^{\mathrm{T}}\boldsymbol{\kappa}^{-2}\bar{\boldsymbol{C}})\boldsymbol{S}^+ - (\boldsymbol{C}^{\mathrm{T}}\boldsymbol{\kappa}^{-1}\bar{\boldsymbol{C}})\boldsymbol{\kappa}^{-1}\boldsymbol{S}^+\}$.

(b) $4\hat{e} = \mu_0^{-1} C^T \kappa^2 \bar{C} + \mu_0^{-2} N \bar{N} \det \kappa^{-1} \{(C^T \cdot \kappa \bar{C})(C^T \kappa^3 \bar{C}) - (C^T \kappa^2 \bar{C})^2\}$, $2\hat{r} = \mu_0^{-2} \det \kappa^{-1} \{(C^T \kappa^3 \bar{C}) \kappa S^+ - (C^T \kappa^2 \bar{C}) \kappa^2 S^+\}$.

9.5

1. (a) When $C \cdot C = 0$, we have $((C \times \mathbf{h}) \cdot (C \times \mathbf{h}))^{1/2} = \mathrm{i}(C \cdot \mathbf{h})$ and $((C \times \mathbf{k}) \cdot (C \times \mathbf{k}))^{1/2} = \mathrm{i}(C \cdot \mathbf{k})$. Then, the roots (9.7.15) become

$$\mu_0 N_1^{-2} = 0, \quad \mu_0 N_2^{-2} = -(\lambda - \nu)(C \cdot \mathbf{h})(C \cdot \mathbf{k}).$$

The second root becomes zero when $C \cdot \mathbf{h} = 0$ or $C \cdot \mathbf{k} = 0$, that is, when C is isotropic in a plane orthogonal to an optic axis ($\mathbf{h}$ or $\mathbf{k}$).

(b) When $C^T \kappa C = 0$, we have $((C \times \mathbf{h})^T \kappa^{-1} (C \times \mathbf{h}))^{1/2} = \mathrm{i}(\lambda \mu \nu)^{1/2}$ $(\mathbf{h}^T \kappa C)$ and $((C \times \mathbf{k})^T \kappa^{-1} (C \times \mathbf{k}))^{1/2} = \mathrm{i}(\lambda \mu \nu)^{1/2}$ $(\mathbf{k}^T \kappa C)$. Then, the roots (9.7.25) become

$$\mu_0 N_1^{-2} = (\lambda - \nu) \lambda \nu (\mathbf{h}^T \kappa C)(\mathbf{k}^T \kappa C), \quad \mu_0 N_2^{-2} = 0.$$

The first root becomes zero when $\mathbf{h}^T \kappa C = 0$ or $\mathbf{k}^T \kappa C = 0$, that is, when in addition to $C^T \kappa C = 0$, C is in a plane conjugate to an optic axis ($\mathbf{h}$ or $\mathbf{k}$) with respect to the Fresnel ellipsoid $x^T \kappa x = 1$.

2. (a) When $C \cdot C = 0$, $\mu_0 N_1^{-2} = 0$ and $\mu_0 N_2^{-2} = -(\nu - \lambda)(C \cdot \mathbf{u})^2 = \lambda \nu C^T \kappa C$.

(b) When $C^T \kappa C = 0$, $\mu_0 N_1^{-2} = (\lambda \nu^{-1})(\nu - \lambda)(C \cdot \mathbf{u})^2 = \lambda C \cdot C$ and $\mu_0 N_2^{-2} = 0$. Both roots are zero when C is isotropic and orthogonal to the optic axis $\mathbf{u}$. In this case $C \cdot C = C^T \kappa C = 0$.

3. (a) Biaxial case: $C \cdot C = 0$, $(C \cdot \mathbf{h})(C \cdot \mathbf{k}) \neq 0$, $N_2^{-2} \neq 0$. Because $((C \times \mathbf{h}) \cdot (C \times \mathbf{h}))^{1/2} = \mathrm{i}(C \cdot \mathbf{h})$ and $((C \times \mathbf{k}) \cdot (C \times \mathbf{k}))^{1/2} = \mathrm{i} C \cdot \mathbf{k}$, we obtain from (9.7.16) and (9.7.17),

$$H_2 = -\mathrm{i} \left\{ \frac{C \times \mathbf{h}}{C \cdot \mathbf{h}} + \frac{C \times \mathbf{k}}{C \cdot \mathbf{k}} \right\},$$

$$D_2 = 2\mathrm{i} N_2 C, \quad H_2 \cdot H_2 = 4, \quad D_2 \cdot D_2 = 0.$$

(b) Uniaxial case: $C \cdot C = 0$, $C \cdot \mathbf{u} \neq 0$, $N_2^{-2} \neq 0$. We obtain from (9.7.28) and (9.7.29),

$$H_2 = C \times \mathbf{u}, \quad D_2 = -N_2 (C \cdot \mathbf{u}) C, \quad H_2 \cdot H_2 = -(C \cdot \mathbf{u})^2,$$
$$D_2 \cdot D_2 = 0.$$

4. Using $\Pi\mathbf{h} = (C\cdot C)\mathbf{h} - (\mathbf{h}\cdot C)$, $\Pi\mathbf{k} = (C\cdot C)\mathbf{k} - (\mathbf{k}\cdot C)C$, and $(C \times \mathbf{h})\cdot(C \times \mathbf{h}) = C\cdot C - (\mathbf{h}\cdot C)^2$, $(C \times \mathbf{k})\cdot(C \times \mathbf{k}) = C\cdot C - (\mathbf{k}\cdot C)^2$, we have

$$\boldsymbol{H}_1 \times \boldsymbol{D}_2 = N_2\left\{ -\frac{C\cdot C - (\mathbf{h}\cdot C)^2}{C\cdot C - (\mathbf{h}\cdot C)^2} + \frac{C\cdot C - (\mathbf{k}\cdot C)^2}{C\cdot C - (\mathbf{k}\cdot C)^2}\right.$$
$$\left. + \frac{(C\cdot C)(\mathbf{h}\cdot\mathbf{k}) - (C\cdot\mathbf{h})(C\cdot\mathbf{k}) - (C\cdot C)(\mathbf{h}\cdot\mathbf{k}) + (C\cdot\mathbf{h})(C\cdot\mathbf{k})}{(C\cdot C - (\mathbf{h}\cdot C)^2)^{1/2}(C\cdot C - (\mathbf{k}\cdot C)^2)^{1/2}}\right\} C = \mathbf{0}.$$

Similarly, $\boldsymbol{H}_2 \times \boldsymbol{D}_1 = \mathbf{0}$.

9.6

1. $N = \mu_0^{1/2}\{\lambda\nu(C\cdot\mathbf{h})(C^{\mathrm{T}}\boldsymbol{\kappa}\mathbf{h})\}^{1/2}$, $\hat{e} = \dfrac{\mu_0}{2\lambda\mu\nu}\left\{\mu\left|\dfrac{C\cdot\mathbf{h}}{C^{\mathrm{T}}\boldsymbol{\kappa}\mathbf{h}}\right| + \lambda\nu\right\}$,

$$2\hat{\mathbf{r}} = \mu_0^{1/2}\overline{\left\{\left(\frac{C\cdot\mathbf{h}}{\lambda\nu C^{\mathrm{T}}\boldsymbol{\kappa}\mathbf{h}}\right)^{1/2}\bar{A}_h \times \boldsymbol{\kappa}^{-1}A_h\right\}}.$$

2. $N = \mu_0^{1/2}\left(\dfrac{\lambda-\mu}{\lambda(\mu-\nu)}\right)^{1/2}$, $2\mu_0^{-1}\hat{e} = \lambda^{-1}(\lambda\alpha_1\bar{\alpha}_1 + \mu\alpha_2\bar{\alpha}_2)$,

$$2\mu_0^{-1/2}\hat{\mathbf{r}} = \left(\frac{\mu-\nu}{\lambda(\lambda-\nu)}\right)^{1/2}(\lambda\alpha_1\bar{\alpha}_1 + \mu\alpha_2\bar{\alpha}_2)\mathbf{u}$$
$$-\left(\frac{\lambda-\mu}{\lambda(\lambda-\nu)}\right)^{1/2}(\mu+\nu)(\bar{\alpha}_1\alpha_2)^{+}\mathbf{s}.$$

Writing $\lambda^{1/2}\alpha_1 = a_1 e^{i\theta_1}$, $\lambda^{1/2}\alpha_2 = a_2 e^{i\theta_2}$, we note that $\hat{\mathbf{r}}$ is parallel to

$$\mathbf{u} - \left(\frac{\lambda-\mu}{\mu-\nu}\right)^{1/2}\frac{\mu+\nu}{2\lambda}\sin 2\phi\cos(\theta_2-\theta_1)\mathbf{s},$$

with $\tan\phi = a_2 a_1^{-1}$. Hence, when α_1 and α_2 are varied (when ϕ, θ_1, θ_2 are varied), $\hat{\mathbf{r}}$ varies from lying along

$$\mathbf{u} + \left(\frac{\lambda-\mu}{\mu-\nu}\right)^{1/2}\frac{\mu+\nu}{2\lambda}\mathbf{s}$$

to lying along

$$\mathbf{u} - \left(\frac{\lambda-\mu}{\mu-\nu}\right)^{1/2}\frac{\mu+\nu}{\lambda}\mathbf{s},$$

and may take any intermediate direction between these two extremes.

3. $N = \mu_0^{1/2}\lambda^{-1/2}(\boldsymbol{C}\cdot\mathbf{u})^{-1}$, $\hat{e} = \mu_0\lambda^{-1}$, $\hat{\mathbf{r}} = \mu_0^{1/2}\lambda^{-1/2}\mathbf{u}$.
4. $N = \mu_0^{1/2}\lambda^{-1/2}(m^2-1)^{-1/2}$, $2\hat{e} = \mu_0\left\{\alpha\bar{\alpha}_1 + \dfrac{\alpha\bar{\alpha}_2 m^2}{(m^2-1)}\right\}$,

$$2\hat{\mathbf{r}} = \mu_0^{1/2}\lambda^{1/2}\left\{\left(\frac{m^2-1}{m^2}\right)^{1/2}\alpha_1\bar{\alpha}_1 + \left(\frac{m^2}{m^2-1}\right)^{1/2}\alpha_2\bar{\alpha}_2\right\}\hat{\mathbf{m}}$$

$$-2\frac{\mu_0^{1/2}\lambda^{1/2}}{(m^2-1)^{1/2}}(\bar{\alpha}_1\alpha_2)^+\hat{\mathbf{m}}\times\hat{\mathbf{n}}.$$

Writing

$$\alpha_1 = a_1 e^{i\theta_1}, \quad \left(\frac{m^2}{m^2-1}\right)^{1/2}\alpha_2 = a_2 e^{i\theta_2},$$

we note that $\hat{\mathbf{r}}$ is parallel to $\hat{\mathbf{m}} - m(m^2-1)^{-1}\sin 2\phi\cos(\theta_2-\theta_1)\hat{\mathbf{m}}\times\hat{\mathbf{n}}$, with $\tan\phi = a_2 a_1^{-1}$. Hence, when α_1 and α_2 are varied (when ϕ, θ_1, θ_2 are varied), $\hat{\mathbf{r}}$ varies from lying along $\hat{\mathbf{m}} + m(m^2-1)^{-1}\hat{\mathbf{m}}\times\hat{\mathbf{n}}$ to lying along $\hat{\mathbf{m}} - m(m^2-1)^{-1}\hat{\mathbf{m}}\times\hat{\mathbf{n}}$, and may take any intermediate direction between these two extremes.

Chapter 10

10.1

1. For purely longitudinal waves: $c_{ijkl}n_k n_l n_j = \rho v^2 n_i$, $t_{ij} = -kc_{ijkl}n_k n_l \sin k(\mathbf{n}\cdot\boldsymbol{x} - vt)$, $t_{ij}n_j = -k\rho v^2 n_i \sin k(\mathbf{n}\cdot\boldsymbol{x} - vt)$. Thus $\boldsymbol{t}_{(\mathbf{n})}\times\mathbf{n} = \mathbf{0}$.
   For purely transverse waves: $c_{ijkl}a_k n_l n_j = \rho v^2 a_i$, $\mathbf{a}\cdot\mathbf{n} = 0$, $t_{ij} = -kc_{ijkl}a_k n_l \sin k(\mathbf{n}\cdot\boldsymbol{x} - vt)$, $t_{ij}n_j = -k\rho v^2 a_i \sin k(\mathbf{n}\cdot\boldsymbol{x} - vt)$. Thus $\boldsymbol{t}_{(\mathbf{n})}\cdot\mathbf{n} = 0$.

10.2

1. For purely longitudinal waves:
   $u_i = n_i \cos k(\mathbf{n}\cdot\boldsymbol{x} - vt)$, $\dot{u}_i = kvn_i \sin k(\mathbf{n}\cdot\boldsymbol{x} - vt)$,
   $t_{ij} = -kc_{ijkl}n_k n_l \sin k(\mathbf{n}\cdot\boldsymbol{x} - vt)$,
   $r_i = -\dot{u}_k t_{ki} = k^2 v c_{ijkl} n_j n_k n_l \sin^2 k(\mathbf{n}\cdot\boldsymbol{x} - vt)$
   $= \rho k^2 v^3 n_i \sin^2 k(\mathbf{n}\cdot\boldsymbol{x} - vt)$, on using the propagation condition for purely longitudinal waves.
   $e = \rho k^2 v^2 \sin^2 k(\mathbf{n}\cdot\boldsymbol{x} - vt)$.
   Thus, $\boldsymbol{r}\cdot\mathbf{n} = ve$ (no mean has been taken).

For purely transverse waves:
$u_i = a_i \cos k(\mathbf{n}\cdot\boldsymbol{x} - vt)$, $\mathbf{a}\cdot\mathbf{n} = 0$, $\dot{u}_i = kva_i \sin k(\mathbf{n}\cdot\boldsymbol{x} - vt)$,
$t_{ij} = -kc_{ijkl}a_k n_l \sin k(\mathbf{n}\cdot\boldsymbol{x} - vt)$,
$r_i = -\dot{u}_k t_{ki} = k^2 vc_{ijkl}a_j a_k n_l \sin^2 k(\mathbf{n}\cdot\boldsymbol{x} - vt)$.
$e = \rho k^2 v^2 \mathbf{a}\cdot\mathbf{a} \sin^2 k(\mathbf{n}\cdot\boldsymbol{x} - vt)$.
Thus

$$\begin{aligned} r_i n_i &= k^2 vc_{ijkl}n_i a_j a_k n_l \sin^2 k(\mathbf{n}\cdot\boldsymbol{x} - vt) \\ &= \rho k^2 v^3 a_j a_j \sin^2 k(\mathbf{n}\cdot\boldsymbol{x} - vt) = ve, \end{aligned}$$

on using the propagation condition for purely transverse waves and the symmetries of c_{ijkl}.

2. Now

$$2\tilde{r}_i^{(\alpha)} = \frac{\omega^2}{v_\alpha} c_{ijkl}a_j^{(\alpha)}a_k^{(\alpha)}n_l,$$

(no sum over α)

$$2\tilde{e}^{(\alpha)} = \rho\omega^2 \boldsymbol{a}^{(\alpha)}\cdot\boldsymbol{a}^{(\alpha)},$$

and hence $\tilde{g}_i^{(\alpha)} = (\rho v_\alpha)^{-1} c_{ijkl}\hat{a}_j^{(\alpha)}\hat{a}_k^{(\alpha)}n_l$, with $\hat{\boldsymbol{a}}^{(\alpha)} = \boldsymbol{a}^{(\alpha)}(\boldsymbol{a}^{(\alpha)}\cdot\boldsymbol{a}^{(\alpha)})^{-1/2}$.

Thus $v_1\tilde{g}_i^{(1)} + v_2\tilde{g}_i^{(2)} + v_3\tilde{g}_i^{(3)} = \rho^{-1}c_{ijjl}n_l$, on using

$$\hat{a}_j^{(1)}\hat{a}_k^{(1)} + \hat{a}_j^{(2)}\hat{a}_k^{(2)} + \hat{a}_j^{(3)}\hat{a}_k^{(3)} = \delta_{jk}.$$

3. With $\boldsymbol{r}_1 = r_1\mathbf{n}$, $\boldsymbol{r}_2 = r_2\mathbf{n}$ $(r_1, r_2 > 0)$, (10.3.7) becomes $\tilde{\boldsymbol{r}} = (\alpha\bar{\alpha}r_1 + \beta\bar{\beta}r_2)\mathbf{n} + (\bar{\alpha}\beta)^+\boldsymbol{r}_{12}$, $\boldsymbol{r}_{12}\cdot\mathbf{n} = 0$.
Writing $\alpha = ae^{i\lambda}$, $\beta = be^{i\nu}$ and defining $a' = ar_1^{1/2}$, $b' = br_2^{1/2}$, and ψ through $\tan\psi = b'a'^{-1}$,

$$\tilde{\boldsymbol{r}} = \frac{(a'^2 + b'^2)}{2(r_1r_2)^{1/2}}\{2(r_1r_2)^{1/2}\mathbf{n} + \sin 2\psi \cos(\nu - \lambda)\boldsymbol{r}_{12}\}.$$

Thus, the direction of $\tilde{\boldsymbol{r}}$ may be any direction in the plane of $\mathbf{n}$ and $\boldsymbol{r}_{12}$, lying between the two directions $2(r_1r_2)^{1/2}\mathbf{n} + \boldsymbol{r}_{12}$ and $2(r_1r_2)^{1/2}\mathbf{n} - \boldsymbol{r}_{12}$.

10.3

1. $\boldsymbol{u} = \mathbf{a}\cos k(\mathbf{n}\cdot\boldsymbol{x} - vt)$, $\dot{\boldsymbol{u}} = kv\mathbf{a}\sin k(\mathbf{n}\cdot\boldsymbol{x} - vt)$, $u_{i,j} = -ka_i n_j \sin k(\mathbf{n}\cdot\boldsymbol{x} - vt)$

(a) $\mathbf{a}\cdot\mathbf{n} = 0$,

$$\begin{aligned} \tilde{r}_i &= \tfrac{1}{2}k^2 v\{-a_i d_{jklp}n_j n_k n_l a_p + d_{ijkl}a_j a_k n_l\}, \\ \tilde{e} &= \tilde{\kappa} + \tilde{\sigma} = \tfrac{1}{4}\rho k^2 v^2 \mathbf{a}\cdot\mathbf{a} + \tfrac{1}{4}k^2 d_{ijkl}a_i a_k n_j n_l. \end{aligned}$$

But, from (10.4.4a), we have $d_{ijkl}a_i a_k n_j n_l = \rho v^2 \boldsymbol{a}\cdot\boldsymbol{a}$. Hence, $\tilde{e} = \frac{1}{2}\rho k^2 v^2 \mathbf{a}\cdot\mathbf{a}$, and $\tilde{\boldsymbol{r}}\cdot\mathbf{n} = v\tilde{e}$.

(b) $\mathbf{a}\cdot\mathbf{l} = 0$,

$$\tilde{r}_i = \tfrac{1}{2}k^2 v d_{ijkl}a_j a_k n_l,$$
$$\tilde{e} = \tilde{\kappa} + \tilde{\sigma} = \tfrac{1}{4}\rho k^2 v^2 \mathbf{a}\cdot\mathbf{a} + \tfrac{1}{4}k^2 d_{ijkl}a_i a_k n_j n_l.$$

But, from (10.4.13a), we have $d_{ijkl}a_i a_k n_j n_l = \rho v^2 \mathbf{a}\cdot\mathbf{a}$. Hence, $\tilde{e} = \frac{1}{2}\rho k^2 v^2 \mathbf{a}\cdot\mathbf{a}$, and $\tilde{\boldsymbol{r}}\cdot\mathbf{n} = v\tilde{e}$.

10.4

1. The propagation condition (10.2.2) with $A = \hat{\mathbf{a}}$ may also be written $\{Q_{ij}(\boldsymbol{k}) - \rho\omega^2\delta_{ij}\}\hat{a}_j = 0$. Taking the derivative with respect to k_l yields

$$\left\{\frac{\partial Q_{ij}}{\partial k_l} - 2\rho\omega\frac{\partial\omega}{\partial k_l}\delta_{ij}\right\}\hat{a}_j + \{Q_{ij} - \rho\omega^2\delta_{ij}\}\frac{\partial \hat{a}_j}{\partial k_l} = 0.$$

Therefore, multiplying by $\hat{a}_i$, we obtain

$$\frac{\partial\omega}{\partial k_l} = \frac{1}{2\rho\omega}\hat{a}_i\frac{\partial Q_{ij}}{\partial k_l}\hat{a}_j. \text{ But } \frac{\partial Q_{ij}}{\partial k_l} = (c_{lijp} + c_{ljip})k_p.$$

Hence,

$$\frac{\partial\omega}{\partial k_l} = \frac{1}{\rho\omega}c_{lijp}\hat{a}_i\hat{a}_j k_p = \frac{1}{\rho v}c_{lijp}\hat{a}_i\hat{a}_j n_p.$$

From (10.3.4) and (10.3.5), because $\hat{\mathbf{a}} = \mathbf{a}(\mathbf{a}\cdot\mathbf{a})^{-1/2}$, we check that $\tilde{r}_l\tilde{e}^{-1} = (\rho v)^{-1}c_{lijp}\hat{a}_i\hat{a}_j n_p$.

2. The propagation condition (10.4.8) with $A = \hat{\mathbf{a}}$ may also be written $\{(k^2\delta_{ip} - k_i k_p)Q_{pr}(\mathbf{k})(k^2\delta_{rj} - k_r k_j) - \rho\omega^2 k^4\delta_{ij}\}\hat{a}_j = 0$. Taking the derivative with respect to k_l, and multiplying by $\hat{a}_i$ yields, on using $\mathbf{k}\cdot\hat{\boldsymbol{a}} = 0$ and the symmetry of $Q_{pr}(\mathbf{k})$, $2k^2(2k_l\hat{a}_p - \hat{a}_l k_p)Q_{pr}\hat{a}_r + k^4\hat{a}_p\ (\partial Q_{pr}/\partial k_l)\hat{a}_r - 4\rho\omega^2 k^2 k_l - 2k^4\rho\omega(\partial\omega/\partial k_l) = 0$. But, multiplying the propagation condition by $\hat{a}_i$ gives $\hat{a}_p Q_{pr}\hat{a}_r = \rho\omega^2$. Thus, $2k^4\rho\omega(\partial\omega/\partial k_l) = -2k^2\hat{a}_l k_p Q_{pr}\hat{a}_r + k^4\hat{a}_p(\partial Q_{pr}/\partial k_l)\hat{a}_r$, and hence

$$\frac{\partial\omega}{\partial k_l} = \frac{1}{\rho v}\{-\hat{a}_l d_{ijkp}n_i n_j n_k\hat{a}_p + d_{lijk}\hat{a}_i\hat{a}_j n_k\}.$$

From the result of Exercise 10.3.1(a), it is easily checked that $\partial\omega/\partial k_l = \tilde{r}_l\tilde{e}^{-1}$.

3. (a) The three possible directions of polarization, which are along the eigenvectors of q_{ik}, are along the principal axes of $\mathscr{E}(\mathbf{n})$.
   (b) The two possible directions of polarization are along the eigenvectors of $q'_{ik} = Q'_{ik}(\mathbf{n})$ given by (10.4.9). Recalling the results of section 5.7, these directions are thus along the principal axes of the elliptical section of the ellipsoid $\mathscr{E}(\mathbf{n})$ by the plane $\mathbf{n}\cdot\boldsymbol{x} = 0$ (see in particular the proof of section 5.7, remark).
   (c) The two possible directions of polarization are along the eigenvectors of $q''_{ik} = Q''_{ik}(\mathbf{n})$ given by (10.4.15). Thus, analogously to case (b), these directions are along the principal axes of the elliptical section of the ellipsoid $\mathscr{E}(\mathbf{n})$ by the plane $\mathbf{l}\cdot\boldsymbol{x} = 0$.
4. The sum of the roots of the secular equation (10.2.4) is the trace of $\boldsymbol{Q}(\mathbf{n})$. Thus, using (10.2.3),

$$\rho \sum_{\alpha=1}^{3} \mathscr{V}_{\alpha}^{2}(n) = \operatorname{tr} \boldsymbol{Q}(\mathbf{n}) = c_{ijil} n_j n_l.$$

Because $n_j n_l + m_j m_l + p_j p_l = \delta_{jl}$, we have

$$\rho \sum_{\alpha=1}^{3} \{\mathscr{V}_{\alpha}^{2}(\mathbf{n}) + \mathscr{V}_{\alpha}^{2}(\mathbf{m}) + \mathscr{V}_{\alpha}^{2}(\mathbf{p})\} = c_{ijij},$$

which does not depend on the choice of the triad **n**, **m**, **p**.
5. From the previous exercise, we have

$$\sum_{\alpha=1}^{3} \mathscr{V}_{\alpha}^{2}(\mathbf{n}) = \frac{1}{\rho} \mathbf{n}^{\mathrm{T}} \boldsymbol{V} \mathbf{n}.$$

Let **s**, **t**, **u** be the unit eigenvectors of $\boldsymbol{V}$ corresponding to λ, μ, ν, respectively. Thus, $\mathbf{n}^{\mathrm{T}} \boldsymbol{V} \mathbf{n} = \lambda(\mathbf{s}\cdot\mathbf{n})^2 + \mu(\mathbf{t}\cdot\mathbf{n})^2 + \nu(\mathbf{u}\cdot\mathbf{n})^2$. Hence

$$\sum_{\alpha=1}^{3} \mathscr{V}_{\alpha}^{2}(\mathbf{n}) = \frac{\gamma}{\rho}$$

may be written

$$(\lambda - \gamma)(\boldsymbol{s}\cdot\mathbf{n})^2 + (\mu - \gamma)(\mathbf{t}\cdot\mathbf{n})^2 + (\nu - \gamma)(\mathbf{u}\cdot\mathbf{n})^2 = 0.$$

This is an equation to be satisfied by the components $\mathbf{s}\cdot\mathbf{n}$, $\mathbf{t}\cdot\mathbf{n}$, and $\mathbf{u}\cdot\mathbf{n}$ of the vector **n**. If γ is not equal to one of the eigenvalues λ, μ, ν, this is the equation of a cone. If $\gamma = \nu$, then $\mathbf{n} = \pm\mathbf{u}$ and the cone degenerates into a single direction. If $\gamma = \lambda$, then $\mathbf{n} = \pm\mathbf{s}$

and the cone again degenerates into a single direction. If $\gamma = \mu$, the equation becomes $(\mathbf{h}\cdot\mathbf{n})(\mathbf{k}\cdot\mathbf{n}) = 0$, where $\mathbf{h}$ and $\mathbf{k}$ are given by (5.6.6). Thus, the cone degenerates into the two planes $\mathbf{h}\cdot\mathbf{n} = 0$, $\mathbf{k}\cdot\mathbf{n} = 0$ of the circular sections of the ellipsoid $\mathbf{x}^T V\mathbf{x} = 1$ (see section 5.6)

10.5

1. For an isotropic material, on comparing (10.7.1) with (10.6.2), we have $d_{11} = d_{33} = \lambda + 2\mu$, $d_{12} = d_{13} = \lambda$, $d_{44} = \mu$. Hence, from (10.7.13) and (10.7.17), $R = 0(0)^{-1}$ (indeterminate), and $S = 1$. Because R is indeterminate, case (b) of (10.7.12) occurs for any direction $\mathbf{n}$.

10.6

1. $\hat{e} = \frac{1}{4}\omega^2\{\rho \boldsymbol{A}\cdot\bar{\boldsymbol{A}} + c_{ijkl}A_i\bar{A}_k S_j \bar{S}_l\}$,
   $\hat{f}_i = \frac{1}{4}\omega^2 c_{ijkl}(A_j\bar{A}_k\bar{S}_l + \bar{A}_j A_k S_l)$,
   $\hat{\mathbf{f}}\cdot\boldsymbol{S} = \frac{1}{4}\omega^2 c_{ijkl}(A_i\bar{A}_k S_j\bar{S}_l + \bar{A}_i A_k \bar{S}_j S_l) = \hat{e}$,
   on using the propagation condition (10.8.2), with $\boldsymbol{Q}(\boldsymbol{S})$ given by (10.8.3).

10.7

1. Taking the + sign in (10.9.7a)

$$\boldsymbol{u} = \{\beta(\tfrac{1}{3}\mathbf{j} + i\mathbf{i}) + \gamma\mathbf{k}\} \exp i\omega\left(3\left(\frac{\rho}{8\mu}\right)^{1/2} y - t\right)\exp\left(-\omega\left(\frac{\rho}{8\mu}\right)^{1/2} x\right).$$

   This is a train of waves propagating along the y-axis with phase speed $\{8\mu(9\rho)^{-1}\}^{1/2}$ and attenuated along the x-axis with attenuation factor $\omega\{\rho(8\mu)^{-1}\}^{1/2}$. If $\gamma\beta^{-1} = a + ib$, the polarization ellipse has $\frac{1}{3}\mathbf{j} + a\mathbf{k}$ and $\mathbf{i} + b\mathbf{k}$ as a pair of conjugate radii. The wave is circularly polarized for $\gamma\beta^{-1} = \pm 2(\frac{2}{3})^{1/2}$.

2. (a) $\hat{e} = \frac{1}{2}\rho\omega^2|\alpha|^2 m^2 \dfrac{\lambda(m^2-1) + 2\mu(m^2+1)}{(\lambda + 2\mu)(m^2 - 1)}$,

$$\hat{\mathbf{f}} = \tfrac{1}{2}\rho\omega^2|\alpha|^2 m \frac{\lambda(m^2-1) + 2\mu(m^2+1)}{(\rho(\lambda+2\mu)(m^2-1))^{1/2}}\hat{\mathbf{m}} = \hat{e}\mathbf{v}_p, \text{ where}$$

$$\mathbf{v}_p = \frac{m\mathbf{m}}{((\lambda+2\mu)(m^2-1)\rho^{-1})^{1/2}} \text{ is the phase velocity.}$$

(b) $\hat{e} = |\beta|^2\hat{e}_1 + |\gamma|^2\hat{e}_2$,
$\hat{\mathbf{r}} = |\beta|^2\hat{\mathbf{r}}_1 + |\gamma|^2\hat{\mathbf{r}}_2 + (\bar{\beta}\gamma)^+\hat{\mathbf{r}}_{12}$,
with

$$\hat{e}_1 = \tfrac{1}{2}\rho\omega^2\frac{m^2+3}{m^2-1}, \quad \hat{e}_2 = \tfrac{1}{2}\rho\omega^2\frac{m^2}{m^2-1},$$

and

$$\hat{\mathbf{r}}_1 = \tfrac{1}{2}\rho\omega^2\left(\frac{\mu}{\rho(m^2-1)}\right)^{1/2}\frac{m^2+3}{m}\hat{\mathbf{m}},$$

$$\hat{\mathbf{r}}_2 = \tfrac{1}{2}\rho\omega^2\left(\frac{\mu}{\rho(m^2-1)}\right)^{1/2} m\hat{\mathbf{m}},$$

$$\hat{\mathbf{r}}_{12} = \rho\omega^2\left(\frac{\mu}{\rho(m^2-1)}\right)^{1/2}\hat{\mathbf{m}}\times\hat{\mathbf{n}} \quad \text{(interaction term).}$$

Writing $\beta = be^{i\lambda}$, $\gamma = ce^{i\nu}$ (see Exercise 10.2.3), we note that $\hat{\mathbf{r}}$ is parallel to $\hat{\mathbf{m}} + (m^2+3)^{-1/2}\sin 2\psi\cos(\nu-\lambda)\hat{\mathbf{m}}\times\hat{\mathbf{n}}$, with $\tan\psi = bc^{-1}m^{-1}(m^2+3)^{1/2}$. Hence, the direction of $\hat{\mathbf{r}}$ may be any direction in the plane of $\hat{\mathbf{m}}$ and $\hat{\mathbf{m}}\times\hat{\mathbf{n}}$, lying between the two directions $\hat{\mathbf{m}} + \{(m^2+3)^{-1/2}\}\hat{\mathbf{m}}\times\hat{\mathbf{n}}$ and $\hat{\mathbf{m}} - (m^2+3)^{-1/2}\hat{\mathbf{m}}\times\hat{\mathbf{n}}$.

3. The constitutive equations are

$$t_{ij} = -p\delta_{ij} + \mu(u_{i,j} + u_{j,i}), \quad u_{i,i} = 0.$$

The propagation condition is

$$iP\boldsymbol{S} + \mu\omega(\boldsymbol{S}\cdot\boldsymbol{A})\boldsymbol{S} = \rho\omega\boldsymbol{A}, \quad \text{with } \boldsymbol{S}\cdot\boldsymbol{A} = 0.$$

Hence, either (a) $\boldsymbol{S}\cdot\boldsymbol{S} = 0$, $\boldsymbol{A} = \alpha\boldsymbol{S}$, $P = -i\alpha\rho\omega$, (α arbitrary); or (b) $P = 0$, $\mu\boldsymbol{S}\cdot\boldsymbol{S} = \rho$, $\boldsymbol{S}\cdot\boldsymbol{A} = 0$ ($\boldsymbol{A}$ arbitrary orthogonal to $\boldsymbol{S}$).

Thus, either

(a) the wave is circularly polarized in the plane of the slowness bivector $\boldsymbol{S}$, which may be any isotropic bivector (solution independent of the shear modulus μ); or

(b) the wave is transverse, given by (10.9.7) as in the compressible case.

10.8

Homogeneous circularly polarized waves have been considered in section 10.7. Here we deal with inhomogeneous waves.

The secular equation must have a double root for given $\boldsymbol{C}$. As in the homogeneous case, this is possible only if either (a) the root (10.10.1) is also a root of the biquadratic (10.10.3), or (b) the biquadratic (10.10.3) has a double root.

Case (a). Insertion of (10.10.1) into (10.10.3) leads to the two possibilities
(i) $C_1^2 + C_2^2 = 0$ and (ii) $(C_1^2 + C_2^2) = RC_3^2$,
where R is given by (10.7.13).

Possibility (i) has been considered (section 10.10.2). The polarization circle lies in the xy-plane.

For possibility (ii), $\rho N^{-2} = \{\frac{1}{2}R(d_{11} - d_{12}) + d_{44}\}C_3^2$, and

$$\boldsymbol{A} = (C_2, -C_1, 0) + \frac{\mathrm{i}}{(S^2 + R)^{1/2}}(SC_1, SC_2, -RC_3),$$

with S given by (10.7.17).

The polarization circle may not lie in the xy-plane, because if either R or C_3 is zero, then we revert to possibility (i).

Case (b). The biquadratic (10.10.3) has the double root

$$2\rho N^{-2} = (d_{11} + d_{44})(C_1^2 + C_2^2) + (d_{33} + d_{44})C_3^2,$$

provided C_i satisfy

$$(d_{11} - d_{44})(C_1^2 + C_2^2) - (d_{33} - d_{44})C_3^2 = \pm 2\mathrm{i}(d_{13} + d_{44})(C_1^2 + C_2^2)^{1/2}C_3.$$

Corresponding to the upper sign (+), we have

$$(C_1^2 + C_2^2)^{1/2} = \mathrm{i}C_3T_\pm,$$

where $T_\pm$ are given by

$$(d_{11} - d_{44})T_\pm = (d_{13} + d_{44}) \pm \{(d_{13} + d_{44})^2 - (d_{11} - d_{44})(\mathrm{d}_{33} - \mathrm{d}_{44})\}^{1/2}.$$

The corresponding amplitude bivector is

$$\boldsymbol{A} = (C_1, C_2, C_3T_\pm).$$

We note $\boldsymbol{A}\cdot\boldsymbol{S} \neq 0$. The circle of polarization may not in general lie in the xy-plane, because $T_\pm \neq 0$ in general, and if $C_3 = 0$, we revert to possibility (i), case (a).

Chapter 11

11.1

1. (a) $\dfrac{\beta'}{\alpha'} = \left(\dfrac{m^2-1}{m^2}\right)\dfrac{\alpha}{\beta}$;

   (b) $\dfrac{\beta'}{\alpha'} = -\left(\dfrac{m^2+1}{m^2}\right)\dfrac{\bar{\alpha}}{\bar{\beta}}$.

2. $\boldsymbol{A}\cdot\boldsymbol{A} = 0$ when

$$\left(\frac{\beta}{\alpha}\right)^2 = \frac{m^2-1}{m^2}, \quad \frac{\beta}{\alpha} = \pm\left(\frac{m^2-1}{m}\right)^{1/2}.$$

$$+\text{ sign: } \boldsymbol{A}_p = \frac{1}{m}\hat{\mathbf{m}} + \mathrm{i}\hat{\mathbf{n}} + \left(\frac{m^2-1}{m}\right)^{1/2}\hat{\boldsymbol{s}},$$

$$-\text{ sign: } \boldsymbol{A}_n = \frac{1}{m}\hat{\mathbf{m}} + \mathrm{i}\mathbf{n} - \left(\frac{m^2-1}{m}\right)^{1/2}\hat{\boldsymbol{s}};$$

$\boldsymbol{A}_p, \boldsymbol{A}_n$ are in planes passing through $\hat{\mathbf{n}}$ of unit normals

$$\frac{1}{m}\hat{\mathbf{s}} \mp \left(\frac{m^2-1}{m}\right)^{1/2}\hat{\mathbf{m}}.$$

Angle with plane of $\boldsymbol{C}$: $\pm\cos^{-1}m^{-1}$.

11.2

1. $p = a\{\omega^+\cos(kx-\omega^+t+\gamma) - \omega^-\sin(kx-\omega^+t+\gamma)\}\exp(-ky+\omega^-t) - \frac{1}{2}\rho k^2a^2\exp 2(-ky+\omega^-t) + p_0$.

11.3

1. For $\boldsymbol{A}_p$, we have $\hat{e} = \frac{1}{2}\rho$, $\hat{d} = \rho\Omega(m^2+3)(m^2-1)^{-1}$,

$$\hat{\mathbf{r}} = \pm\left(\frac{\mu\rho\Omega}{m^2-1}\right)^{1/2}\left\{\left(m+\frac{1}{m}\right)\sin\left(\frac{\pi}{4}+\frac{\delta}{2}\right)\hat{\mathbf{m}} + 2\cos\left(\frac{\pi}{4}+\frac{\delta}{2}\right)\hat{\mathbf{n}} + \left(\frac{m^2-1}{m}\right)^{1/2}\cos\left(\frac{\pi}{4}+\frac{\delta}{2}\right)\hat{\mathbf{s}}\right\}.$$

For A_n, we have $\hat{e} = \frac{1}{2}\rho$, $\hat{d} = \rho\Omega(m^2+3)(m^2-1)^{-1}$,

$$\hat{\mathbf{r}} = \pm\left(\frac{\mu\rho\Omega}{m^2-1}\right)^{1/2}\left\{\left(m+\frac{1}{m}\right)\sin\left(\frac{\pi}{4}+\frac{\delta}{2}\right)\hat{\mathbf{m}} + 2\cos\left(\frac{\pi}{4}+\frac{\delta}{2}\right)\hat{\mathbf{n}} - \frac{(m^2-1)^{1/2}}{m}\cos\left(\frac{\pi}{4}+\frac{\delta}{2}\right)\hat{\mathbf{s}}\right\}.$$

2. The vector $m\cos(\frac{1}{4}\pi+\frac{1}{2}\delta)\hat{\mathbf{m}} - \sin(\frac{1}{4}\pi+\frac{1}{2}\delta)\,\hat{\mathbf{n}}$ is along $\boldsymbol{K}^+$.
   Inserting $\hat{r}_x = m\cos(\frac{1}{4}\pi+\frac{1}{2}\delta)$, $\hat{r}_y = -\sin(\frac{1}{4}\pi+\frac{1}{2}\delta)$, $\hat{r}_z = 0$ into (11.5.22), we obtain

   $$0 \leqslant -4(m^2+1)\cos^2\delta - (1+3m^2)(1-\sin\delta)^2 - (3+m^2)(1+\sin\delta)^2,$$

   which is never satisfied.
3.

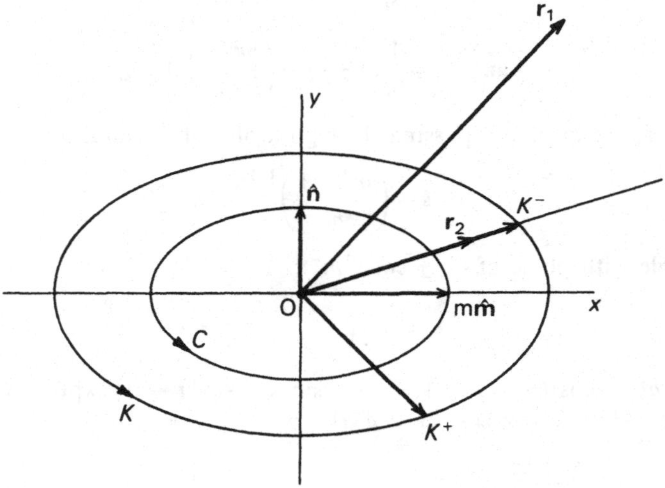

Energy flux for zero pressure wave. $\delta = 30°$, $m = 2.22$.

Bibliography

References

Ahlfors, L.V. (1953) *Complex analysis*, McGraw-Hill, New York.

Azzam, R.M.A. and Bashara, N.M. (1977) *Ellipsometry and Polarised Light*, North-Holland, Amsterdam.

Born, M. and Wolf, E. (1980) *Principles of Optics*, 6th edn, Pergamon, Oxford.

Boulanger, Ph. and Hayes, M. (1990) Electromagnetic plane waves in anisotropic media: an approach using bivectors. *Philosophical Transactions of the Royal Society of London*, A, **330**, 335–93.

Brady, P. (1989) *Bivector analysis*. Minor M.Sc. thesis, University College Dublin, unpublished, 50 pp.

Chadwick, P. (1989) Wave propagation in transversely isotropic elastic media. I. Homogeneous plane waves. *Proceedings of the Royal Society of London*, A, **422**, 23–66.

Fedorov, F.I. (1968) *Theory of Elastic Waves in Crystals*, Plenum Press, New York.

Gibbs, J.W. (1881, 1884) *Elements of Vector Analysis*, privately printed. (Published in *Scientific Papers*, Vol. 2, Dover, New York (1961), pp. 17–90.)

Halberstam, H. and Ingram, R.E. (eds) (1967) *Mathematical Papers of Sir William Rowan Hamilton*, Vol. **3**, University Press, Cambridge.

Hamilton, W.R. (1853) *Lectures on Quaternions*, Hodges and Smith, Dublin.

Hayes, M. (1984) Inhomogeneous plane waves. *Archive for Rational Mechanics and Analysis*, **85**, 41–79.

Hayes, M. (1987) Inhomogeneous electromagnetic plane waves in crystals. *Archive for Rational Mechanics and Analysis*, **97**, 221–60.

Hayes, M. and Rivlin, R.S. (1962) A note on the secular equation for Rayleigh waves. *Journal of Applied Mathematics and Physics* (*ZAMP*), **13**, 80–3.

Love, A.E.H. (1927) *A Treatise on the Mathematical Theory of Elasticity*, 4th edn, University Press, Cambridge.

MacCullagh, J. (1847) On total reflexion. *Proceedings of the Royal Irish Academy*, **3**, 49–51.

Milne, E.A. (1948) *Vectorial Mechanics*, Methuen, London.

Musgrave, M.J.P. (1970) *Crystal Acoustics*, Holden-Day, San Francisco.

Shurcliff, W.A. (1962) *Polarized Light*, Harvard University, Cambridge, Massachusetts.

Stokes, G.G. (1852) On the composition and resolution of streams of polarized light from different sources. *Transactions of the Cambridge Philosophical Society*, **9**, 399–416.

Synge, J.L. (1964) The Petrov classification of gravitational fields. *Communications of the Dublin Institute for Advanced Studies*, A, **15**, 51 pp.

Further reading

Chapter 1

A good text is

Clement-Jones, A. (1912) *An Introduction to Algebraical Geometry*, Clarendon Press, Oxford.

Chapter 2

Airy's decomposition is mentioned in

Preston, T. (1901) *Theory of Light*, Macmillan, London.

Bivectors and their associated ellipses are considered in

Lindell, I.V. (1988) Complex vectors and dyadics for electromagnetics. *Helsinki University of Technology, Faculty of Electrical Engineering, Electromagnetics Laboratory Report Series*, **36**, 50 pp.

Extensive use is made of bivectors in

Stone, J.M. (1963) *Radiation and Optics*, McGraw-Hill, New York.

Chapters 3 and 4

The classic text on matrices is

Gantmacher, F.R. (1959) *The Theory of Matrices*, 2 vols, Chelsea, New York.

Hamilton's cyclic form is mentioned in

Joly, C.J. (1905) *A Manual of Quaternions*, Macmillan, London.

Chapter 5

A useful text is

Bell, R.J.T. (1938) *Coordinate Solid Geometry*, Macmillan, London.

Chapter 6

General properties of waves are covered very well in the texts

Lighthill, M.J. (1978) *Waves in Fluids*, University Press, Cambridge.
Whitham, G.B. (1974) *Linear and Nonlinear Waves*, John Wiley, New York.

Chapter 7

Good treatments may be found in

van de Hulst, H.C. (1981) *Light Scattering by Small Particles*, Dover, New York.
Kliger, S., Lewis, J.W. and Randall, C.E. (1990) *Polarized Light in Optics and Spectroscopy*, Academic Press, London.
Ramachandran, G.N. and Ramesehan, S. (1961) Crystal optics, in *Crystal Optics, Diffraction*, (ed. S. Flügge), *Handbuch der Physik* **25**(1), Springer, Berlin.
Simmons, J.W. and Guttmann, M.J. (1970) *States, Waves and Photons: a Modern Introduction to Light*, Addison-Wesley, Reading, Massachusetts.

Chapter 8

For waves in dissipative media the reader may consult

Buchen, P.W. (1971) Waves in linear viscoelastic media. *Geophysical Journal of the Royal Astronomical Society*, **23**, 531–42.

The material in the text is based upon

Hayes, M. (1977) A note on group velocity. *Proceedings of the Royal Society of London*, A, **354**, 533–35.
Hayes, M. (1980) Energy flux for trains of inhomogeneous plane waves. *Proceedings of the Royal Society of London*, A, **370**, 417–29.

Chapter 9

There is an excellent treatment of homogeneous electromagnetic waves in the texts of Born and Wolf (1980) and of Stone (1963).

A good account of homogeneous electromagnetic waves and homogeneous elastic waves in anisotropic media may be found in

Sirotine, Y. and Chaskolskaia, M. (1984) *Fondements de la Physique des Cristaux* (traduit du russe), Editions MIR, Moscow.

Chapter 10

There is an elegant treatment of elastic waves in anisotropic media in

Schouten, J.A. (1989) *Tensor Analysis for Physicists*, 2nd edn, Dover, New York.

Chapter 11

The treatment in the text is based upon

Boulanger, Ph. and Hayes, M. (1990) Inhomogeneous plane waves in viscous fluids. *Continuum Mechanics and Thermodynamics*, **2**, 1–16.

Index

Printed in the United States
by Baker & Taylor Publisher Services